SORGHUM BIOCHEMISTRY: AN INDUSTRIAL PERSPECTIVE

SORGHUM BIOCHEMISTRY: AN INDUSTRIAL PERSPECTIVE

C.V. RATNAVATHI
ICAR-Indian Institute of Millets Research, Rajendranagar, Hyderabad, India

J.V. PATIL
College of Agriculture, Pune, Maharashtra, India

U.D. CHAVAN
Mahatma Phule Krishi Vidyapeeth, Rahuri, Maharashtra, India

ELSEVIER

AMSTERDAM • BOSTON • HEIDELBERG • LONDON
NEW YORK • OXFORD • PARIS • SAN DIEGO
SAN FRANCISCO • SINGAPORE • SYDNEY • TOKYO

Academic Press is an imprint of Elsevier

Academic Press is an imprint of Elsevier
125 London Wall, London EC2Y 5AS, UK
525 B Street, Suite 1800, San Diego, CA 92101-4495, USA
50 Hampshire Street, 5th Floor, Cambridge, MA 02139, USA
The Boulevard, Langford Lane, Kidlington, Oxford OX5 1GB, UK

British Library Cataloguing-in-Publication Data
A catalogue record for this book is available from the British Library

Library of Congress Cataloging-in-Publication Data
A catalog record for this book is available from the Library of Congress

ISBN: 978-0-12-803157-5

For Information on all Academic Press publications
visit our website at https://www.elsevier.com/

Publisher: Nikki Levy
Acquisition Editor: Nina D. Bandeira
Editorial Project Manager: Ana Claudia Garcia
Production Project Manager: Julie-Ann Stansfield
Designer: Mark Rogers

Contents

Biography

C.V. RATNAVATHI, PhD

Principal Scientist, Plant Biochemistry and Joined Agricultural Research Service in 1985 and working on Sorghum since 1989.

Dr. Ratnavathi has a PhD in Biochemistry from Osmania University, and currently serves as the Principal Investigator on the NASF project on sweet sorgum for efficient bioethanol production, while also leading a research project from the DBT on the therapeutic properties of sorghum. She has led several other externally funded projects like NATP on value addition to sorghum and was instrumental in establishing a food processing laboratory under the NAIP Millet Value chain project. She has developed 30 sorghum recipes and 10 semiprocessed products, and has published numerous journal articles, books, and book chapters. She is recognized by the CODEX Committee for her research work on aflatoxins in sorghum. Dr. C.V. Ratnavathi has been awarded the fellowship of the Andhra Pradesh Academy of Sciences, Guntur.

J.V. PATIL, PhD

Former Director, ICAR-Indian Institute of Millets Research (IIMR) (Formerly Directorate of Sorghum Research), Indian Council of Agricultural Research (ICAR), Hyderabad, India. He is presently Associate Dean and Principal, College of Agriculture, Pune, Maharashtra, India.

Dr. Jagannath Vishnu Patil is a leader in sorghum and pulses research and has developed improved varieties/hybrids of it. He has also conceptualized innovative technologies for germplasm conservation and productivity enhancement in crops. He has been elected a Fellow of the Indian Society of Genetics and Plant Breeding, Indian Society for Pulses Research and Development, and National Academy of Agricultural Sciences. He is instrumental for elevating the Directorate of Sorghum Research (DSR) to the Indian Institute of Millet Research (IIMR).

Under the leadership of Dr. J.V. Patil, IIMR has been awarded Centre of Excellence for value addition and processing of Sorghum. He has successfully demonstrated the production to consumption value chain in sorghum.

U.D. CHAVAN, PhD

Professor and Senior Cereal Food Technologist, Department of Food Science and Technology, Mahatma Phule Krishi Vidyapeeth, Rahuri, Ahmednagar, Maharashtra, India.

Dr. Chavan worked as the Senior Research Assistant in the Department of Biochemistry and Food Science and Technology at Mahatma Phule Krishi Vidyapeeth, Rahuri from 1988 to 2000. He received his Ph.D. degree in Food Science from Memorial University of Newfoundland, St. John's, Canada in 1999. During his PhD, he worked as a Technician/Research Associate at Atlantic Cool Climate Crop Research Center and Agriculture and Agri-Food Canada. He has received several awards and fellowships, including the International Scholar Award for his PhD work at Memorial University and a Lifetime Achievement Award for his contribution to postharvest technology. He has written 138 research papers, 140 popular articles, and 42 books in multiple languages. He has done International training on "Global Nutrition 2002" at Uppsala University, Uppsala, Sweden in 2002. He has also contributed in the development of crop varieties in wheat (one) and Sorghum (eleven). He has been given nine recommendations in the field of Food Science and Technology. He is handling five National and International Research Projects on Sorghum. He has guided 25 MSc Students in the Discipline of Food Science and Technology and Biochemistry. Now he is working as a Professor in the Department of Food Science and Technology, as well as a Senior Cereal Food Technologist at the Sorghum Improvement Project and a Foreign Student Advisor at Mahatma Phule Krishi Vidyapeeth, Rahuri, Ahmednagar, Maharashtra, India.

Introduction

Sorghum is an important food and fodder crop of the semiarid tropic regions (14–24°N to 70–82°E). India is the third largest producer (8.06 MT) of sorghum in the world after United States and Nigeria (FAO Report, 2002), and has the largest area under cultivation for this crop (8.5 Mha) covering the states of Maharashtra, Karnataka, Madhya Pradesh, Andhra Pradesh, Rajasthan, Gujarat, Tamil Nadu, and Uttar Pradesh. The rainy season sorghum is often damaged due to grain mold (GM), so fetches lower prices, but has the greatest potential for its use in the nonfood and industrial sectors for the production of various value-added products.

At present, most of the sorghum produced in India is consumed as a human food in the form of roti or chapatti (unleavened flat bread). Alternatively, the sweet sorghum with its juicy sweet stalk has enormous potential as a bioenergy crop. Sorghum is known to have a highest dry matter productivity rate (50 $g/m^2/day$) in a number of locations across the globe and is reported to surpass the productivity of sugarcane, another C_4 plant (Somani et al., 1997).

Alternative uses of sorghum encompass the utilization of grain and sweet stalk in food and nonfood sectors for the production of commercially valued products, such as alcohol (potable and industrial grade), syrups (natural and high fructose), glucose (liquid and powder), modified starches, maltodextrins, jaggery, sorbitol, and citric acid (downstream products from starch).

Global and domestic demand for animal and poultry feed are rapidly rising. Sorghum can act as a prominent source of poultry and animal feed (Kleih et al., 2000). Similarly, demand for starch and industrial potable alcohol are continuously increasing. In comparison to other raw materials, sorghum ensures itself a place as a potential alternative in this sector. However, sorghum always faces stiff competition from maize. Apart from this, demand can be created for sorghum in beverage industries as malt and adjunct (unmalted cereal component used for beer preparation). In this book we detail the gradual shift from traditional uses to the envisaged use of sorghum as value-added food and nonfood industrial products.

Sweet sorghum (*Sorghum bicolor* (L.) Moench), a C4 graminaceous crop which has sugar-rich stalks and which is a water-use-efficient crop, has a very good potential as an alternative feedstock for ethanol production. It is the only crop that provides both grain and stem that can be used for sugar, alcohol, syrup, jaggery, fodder, fuel, bedding, roofing, fencing, paper, and chewing. Sweet sorghum juices usually contain approximately 16–18% fermentable sugar, which can be directly fermented into ethanol by yeast. Technical challenges of using sweet sorghum for biofuels are a short harvest period for the highest sugar content and fast sugar degradation during storage. Evaluation of 160 sweet sorghum genotypes was done for juice extractability. Sweet sorghum juice mainly contains sucrose, glucose, and fructose. While sucrose is the predominant sugar during the whole developmental stages, it constituted only about 50% of the soluble sugar at the boot stage, glucose and fructose making the remainder. Some studies have been carried out on the genetic potential of sweet sorghum for higher green cane yield and juice extractability, whereas some studies have been purely on the shelf life of the juice and how to preserve the juice at low cost against bacterial contamination. Studies have also involved testing the potential of genotypes for superior yields of cane during the post rainy and summer seasons to make the cane available for the maximum period in a year. Fermentation studies have also been carried out using *Saccharomyces cerevisiae*, the most commonly used species of yeast in alcohol fermentation. Genotypic variation for ethanol production from sweet sorghum juice has also been studied. A pilot study was carried out for the evaluation of ethanol production from sweet sorghum stalk juice.

GM is one of the major biotic constraints of sorghum for feed and food production. The principal GM fungi in India are *Fusarium moniliforme, Curvularia lunata, Phoma sorghina, Alternaria alternate, Exserohilum, Gonatobotrytis* sp., and *Aspergillus* spp. GM is the result of a complex of fungus–host interactions, which leads to the complete deterioration of sorghum grain. Infestation of sorghum grain by storage fungi results in varying degrees of damage, including (1) discoloration of the kernel, (2) reduction in kernel germination, (3) heating, (4) mustiness, and (5) production of mycotoxins. The occurrence of *Aspergillus* is less compared to *Fusarium*.

The incidence of GM is moderate to low in *Kharif*. The disease intensity was medium in Andhra Pradesh, Gujarat, and Tamil Nadu, whereas it was low in Maharashtra, Karnataka, and Rajasthan. The natural variability in *Aspergillus* and *Fusarium*, the toxicity and potential of toxin production in sorghum, and the extent of damage to the grain in detail were given in this report. In sorghum, other toxins detected are *Tricothecenes*, Zearalenone, and Fumonisin B_1 and B_2 toxins produced by the fungus *Fusarium*. Some of the fusarial toxins—such as zearalenone—exhibit uterotrophic activity. However, *Phoma sorghina*, a widely distributed GM fungus, is known to produce tenuazonic acid and may be responsible for onyalai, a human disorder prevalent in Africa. Onyalai is diagnosed by hemorrhagic vesicles in the mouth that appear after the ingestion of phoma-infected grain.

References

FAO, 2002. HTM Published in 2002 by the Food and Agriculture Organization of the United Nations, Viale delle Terme di Caracalla Rome, Italy. <www.fao.org/docrep/005/y7352e/y7352e00>. ISBN 92-5-104815-0.

Klieh, U., Bala Ravi, S., Dayakar Rao, B., Yoganand, B., 2000. Industrial Utilization of sorghum in India. International Crop Research Institute for Semi-Arid Tropics, Hyderabad, India.

Somani R.B., A. Almodares, R.B. Pandrangi and A. Sepahi, 1997. Potential of sweet sorghum in India and Iran for alcohol and sweetners production. First International Sweet sorghum Conference, China.

CHAPTER

1

Sorghum Grain Quality

C.V. Ratnavathi and V.V. Komala

ICAR-Indian Institute of Millets Research, Rajendranagar, Hyderabad, India

OUTLINE

Sorghum Biochemistry: An Industrial Perspective.
DOI: http://dx.doi.org/10.1016/B978-0-12-803157-5.00001-0

Abstract

Sorghum has been an important staple in the semiarid tropics of Africa and Asia for centuries. Sorghum is also a rich source of various phytochemicals including tannins, phenolic acids, anthocyanins, phytosterols, and policosanols apart from proximate composition, vitamins, and minerals. These phytochemicals have significant impact on human health. Sorghum is consumed as food in various forms as roti, tortilla, injera, kisra, tuwo, ugali, bogobe, sankati, ambali, edi, couscous, wowoto, noodles, soru, burkutu, busa, ting, and obhshera in different parts of the world. Since sorghum flour is a gluten-free food, it is a safe alternative for those with celiac disease. Apart from the traditional products like (bhakri, bhatwadi, papad) non traditional foods like popped gains, kurdai, biscuits, flakes, upma, idli, dosa, utappa, chiwada, chakli, ambali, shankarpale, and cakes are also prepared from sorghum. Alternative uses of sorghum include in nonfood sectors for the production of commercially valued products.

1.1 INTRODUCTION

Sorghum (*Sorghum bicolor* (L.) Moench) is a genus of numerous species of grasses, one of which is grown for grain and many of which are used as fodder plants, either cultivated or as part of pasture. The plants are cultivated in warmer climates worldwide. Species are native to tropical and subtropical regions of all continents in addition to the southwest Pacific and Australasia. Sorghum is in the subfamily *Panicoideae* and the tribe of *Andropogoneae* (the tribe of big bluestem and sugar cane). Sorghum is divided in 29 species and two hybrids as follows:

1.1.1 Sorghum Species

Sorghum almum; Sorghum amplum; Sorghum angustum; Sorghum arundinaceum; S. bicolor—cultivated sorghum, often individually called sorghum. Also known as durra, jowari, or milo; *S. bicolor* subsp. *drummondii*—Sudan grass; *Sorghum brachypodum; Sorghum bulbosum; Sorghum burmahicum; Sorghum ecarinatum; Sorghum exstans; Sorghum grande; Sorghum halepense*—Johnson grass; *Sorghum interjectum; Sorghum intrans; Sorghum laxiflorum; Sorghum leiocladum; Sorghum macrospermum; Sorghum matarankense; Sorghum nitidum; Sorghum plumosum; Sorghum propinquum; Sorghum purpureosericeum; Sorghum stipoideum; Sorghum timorense; Sorghum trichocladum; Sorghum versicolor; Sorghum verticiliflorum; Sorghum vulgare var. technicum*—broomcorn and two hybrids *Sorghum × almum; Sorghum × drummondii.*

One species, *S. bicolor*, is an important world crop, used for food (as grain and in sorghum syrup or "sorghum molasses"), fodder, the production of alcoholic beverages, and biofuels. Most varieties are drought- and heat-tolerant and are especially important in arid regions, where the grain is one of the staples for poor and rural people. These varieties form important components of pastures in many tropical regions. *Sorghum bicolor* is an important food crop in Africa, Central America, and South Asia and is the fifth most important cereal crop grown in the world.

Sorghum (*S. bicolor* (L.) Moench) is one of the major cereal crops among all the millets grown mostly in arid land. It is grown as a rainfed crop in both *Kharif* (mansoon) and *Rabi* (winter). It is mainly grown in the Deccan plateau, Central and Western India apart from a few patches in northern India. Almost the entire grain produced is used for human consumption in India. It is nutritionally superior to other fine cereals such as rice and wheat, and hence, it is known as a nutritious cereal. Nutritionally sorghum grain contains 4.4 to 21.1% protein, 2.1 to 7.6% fat, 1.0 to 3.4% crude fiber, 57.0 to 80.6% total carbohydrates, 55.6 to 75.2% starch, and 1.3 to 3.5% total minerals (ash). Sorghum also provides 350 Kcal energy, calcium, phosphorus, potassium, carotene, and thiamin as well as antioxidants through phenolics and various types of tannins. It is mainly consumed as *Bhakri* (*roti*), that is, unleavened pancake in various states of India. Apart from the traditional products like *bhakari, bhatwadi, papad, popped grains, kurdai, high fiber cookies, biscuits, flakes, thalipeeth, upama, rawa idali, dosa, uttappa, chiwada, chakali, papadi, ambil, shankarpale, cookies, and cakes* are also prepared from sorghum and consumed as snack food items. Sorghum grains are not only a good source of nutrients but also contain special constituents such as phyto-chemicals and dietary fiber, as well as resistant starch, which are essential to human nutrition. However, the grain sorghum consumption has remained restricted to the poorer sections of society due to poor nutritional quality of grain and inferior quality of the products, such as Bhakari, as well as the very low price of this food item as compared to the other cereal grains such as wheat, rice, and other millets.

Sorghum is of African origin. A large variety of wild and cultivated sorghum are grown in the tropics and subtropics of the world. In India, sorghum constitutes an important article of food, after rice and wheat. The sorghum grain is small and rounded, varying in color from off-white to white to varying shades of red, yellow, or brown. The grain size varies, the weight ranging from 7.0 to 61 g/1000 grains, with most sorghums weighing 20–30 g/1000 grains. The chemical composition of grain sorghum is similar to that of maize. Generally, sorghum has more protein than maize, a lower fat content, and about the same amount and proportions of carbohydrate components. The proximate analysis of Indian sorghum grain indicates the percentages of moisture 11.9, protein 10.4, fat 1.9, fiber 1.6, carbohydrates 72.6, and minerals 1.6 respectively; minerals present in the grain are calcium, magnesium, potassium, and iron.

In comparison with maize, sorghum grain contains approximately the same quantities of riboflavin and pyridoxine but more pantothenic acid, nicotinic acid, and biotin. Nicotinic acid occurs in the grain in an available form. Starch is the major carbohydrate of the grain. The other carbohydrates present are simple sugars, cellulose, and hemicelluloses. The amylose content of starch varies from 21 to 28%. Starch from waxy varieties contains little amylose. Both waxy and regular starches contain free sugars up to 1.2%—sucrose being the major constituent (0.85%) followed by glucose (0.09%), fructose (0.09%), maltose, and stachyose. Sorghum grain contains no detectable amount of glucoside, but on germination a three day-old seedling gives 3.5% dhurrin, which leads to the poisoning of animals consuming such sorghum seedlings.

The percentage of different protein fractions to the total protein of sorghum grown in India is albumin 5%, globulin 6.3%, prolamin 46.4%, and glutelin 30.4%. Prolamin and glutelin are principally present in the endosperm. Amino acid analysis of various protein fractions shows that there is better distribution of all essential amino acids in globulins than in prolamins. Sorghum protein is superior to wheat protein in biological value and digestibility. A vegetarian diet based on some varieties of sorghum is somewhat better than a rice-based diet. Sorghum lipids mostly consist of triglycerides, which are rich in the unsaturated fatty acids, oleic and linoleic, their percentage being 33 and 47%, respectively.

Currently, the demand of sorghum for animal feed is the driving force for sorghum production around the world (FAO, 1995). Sorghum grain has been regarded as a nutritionally inferior grain compared with other cereals, which may be one reason why its demand as human food has declined over the years; other factors that contribute to lower human consumption include cheap imports of corn to countries where sorghum is the staple food, complex processing techniques, and long preparation time for sorghum products. Even in the feed industry, sorghum is confronted with the same dilemma: somewhat lower feed value compared with other feed grains, primarily corn. Although slightly less nutritious, sorghum is still widely used as an animal feed due to its lower cost.

Sorghum is an important crop for food and fodder in the semiarid tropic regions (14–24°N latitude and 70–82° longitude) in India. It is mainly cultivated in the states of Maharashtra, Karnataka, Madhya Pradesh, Andhra Pradesh, Rajasthan, Gujarat, Tamil Nadu, and Uttar Pradesh and is mainly grown in Kharif (rainy), Rabi (post rainy), and summer seasons. India is the second largest producer of sorghum worldwide and has the largest area of sorghum under cultivation. Sorghum is grown both as a rainy season (June–October) and post rainy season (September–January) crop. The rainy season sorghum is often damaged due to deterioration and fetches lower prices and hence is less profitable. This molded grain has promoted its use in the nonfood and Industrial sector.

The superiority of sorghum over other crops lies in its ability to produce grain with a relatively limited supply of water. The crop is unique in that it can remain dormant during stress periods and renew growth when conditions are more favorable. Currently sorghum is a dietary staple for more than 500 million people in over 30 countries. On a worldwide basis, an average of 50% of the sorghum grown is used for human consumption. In addition to human consumption, sorghum is a popular component in animal feed, for both roughage and grain, because it is much less expensive to produce than other crops such as corn. Sorghum is also used in industrial waxes, packing material, wallboard, ethanol, beverages, and brewing. Despite its worldwide economic importance, sorghum is less well characterized at the genetic and molecular levels than many other cereal crops. Sorghum is, however, considered a good model species for drought tolerance, especially in grasses, and many researchers are beginning to realize its importance to dry-land agriculture. Today's agriculturalists are faced with a unique challenge of how to cope with the loss of arable land and plateaus in yield increases in the face of a growing world population. For at least the past century, the population increase has been exponential. From the present population of approximately 6.5 billion, the world population is projected to reach 8.3 billion by 2025; to meet the projected food demand for this increasing population, researchers and producers agree on the necessity of an increase in the average yield of cereal crops.

At present most of the sorghum produced in India is consumed as a human food in the form of Roti or Chapatti (unleavened flat bread). On the other hand, a different sorghum called sweet sorghum with its juicy sweet stalk proved its potential as a bioenergy crop. Sorghum is known to have a highest productivity rate of about 50 g dry matter/m^2/day in a number of locations around the world, and also it has been reported to surpass sugarcane, another C4 plant. Sweet sorghum also produces panicle with good quality grain. Grain sorghum and sweet sorghum offer themselves as a very good raw material for alternate uses.

1.1.2 Grain Quality

Grain quality is a nebulous term that means different things to different people. Grain quality largely depends on the grain type and its end use. It includes a range of properties that can be defined in terms of *physical* (moisture content, kernel size), *sanitary* (fungi and mycotoxin count), and *intrinsic* (fat content, protein content, hardness, starch content) quality characteristics. The quality properties of a grain are affected by its genetic traits, the growing period, timing of harvest, grain harvesting and handling equipment, drying system, storage management practices, and transportation procedures.

1.1.3 Sorghum Kernel Structure

Sorghum kernels are generally spherical in shape and come in different sizes and colors. Typical sorghum seeds are usually 4 mm long, 2 mm wide, and 2.5 mm thick with color ranging from black, red, purple, brown, yellow to white. Fig. 1.1 illustrates the sorghum grain structure (Rooney and Miller, 1982). The basic anatomical components are pericarp (outer layer), germ (embryo), and endosperm (storage tissue). The distribution of these components differs among varieties and environment, with an average of 8, 82, and 10% for pericarp, endosperm, and germ, respectively (Hubbard et al., 1950). The pericarp has a thickness from 8 to 160 μm and is divided into three layers: epicarp, mesocarp, and endocarp (Earp and Rooney, 1982). Immediately underneath the pericarp is a layer known as the seed coat or testa. Like the pericarp, the thickness of the testa also varies from 8 to 40 μm depending on sorghum genotype (Earp and Rooney, 1982). This layer is

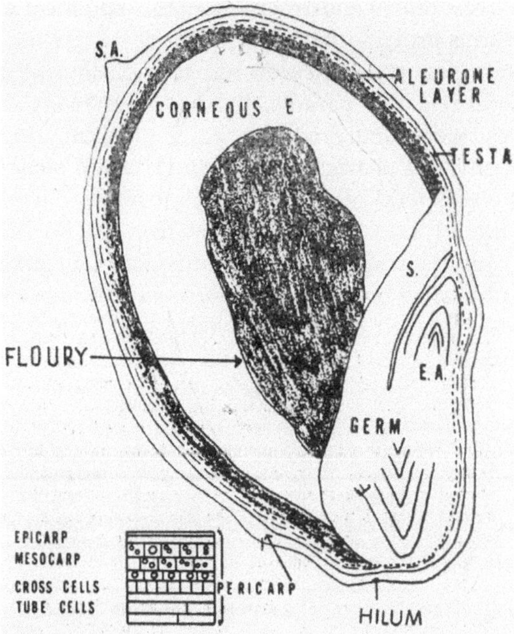

FIGURE 1.1 Structure of the grain sorghum showing the pericarp, endosperm (aleurone layer, corneous and floury), and germ (scutellum (S) and embryonic axis (EA)).

responsible for storing tannins and pigments in some of the sorghum genotypes. Tannins are polyphenolic compounds that negatively affect the nutritional quality of sorghum by binding and precipitating proteins, thus rendering the protein indigestible (Butler et al., 1984). In addition, tannins also can interact with various digestive enzymes and reduce the activity of these enzymes (Maxson et al., 1973; Griffiths and Mosely, 1980; Butler et al., 1984).

Endosperm, the major storage tissue, is composed of the aleurone layer, peripheral endosperm, vitreous (hard) endosperm, and floury (soft) endosperm. The aleurone layer is a single layer of cells that is located right under the seed coat. This layer is rich in proteins and enzymes, oils, vitamin B complexes, and minerals. Rooney and Miller (1982) provided a detailed picture and description of the sorghum endosperm. The location of peripheral endosperm is not well defined but is usually located right under the aleurone layer with thickness between two to six block cells. Right under the peripheral endosperm is the vitreous endosperm followed by the floury endosperm. Vitreous and floury endosperm can be distinguished from one another based on the packing and shape of the starch granules, and the distribution of the protein matrix. Vitreous endosperm contains polygonal-shaped starch granules surrounded by a continuous protein matrix, while floury endosperm contains spherical starch granules with a discontinuous protein matrix.

Indian sorghum genotypes include 160 germplasm lines, 200 elite SPV genotypes, and 60 released parents, hybrids, and many other varieties. These genotypes were analyzed for both physical, and biochemical characters. The physical characters include (1) grain size, (2) 100 grains weight, and (3) endosperm texture. The proposed biochemical characters to be evaluated are (1) %starch, (2) %protein, (3) %fat, (4) %in vitro protein digestibility, (5) phytic acid content, (6) amylose, (7) starch digestibility, and (8) dietary fiber. The progress of analysis of the lines and distribution of the genotypes analyzed for each physical and chemical parameter is presented in Table 1.1.

TABLE 1.1 Summary Table Physico-Chemical Parameters Analyzed

S. no.	Parameters	Genotypes analyzed	Range	Genotypes distribution
PHYSICAL PARAMETERS				
1.	Grain size	430	2 mm to 4 mm	3.35 mm (272), 4 mm (20), below 3.35 mm (138)
2.	Grain weight	430	1.22 g to 6 g	3 g above (207), 3.9 g above (70), below 3 g (153)
3.	Endosperm texture	430	1 to 9	5 (218), 3 (108), 7 (81), 9 (9), 1 (2)
BIOCHEMICAL (INTRINSIC) PARAMETERS				
1.	Protein (%)	284	7.18 to 18.93	Between 9% to 12% (160), Between 12% to 15% (62), Above 15% (12)
2.	Fat (%)	70	1.94 to 4.27%	Between 2.5% to 3.5% (56), Above 3.5% (3)
3.	Starch (%)	268	30 to 80.75%	Above 68% (32)
4.	Protein digestibility (%)	108	20.37 to 70.57%	Above 50% (24), Above 60% (13), Above 70% (17)
5.	Phytic acid (mg/100 g)	59	2.4 to 6.70	Between 3 mg/100 g to 5 mg/100 g (42), Above 5 mg/100 g (14)
6.	Amylose (%)	180	8.35 to 29.05	Below 10% (11), below 20% (56), above 20% (113)
7.	Total phenols (mg/gm)	100	0.5-3	Below 1 mg/g (25), Between 1 mg/g to 2 mg/g (63), Above 2 mg/g (12)

TABLE 1.2 Genotypes Identified With Floury Endosperm

S. no.	Genotype	Grain size (μm)	Grain weight (g)	Endosperm texture	Suitable product
1.	GGUB-40	3.35	3.87	7	Flour
2.	ELG-2	3.35	3.53	7	Flour
3.	EC-4	3.35	3.45	7	Flour
4.	E-71	3.35	3.53	7	Flour
5.	EP-2	3.35	3.99	7	Flour
6.	EP-3	3.35	3.54	7	Flour
7.	EP-4	3.35	3.97	7	Flour
8.	EP-29	3.35	4.25	9	Flour
9.	EP-30	3.35	3.64	7	Flour
10.	EP-38	3.35	3.67	7	Flour
11.	EP-39	3.35	3.88	7	Flour
12.	EP-44	3.35	4.07	7	Flour
13.	SPV-1259	3.35	3.64	7	Flour
14.	SPV-1626	3.35	4.22	7	Flour
15.	SPV-1759	3.35	3.63	7	Flour
16.	SPV-1802	3.35	4.01	7	Flour
17.	DSV-5	3.35	3.35	9	Flour
18.	CSV-216 R	3.35	5.03	7	Flour

1.1.4 Physical Characters

The physical characters include (1) grain size, (2) 100 grains weight, (3) endosperm texture, and (4) grain hardness.

Out of 420 genotypes analyzed, 18 superior genotypes for grain size and weight and floury endosperm were selected (Table 1.2). These genotypes include 12 germplasm lines, four SPV lines, and two cultivars. These genotypes yield flour mostly suited for roti and other bakery products like cake and biscuit. Eleven genotypes were superior in grain size and weight having corneous endosperm, and these genotypes were suitable for coarse rawa and poha preparation. Among these, three lines are elite SPV lines and eight are germplasm lines (Table 1.3). Twenty-five superior genotypes with grain size, grain weight, and intermediary endosperm were identified for the preparation of fine semolina, which is further used in preparation of cold extruded product

TABLE 1.3 Genotypes With Corneous Endosperm

S. no.	Genotype	Grain size (μm)	Grain weight (g)	Endosperm texture	Suitable product
1.	EB-13	3.35	3.43	3	Coarse rawa
2.	GGUB-5	3.35	3.39	3	Coarse rawa
3.	GGUB-9	3.35	3.94	3	Coarse rawa
4.	EC-19	3.35	3.51	3	Coarse rawa
5.	EP-6	4.00	5.02	3	Coarse rawa
6.	EP-36	3.35	3.86	3	Coarse rawa
7.	EP-101	3.35	3.96	3	Coarse rawa
8.	EP-109	3.35	3.80	3	Coarse rawa
9.	SPV-1253	3.35	3.44	3	Coarse rawa
10.	SPV-1383	3.35	4.21	3	Coarse rawa
11.	SPV-1808	3.35	3.79	3	Coarse rawa

(pasta and vermicelli). These genotypes include 13 genotypes from germplasm pool and seven from SPV elite genotypes, and five genotypes are released cultivars. The SPV lines are SPV 727, SPV 1155, SPV 1763, SPV 1764, SPV 1766, SPV 1787, and SPV 1809. The released genotypes are PKV kranti, CSV 18 R, CSV 22R, Dagadi, and M 35-1 (Table 1.4).

1.1.5 Biochemical Parameters

The proposed biochemical characters to be evaluated are (1) percent protein, (2) percent starch, (3) percent fat, (4) percent in vitro protein digestibility, (5) phytic acid content, (6) amylose, (7) starch digestibility, and (8) dietary fiber.

1.1.6 Protein Estimation

Protein content was estimated by colorimetric determination of total Kjeldhal nitrogen using salicylate (Willis et al., 1996).

Protein content present in the sample was digested with acid and converted to ammonia also called as Kjeldhal nitrogen. Salicylate nitroprusside reagent reacts with hypochlorite reagent to form a green colored complex, and the absorbance was read at 685 nm. The amount of protein was calculated by multiplying the amount of nitrogen obtains with a factor 6.25 ($N \times 6.25$).

TABLE 1.4 Suitable Genotypes Identified for Rawa

S. no.	Genotype	Grain size (μm)	Grain weight (g)	Endosperm texture	Suitable product
1.	EC-33	3.35	4.06	5	Rawa & Flakes
2.	EP-8	3.35	4.31	5	Rawa & Flakes
3.	EP-15	4.00	4.63	5	Rawa & Flakes
4.	EP-18	3.35	4.30	5	Rawa & Flakes
5.	EP-19	3.35	4.18	5	Rawa & Flakes
6.	EP-20	4.00	4.61	5	Rawa & Flakes
7.	EP-21	3.35	4.23	5	Rawa & Flakes
8.	EP-25	4.00	4.21	5	Rawa & Flakes
9.	EP-27	3.35	4.06	5	Rawa & Flakes
10.	EP-31	4.00	4.29	5	Rawa & Flakes
11.	EP-98	3.35	4.05	5	Rawa & Flakes
12.	EP-99	3.35	5.99	5	Rawa & Flakes
13.	EP-100	4.00	4.73	5	Rawa & Flakes
14.	SPV-727	3.35	3.95	5	Rawa & Flakes
15.	SPV-1155	3.35	4.03	5	Rawa & Flakes
16.	SPV-1763	3.35	4.01	5	Rawa & Flakes
17.	SPV-1764	3.35	3.90	5	Rawa & Flakes
18.	SPV-1766	3.35	4.30	5	Rawa & Flakes
19.	SPV-1787	3.35	3.90	5	Rawa & Flakes
20.	SPV-1809	3.35	4.07	5	Rawa & Flakes
21.	CSV-22R	3.35	5.01	5	Rawa & Flakes
22.	PKV Kranti	3.35	3.96	5	Rawa & Flakes
23.	DAGADI	3.35	4.38	5	Rawa & Flakes
24.	CSV-18 R	3.35	4.94	5	Rawa & Flakes
25.	M35-1	3.35	4.26	5	Rawa & Flakes

Among 284 Sorghum genotypes, the total protein content of the sorghum ranged from 7.18% for line SPV-1288 to 18.93% for line SPV-1293. Out of 284 genotypes, 160 genotypes ranged between 9 to 12%, 62 genotypes ranged between 12 to 15%, and 12 genotypes were above

15% and the genotypes are listed here: SPV-621, SPV-1259, SPV-1293, SPV-1758, IS 602, IS 348, IS8348, IS33648, IS23992, IS20740, IS12937, IS 9108, and IS4060 (Table 1.1).

1.1.7 Protein Body and Protein Matrix Characteristics of Sorghum

Sorghum storage proteins are accumulated inside organelles known as protein bodies. Sorghum protein bodies are circular in shape with a diameter of 0.3 to 3.0 μm depending on their location within the kernel (Rooney and Miller, 1982; Taylor et al., 1985). The protein bodies in the peripheral and vitreous endosperm average 0.3 to 3.0 μm in diameter, whereas the ones in the floury endosperm range from 0.3 to 1.5 μm in diameter (Rooney and Miller, 1982). It has been observed that the protein bodies are encapsulated by protein matrix. This matrix protein consists mainly of glutelins and small amounts of albumins or globulins (Seckinger and Wolf, 1973). Kafirins comprise by far the majority of the protein inside protein bodies accompanied by a small amount of glutelins and minute amounts of albumins and globulins (Taylor et al., 1984). Using immunocytochemistry and transmission electron microscopy, it was shown that β- and γ-kafirins are located on the core and the periphery of the protein bodies, while the α-kafirin, which makes up about 80% of the total kafirin, is located in the interior (Shull et al., 1992). This distribution is similar to that of zein distribution in corn protein bodies.

Recently, Oria et al. (2000) identified a unique structure of protein bodies from a high protein digestibility sorghum mutant genotype. These protein bodies were irregular in shape and had crevices that reached the center of the structure forming lobes. In the mutant, α-kafirin was still the major protein of the protein bodies, with β-kafirin mainly found distributed in the lobes and γ-kafirin concentrated at the base of the crevices. This structure resulted in the rapidly digesting α-kafirin protein exposed to proteases and is the basis for the high protein digestibility of this genotype.

1.1.8 Digestibility of Uncooked Sorghum Proteins

It is well accepted that sorghum has lower protein digestibility compared to other cereal grains (MacLean et al., 1981; Hamaker et al., 1987). The low protein digestibility characteristic is more prominent in cooked than in uncooked sorghum (Axtell et al., 1981; Hamaker et al., 1987). Although protein digestibility of uncooked sorghum is only slightly lower than corn, it is still considered to affect its feed grain value especially for nonruminant animals. Ruminants, likes cattle and sheep, have microfloras in their rumen that have the ability to convert poor quality protein and nonprotein nitrogen into quality protein. As for nonruminants, like pigs and poultry, the quality and quantity of the protein from cereal-based diets is important because it contributes approximately 30–50% of the dietary protein required by the animals (Bramel-Cox et al., 1995).

One of the first factors identified as a cause of low protein digestibility in sorghum was polyphenolic compounds known as tannins. Tannins can bind proteins and render them indigestible by proteases (Butler et al., 1984). Tannins also interact with protein digesting enzymes (proteases) and inhibit their activity (Weaver et al., 1998), indicating that tannin is not the only factor that affects protein digestibility of sorghum. Hamaker et al. (1987) proposed that the comparably low protein digestibility of low tannin sorghum genotypes is due to the composition and structure of protein bodies and accessibility to interior proteins by digestive proteases. Additional detailed studies on protein bodies indicated that the γ- and possibly β-kafirins, which are high in cysteine, form disulfide bonded polymers around the periphery of the protein body, thus limiting the access of the enzyme to the α-kafirin in the interior of the protein bodies (Oria et al., 1995a,b; Weaver et al., 1998). This theory is supported by recent findings by Oria et al. (2000) for the mutant genotype with high protein digestibility. In this mutant genotype, the exposed α-kafirin, the major storage protein, and the position of the β- and γ-kafirins at the base of the highly folded structure allowed proteases to reach and hydrolyze α-kafirin directly, thus giving higher protein digestibility value.

1.1.9 Determination of Protein Digestibility

The samples are estimated for protein digestibility by turbidity assay method (Hamaker and Bugusu, 2003). Digestion of proteins present in sorghum flour was carried out by pepsin followed by extraction and removal of digested protein with phosphate buffer (pH 7, 0.1 M). Undigested protein present in the sample was extracted using borate buffer (pH 10, 0.0125 M) containing 1% SDS and 2% mercaptoethanol for 1 h. Reaction of trichloro acetic acid (72%) with extracted undigested protein results in turbidity; the turbidity of the sample was measured at 562 nm. Corresponding absorbance indirectly gives the percentage of digested protein present in the sample.

1.1.10 Variation in Protein Digestibility

The samples used in this study were specifically selected to show a wide range of protein digestibility. Table 1.1 lists the percentage protein digested after 1 h pepsin treatments. Protein digestibility content varied from 3.37% for line SPV37 to 70.57% for the line SPV1775. It was speculated that the amount of sulfur-containing amino acids and degree of disulfide-mediated polymerization at the periphery of protein bodies differed among genotypes, thus affecting the digestion rate based on the extent of disulfide-bonded protein present.

Among 108 genotypes, the protein digestibility content of the sorghum ranged from 20.37% for line SPV-265 to 70.57% for the line SPV1775. Out of 108 genotypes, 24 genotypes ranged above 50%, 13 genotypes ranged above 60%, and 17 genotypes ranged above 70% content of protein digestibility (Table 1.1).

1.1.11 Starch

Starch is the storage form of energy in cereals and usually makes up between 60 to 75% of the total weight of cereal grains. It is found in plants in the form of granules. Starch granules are made up of two distinct components of glucose polymers: amylose and amylopectin.

1.1.12 Amylose

Before 1950, amylose was believed to be a completely linear polymer of D-glucopyranosyl units. However, in the early 1950s, a study (Peat et al., 1952) showed that amylose is not completely hydrolyzed into maltose when treated with crystalline β-amylase, indicating that amylose is not entirely linear. Banks and Greenwood (1966) concluded the presence of α-(1→6) side-chains with considerable length using bacterial pullulanase as the debranching enzyme. In a more recent study, Cura et al. (1995) confirmed that the branching linkages in amylose are indeed that of α-(1→6). Takeda et al. (1987) found, depending on the source, three to 10 branch points per amylose molecule. It is now widely accepted that amylose consists of the linear polymer of (1→4) linked α-D-glucopyranosyl units with lightly branched side chains joined by (1→6)-α-linkage. The side chains are either very long or very short and are located far from one another. Since the building block of amylose is the chair ($^4C_1$) conformer of the glucose molecule, a helical twist is imparted on amylose. In the interior of the helix, starch hydroxyl groups are hydrogen bound to each other, which both stabilizes the helix and makes it relatively hydrophobic. The molecular weight of amylose depends on the botanical source of the starch and the extraction methods (Ong et al., 1994). The average molecular weight of amylose is about 1.3 × 106 to 9 × 106 Daltons (Ong et al., 1994). By making several assumptions, Buléon et al. (1998) calculated the number of amylose chains inside a single granule to be approximately 1.8 × 109. Although the structure of amylose is well established, its location is still not well understood. Two theories have emerged: one states that amylose is found in bundles between amylopectin molecules (Nikuni, 1978; Blanshard, 1986; Zobel, 1992), while the other theory declares that amylose is arbitrarily scattered among amylopectin molecules (Jane et al., 1992; Kasemsuwan and Jane, 1994).

1.1.13 Amylose Estimation

The samples were estimated for amylose by the assay method reported by Williams et al. (1958). Amylose in flour is released by treatment with alkali. By the addition of Tri-iodide ion, amylose produces a blue color and the absorbance of the blue color is measured.

Among 180 genotypes, the total amylose content of the sorghum ranged from 8.35% for line ICSV 745 to 29.05% for line SPV 1730. Out of 180 genotypes, 11 genotypes were below 10%, 56 genotypes were below 20%, and 113 genotypes ranged above 20% of amylose content (Table 1.1).

1.1.14 Amylopectin

Amylopectin, the counterpart of amylose, is the major component of starch by weight and one of the largest molecules found in nature. It also is composed of linear chains of (1→4) linked α-D-glucopyranosyl units but with a much greater extent of α-(1→6) branching than amylose.

These branch points make up approximately 4 to 6% of total linkages (Hood, 1982). Peat et al. (1952) proposed that amylopectin consists of three different types of chains. The A chains, also known as the unbranched chains, are the linear segments joined to other chains by a single (1→6)-α-linkage. The B chains are those connected to other chains via α-(1→6) linkages and also carry one or more A or B chains attached to them. The C chain is the single, central chain that carries the only reducing group of the amylopectin molecule.

The linear glucopyranosyl units are connected by α-(1→4) linkages, and the branch point is connected by α-(1→6) linkages. Various models have been proposed as the structure of amylopectin since 1940, but the widely accepted model used until today is the "cluster" model proposed by French (1972) and modified by Robin et al. (1974). A further refinement of the "cluster" model of amylopectin structure with its chains arrangement was proposed by Hizukuri (1986). Lineback (1984) proposed that amylopectin is the molecule responsible for the crystallinity of the starch granule based on the fact that waxy starch, which is mainly amylopectin, showed the same degree of crystallinity as normal starch. Physicochemical studies showed that the relative molecular weight (MW) of amylopectin, depending on the sources, is between 10^7 and 5×10^8 daltons. Buléon et al. (1998) estimated the number of amylopectin chains in a single starch granule to be 5.4×10^7.

1.1.15 Starch Granule Structure

The size, shape, and composition of starch granules differ depending on the botanical sources and the maturity of the plant. Due to its complex organization, the accurate internal structure of starch granule has not yet been identified. However, decades of various investigations have provided scientists with enough information to provide a scientific speculation of the granule organization. It now is believed that starch granules are made up of alternating layers of crystalline and semicrystalline regions around a nucleus called the hilum (French, 1984; Gallant et al., 1992). According to Zobel (1988), native starch granules have crystalline regions between 15 and 45% of the total starch. Cereal starches give an A type crystalline pattern when observed using X-ray diffraction, which means the crystallites are densely packed (Imberty et al., 1987).

Several studies looking at the crystalline layers have suggested that these layers consist of alternating amorphous and crystalline lamellae with thickness of approximately 10 nm (Hizukuri, 1986; Oostergetel and van Bruggen, 1989; Jenkins et al., 1993). These crystalline and amorphous lamellae represent the side chain clusters and branching portions, respectively, of the amylopectin model suggested by French (1972). Gallant et al. (1997) suggested a more detailed description of the arrangement of the amylopectin side chain cluster. Each cluster can carry from 9 to 17 double helical side chains, and the crystalline lamellae are not continuous, indicated by the amorphous regions between clusters.

Studies using scanning electron microscopy, transmission electron microscopy, and atomic force microscopy indicated that the lamellae layers of the amylopectin are arranged into larger, almost spherical structure called "blocklets." The blocklets' diameter ranges from 20 to 500 nm based on their location in the granule and the botanical source. The blocklets found in the semicrystalline layers are usually much smaller than the ones found in the crystalline layer, which indicates a less crystalline organization in the former (Gallant et al., 1992).

It is suggested that the lower level of crystallinity in the semicrystalline region is due to the interaction of the amylopectin with amylose (Gallant et al., 1997).

Grain source starches have been shown to have smaller granules compared to tuber starches (Lineback, 1984). Normal sorghum starch contains, on average, 25 and 75% of amylose and amylopectin, respectively (Miller and Burns, 1970; Ring et al., 1982). Waxy varieties of sorghum are essentially 100% amylopectin, but some can contain amylose up to 5.8% (Miller and Burns, 1970). Heterowaxy sorghum lines, which contain 25% waxy and 75% normal phenotype seeds, can have amylose ranging from 15 up to 22% (Ring et al., 1982). Miller and Burns (1970) reported the size of sorghum granules ranging from 4 to 25 microns with the majority being 15 to 20 microns. In addition to their sizes, the shapes of sorghum starch granules vary from spherical to polygonal based on their location within the sorghum.

In sorghum, nearly all starch granules are found in the endosperm fraction (some sorghum phenotypes also have a starchy mesocarp). The vitreous endosperm portion has a continuous protein matrix that engulfs starch granules and protein bodies. The starch granules in the vitreous regions are generally polygonal in shape and contain dents due to the tight packing with the protein bodies and among the starch granules themselves (Rooney and Miller, 1982). Hoseney et al. (1974) believed that this arrangement is the reason for the hardness and translucent appearance of the vitreous endosperm. On the other hand, floury endosperm usually contains round, large, and loosely packed starch granules with a lot of intergranular air spaces within the thin layer of discontinuous protein matrices and occasional protein bodies. As light passes through the floury endosperm, the air spaces diffract and diffuse the light to give the opaque or chalky appearance (Hoseney et al., 1974). The endosperm texture is determined often from the ratio between the vitreous to the floury endosperm.

The physical quality of grain deteriorates due to mold; the chemical quality does not always get affected. The chemical analysis of %starch and protein for the advanced sorghum varieties and hybrids was undertaken and results are presented below. The grain samples from Akola were collected during different years for the study of grain quality.

1.1.16 Starch Estimation

Starch content was estimated by colorimetric determination of sugars obtained on hydrolysis of starch by glycosidase enzyme (Southgate, 1976; Dubois et al., 1956).

Starch comprises of amylase and amylopectin. On hydrolysis by glycosidase enzyme, starch is converted into maltose and finally to glycols. The sugars thus obtained were analyzed quantitatively using phenol–sulphuric acid reagent.

Among 268 genotypes, the total starch content of the sorghum ranged from 30% for line SPV-1284 to 80.75% for line IMS 9B. Out of 268 genotypes, 32 genotypes ranged above 68% of starch content (Table 1.1).

1.1.17 Starch Digestibility of Raw Sorghum Grain

Based on in vivo feeding studies and in vitro enzymatic studies, it is clear that starch in sorghum is poorly digested as compared to corn and also is digested at different rates among cultivars (Hibberd et al., 1982 (as indicated by difference in dry matter disappearance and gas production); Elmalik et al., 1986; Knabe, 1990 (as indicated by difference in feed efficiency); Wester et al., 1992). Various factors have been suggested as the cause of the variation in starch digestibility. Tannin was one of the first factors identified as the cause of the lower starch digestibility. Numerous studies have shown that various animal species fed sorghum grain with comparably high concentration of tannin performed more poorly than those consuming low tannin sorghum (Maxson et al., 1973; Cousins et al., 1981; Elkin et al., 1996; Pour-Reza and Edriss, 1997; King et al., 2000). In addition to binding protein and reducing its digestibility, tannin also has the ability to bind to starch granules (Davis and Hoseney, 1979). Davis and Hoseney (1979) suggested that starch-tannin complexes are resistant to amylases. Tannin can also interact with digestive enzymes including amylases, thus inhibiting the activity of the amylase in breaking down starch components into simple sugars for energy use (Maxson et al., 1973; Longstaff and McNab, 1991). Because some sorghum genotypes carry higher tannin than other cereals, the lower starch digestibility in high tannin sorghum is understandable. However, sorghum cultivars with similar or no tannin contents also exhibit significant differences in starch digestibility (Elmalik et al., 1986; Knabe, 1990 (as indicated by differences in fed efficiency); Elkin et al., 1996 (as indicated by differences in total metabolizable energy)). These results suggest that other factors besides tannin contribute to the inhibition of starch digestion.

1.1.18 Starch Digestibility

Endosperm characteristics of the sorghum kernel have been pointed out as one of the responsible factors for digestibility. Various researchers have found that sorghum cultivars with floury endosperm are more digestible than the ones with vitreous endosperm (Miller et al., 1972; Elmalik et al., 1986). Researchers are now looking at the endosperm texture in more detail to try to identify the component in the endosperm that causes the difference in starch digestibility. One prevailing theory is that arrangement, composition, and digestibility of the protein around starch granules play a major role in starch digestibility (Rooney and Pflugfelder, 1986). As described previously, sorghum starch granules are surrounded by protein bodies and embedded in a protein matrix to different degrees depending on their location. Rooney and Pflugfelder (1986) suggested that endosperm sections with dense packing are less accessible to starch degrading enzymes (amylases). This theory supports the abovementioned endosperm texture studies because the vitreous endosperm is usually densely packed, thus reducing the accessibility of enzyme, while floury endosperms are more loosely packed and more easily accessible to enzymes. Lichtenwalner et al. (1978) showed that samples treated with the protease enzyme, Pronase, had a significant increase in starch digestibility. Based on these observations, it is clear that getting through the protein barrier is a critical step for enzymes to digest starch. As explained earlier, protein digestibility varies between cereals and within sorghum cultivars due to the protein body composition and arrangement. One hypothesis tested in this thesis is that differences in protein digestibility might be the basis for the significant variation in starch digestibility observed in the animal feeding studies and could provide a strategy toward developing high digestibility sorghum genotypes.

Waxy sorghum genotypes, which carry a higher amount of amylopectin fraction of starch, have been shown to have higher nutritive value than nonwaxy genotypes. Feeding studies done on steers and rats have shown that animals fed waxy sorghum gain better feed efficiency compared to those fed nonwaxy sorghum (Sherrod et al., 1969; Elmalik et al., 1986). In vitro studies also showed that waxy sorghum has higher starch digestibility than nonwaxy sorghum (Lichtenwalner et al., 1978; Wester et al., 1992). It was proposed that starch granules with higher amounts of amylopectin were more susceptible to enzymatic degradation (Sherrod et al., 1969; Sullins and Rooney, 1974). Another theory attributed the higher nutritive value in waxy sorghum to the smaller proportion of peripheral endosperm, which allows for easier access of starch-digesting enzymes to starch granules (Sullins and Rooney, 1974).

Even though several factors have been proposed, the reason for low starch digestibility in sorghum is not well understood, especially in low tannin or tannin-free sorghum. Additional studies are required to further understand and improve sorghum starch digestibility through plant breeding. Feed efficiency was improved by 5 percent with the mixture of sorghum and corn (Michael and Brouk, 2012). On a larger scale, the improvements in sorghum starch digestibility might save the overall feed industry billions of dollars around the world. This large economic benefit should be enough incentive for the feed industry and researchers to pursue improvement in starch digestibility of sorghum.

Previous studies have shown that starch digestibility of sorghum varies not only compared to other grains but also among sorghum genotypes (Elmalik et al., 1986; Hibberd et al., 1982; Wester et al., 1992). Therefore, to improve the starch digestibility of sorghum, it is essential to first understand the cause of differences in starch digestibility among sorghum genotypes. Researchers have suggested tannins as one of the components that significantly reduce starch digestibility (as indicated by differences in metabolizable energy) (King et al., 2000; Elkin et al., 1996; Cousins et al., 1981). Tannins are polyphenolic compounds that have the ability to interact with protein, starch, and digestive enzymes (Maxson et al., 1973; Butler et al., 1984; Longstaff and McNab, 1991). The interaction of tannins with starch digestive enzymes (amylases) has been proposed as one of the reasons for the reduction of starch digestibility in high tannin sorghum (Maxson et al., 1973; Longstaff and McNab, 1991). Davis and Hoseney (1979) showed that tannins also bind starch granules and suggested the possibility that the starch-tannin complex reduced the activity of starch degrading enzymes. Although the negative effect of tannin on starch digestibility is clear, comparison of sorghum free of tannin and with tannin content also demonstrated significant variation in the starch digestibility, suggesting that tannin is not the only factors that negatively influenced starch digestibility (Elmalik et al., 1986; Knabe, 1990; Elkin et al., 1996 (as indicated by differences in total metabolizable energy)). Moreover, by far the majority of sorghum cultivars now grown in the USA, as well as in Africa and Asia, are of the low tannin or tannin-free phenotypes. This puts an importance on identifying the factor or factors that reduce starch digestibility in these widely grown sorghums.

One prevailing theory for the variation in starch digestibility among genotypes with similar tannin content or tannin-free is the inaccessibility of enzyme to starch granules (Sullins and Rooney, 1974; Rooney and Pflugfelder, 1986). In the sorghum endosperm, starch granules are surrounded by protein bodies and protein matrix. The texture of the endosperm is affected by the arrangement of the granules with regard to protein bodies and protein matrices (Rooney and Miller, 1982). In vitreous endosperm, starch granules are packed tightly within a protein matrix that is embedded with protein bodies. In floury endosperm, starch granules are loosely packed and surrounded by less protein bodies and discontinuous protein matrix. Rooney and Pflugfelder (1986) believed that the dense packing of the vitreous endosperm inhibits the accessibility of amylases to the starch granules, while the granules in the floury endosperm are more available to the enzyme. Digestibility studies have shown that sorghum with floury endosperm is more digestible than the one with vitreous endosperm (Miller et al., 1972; Elmalik et al., 1986). These studies supported the enzyme accessibility theory.

As mentioned previously, tannins have the ability to bind protein and render the protein indigestible by proteases (Butler et al., 1984). In addition, tannin also can reduce protein digestibility by binding to the proteases and reduce the activity. As with starch digestibility, studies examining protein digestibility in sorghum grain have found that proteins are digested at different rates among tannin-free or low-tannin genotypes (Axtell et al., 1981; Elkin et al., 1996). It is believed that the structure and composition of the protein within the protein bodies is responsible for the differences in protein digestibility among the tannin-free sorghum genotypes. As described previously, protein bodies consist mainly of kafirins. The true storage protein, α-kafirin, is located in the interior of the protein bodies and surrounded by β- and γ-kafirin on the periphery (Shull et al., 1992). Studies have shown that protein bodies are digested from the outer layer first and digestion continues gradually toward the interior (Oria et al., 1995a). It is postulated that the β- and γ-kafirins form disulfide linkages into polymers that are more resistant to enzyme digestion around the periphery of the protein bodies, thus limiting the accessibility of protease to α-kafirin (Oria et al., 1995b). Variation in protein digestibility among normal sorghum genotypes might be due to the different extent of disulfide linkages formed around the protein bodies. As for a high protein digestibility mutant genotype identified at Purdue, Oria et al. (2000) showed that the protein bodies were folded and the γ- and β-kafirins were located at the base of the folds. Thus, this structure exposed the α-kafirins and allowed faster protein digestion.

Based on the differences in protein digestibility and the suggestion regarding the accessibility of the enzyme, it is reasonable to hypothesize that faster digesting proteins will expose more of the starch granules and consequently lead to faster starch digestion. This study to determine whether sorghum with higher protein digestibility will lead to a concomitant higher starch digestibility.

1.1.19 Starch Digestibility With and Without Protease Pretreatment

The samples are estimated for starch digestibility, which was modified from Zhang (1997). The starch digestibility determination was based on the amount of reducing sugar released after α-amylase treatment. Initially sorghum flour was subjected to protease (Pepsin) pretreatment in order to remove the protein allowing the easy access of α-amylase to starch. Digestion of proteins present in sorghum flour was carried out by pepsin followed by α-amylase digestion. The digested sample was analyzed for the reducing sugar content using the Dinitrosalicylic acid (DNS) method. The intensity of the dark red color was measured at 510 nm and the amount of reducing sugars present in the sample was calculated using the standard graph. Corresponding absorbance directly gives the percentage of starch digested in the sample.

1.1.19.1 *Starch Digestibility Without Protease Pretreatment of Sorghum Genotypes*

Starch digestibility of the sorghum samples without pepsin pretreatment ranged from 96 mg (glucose)/h for the line SPV 839 to 258.80 mg (glucose)/h for the line SPV-1231. These values were determined after 1 h of amylase digestion. High values were observed in SPV-1231, SPV-1294, and SPV-711. A low value was observed in SPV 839.

1.1.19.2 *Starch Digestibility With Protease Pretreatment of Sorghum Genotypes*

Starch digestibility of the sorghum samples with pepsin pretreatment ranged from 117.50 mg (glucose)/h for the line SPV 1253 to 293.50 mg (glucose)/h for the line SPV-1231.

1.1.20 Influence of Protein Predigestion on Starch Digestibility of Sorghum Genotypes

Starch digestibility values after 1 h pepsin pretreatment represent more closely the digestive system in which proteins are largely digested prior to and during starch digestion. Comparison of profiles without and with protease predigestion also provides information on the effect of protein digestibility differences on starch digestibility. Based on the protein digestibility values, 1 h was selected as the digestion time, because the effect of the protein digestibility has been observed more prominently than starch digestibility. We speculated that digesting the protein for too long might remove too much of the protein from the samples, and thus the effect of protein on starch digestibility may have become unobservable.

Starch digestibility values from this experiment showed considerable differences among the samples, thus confirming the results from other studies (Breuer and Dohm, 1972 (as indicated by difference in feed efficiency); Elmalik et al., 1986; Wester et al., 1992; Elkin et al., 1996 (as indicated by differences in total metabolizable energy)). After pepsin pretreatment, it was noted that the overall starch digestibility increased significantly. This trend supports the idea that removal of the protein allowed for easier enzyme accessibility and thus increased the starch digestion rate.

Starch digestibility values without pepsin treatment appeared to mirror the rank order of samples based on protein digestibility values. The reason for this trend is unclear from the present study. Digesting away the protein with pepsin before α-amylase treatment did increase the overall starch digestibility of sorghum flour samples. This result supports the idea that removing the protein allowed for better accessibility of amylase to starch granules. There was a prominent trend that after pepsin treatment, sorghum lines with higher protein digestibility had better starch digestibility than lines with lower protein digestibility. This trend is apparently due to its different starch composition. This trend suggests that the intrinsic properties of starch (composition, crystallinity, granule structure) may contribute to the overall starch digestibility.

Overall, starch digestibility was shown to be affected by various factors including protein digestibility, starch composition, starch granule structure, and perhaps some other factors not observed in this study.

1.1.21 Fat Estimation

Finely powdered grains were subjected to defatting (AOAC, 1995) by soxhlets (pelican instruments) apparatus. Fat was extracted in soxhlets apparatus using n-hexane as a solvent. The solvent is evaporated and the fat is extracted.

Among 70 genotypes, the total fat content of the sorghum ranged from 1.94% for line GJ-9 to 4.27% for line 2219 B. Out of 70 genotypes, 56 genotypes ranged between 2.5 and 3.5%, and three genotypes were above 3.5% of fat content and are listed here: 2219 B, SPH 1679(2), and SPH 1704 (Table 1.1).

1.1.22 Tannins and Phenols of Sorghum Grain

All sorghums contain phenols, which can affect the color, appearance, and nutritional quality of grain and sorghum products. The phenolic compounds can be divided into three basic groups: phenolic acids, flavonoids, tannins. All sorghum contain phenolic acids and most contain flavonoids. Only the brown, high-tannin, bird-resistant sorghums contain condensed tannins.

Many phenolic acids inhibit the growth of microorganisms and may impart resistance to grain moulds before and later grain maturity. Phenolic acids apparently do act adversely affect the nutritional quality of sorghum grain, but they may form undesirable colors under some food processing conditions, such as alkaline conditions used in the making of tortillas.

Flavonoids are the largest group of phenols in the plant kingdom. Flavonoid compounds consist of two distinct units: A C6–C3 fragment from cinnamic acid forms the β-ring and a C6 fragment from malonyl-CoA forms the A-ring. The major groups of flavonoids are the flavonols and flavans. The major group of flavonoids in sorghum are the flavans. Flavan-3-en-3-ols (double bond between C5 and C4 hydroxyl at C3) are called anthocyanidins and are the major flavans in sorghum. Anthocyanidins are found in their ionized form flavylium. The flavyliumion is primarily responsible for the intense red pigmentation of anthocyanidins in an acid medium. Flavan-3-ols (hydroxyl at C3) are called catechins, and 4-deoxyleuco-anthocyanidinsflavan-3,4-diols (hydroxyl at C3 and C4) are called leucoanthocyanidins. When treated with mineral acid, the colorless leucoanthocyanidins produce red anthocyanidins. In plants, anthocyanidins and leucoanthocyanidins exist often as glucosides at the 3 or 7 position and are called anthocyanidins and leucoanthocyanidins, respectively. Anthocyanidins are the major pigments in many flowers, stalks, and leaves. The color depends on the pH and substitution on the β-ring. Many plant colors in the orange to blue region are caused by copigmentation of anthocyanidins with metal ion and other phenolic compounds.

Anthocyanidins are very unstable in acid medium and are readily converted to their corresponding anthocyanidin in even slightly acidic solvents. This makes it difficult to determine whether a pigment is an anthocyanin or anthocyanidin. Both types of pigments have been reported in sorghum. Luteolindin and apigeninidin are actually 3-deoxyanthocyanidins and would be produced from a flavan-4-ol rather than a flavan-3,4-diol. The flavan-4-ol apiforol, a precursor for apigeninidin, has been found in sorghum leaf tissue and grains. The pericarp color of sorghum appears to be due to a combination of primarily anthocyanin and anthocyanidins pigments and other flavonoidin compounds. There appears to be a good deal of variation between sorghum of the same genetic pericarp color. The most abundant polyphenols are the condensed tannins, lignins, catechol melanins, and flavolans found in virtually all families of plants and comprising up to 50% of the dry weight of leaves.

Some polyphenols produced by plants in case of pathogen attacks are called phytoalexins. Such compounds can be implied in the hypersensitive response of plants. High levels of polyphenols in some woods can explain their natural preservation against rot.

1.1.23 Extraction of Polyphenols

Sorghum samples were ground on a UDY mill, to pass a 0.4-mm mesh. The sorghum flour was defatted and extracted with acidified methanol. The extract was centrifuged and supernatant was used for the estimation of polyphenols.

1.1.24 Prussian Blue Assay

Total phenols in grain sorghum were estimated by the method reported by Price and Butler (1977).

Phenols reduce the ferric ion to ferrous ion, and it will develop a deep blue color called Prussian blue, which is measured calorimetrically at 720 nm. The resultant color may be yellow, green, or deep blue, depending on the concentration of total phenols present in the sample.

Among 100 SPV Sorghum genotypes, total polyphenol content of the sorghum ranged from 0.5 mg/g for line SPV- 655 to 3.02 mg/g for line SPV- 281. Out of 100 genotypes, 25 genotypes were below 1 mg/g, 63 genotypes ranged between 1 mg/g to 2 mg/g, and 12 genotypes ranged above 2 mg/g content of polyphenols (Table 1.1).

1.1.25 Phenolic Compounds and Antioxidant Activity of Sorghum Grains of Varying Genotypes

Numerous methods are used to evaluate antioxidant activities of natural compounds in foods or biological systems with varying results. Two free radicals that are commonly used to assess antioxidant activity in vitro are 2,2-azinobis (3-ethylbenzothiazoline-6-sulfonic acid) (ABTS) and 2,2-diphenyl-1-picrylhydrazyl (DPPH). However, both of these radicals are foreign to biological systems. The ABTS assay measures the relative ability of antioxidants to scavenge the ABTS generated in aqueous phase, as compared with a Trolox (water soluble vitamin E analogue) standard. The ABTS is generated by reacting with a strong oxidizing agent (eg, potassium permanganate or potassium persulfate) with the ABTS salt. The reduction of the blue-green ABTS radical by hydrogen-donating antioxidants is measured by the suppression of its characteristic long wave absorption spectrum. The method is usually expressed as Trolox equivalent antioxidant capacity (TEAC). The method is rapid and can be used over a wide range of pH values, in both aqueous and organic solvent systems. It also has good repeatability and is simple to perform; hence, it is widely reported. The method, however, has not been correlated with biological effects; hence, its actual relevance to in vivo antioxidant efficacy is unknown.

The DPPH is a stable free radical with an absorption band at 515 nm. It loses this absorption when reduced by an antioxidant or a free radical species. The DPPH method is widely used to determine antiradical/antioxidant activity of purified phenolic compounds as well as natural plant extracts. It has been found that most phenolic antioxidants react slowly with DPPH, reaching a steady state in 1–6 h or longer. This suggests that antioxidant activity using DPPH should be evaluated over time. The method also has good repeatability and is used frequently. However, like ABTS, it has limited, if any, relevance to biological systems. Color interference of DPPH with samples that contain anthocyanin leads to underestimation of antioxidant activity.

1.1.26 Phytic Acid Estimation

Phytic acid is estimated by colorimetric determination of ferric ions (Wheeler and Ferral, 1971).

Phytic acid and phytic acid chelates react with ferric chloride and forms ferric phytate. The available ferric ion after reaction is determined by developing blood-red color with potassium thiocyanate.

Among 59 genotypes, the total phytic acid content of the sorghum ranged from 2.40 mg/g for line RS 29, CSH-16, CSH-17, and CSV-13 to 6.70 mg/g for line PVK-809. Out of 59 genotypes, 42 genotypes ranged between 3 mg/100 g to 5 mg/g and 14 genotypes ranged above 5 mg/g of phytic acid content (Table 1.1).

1.1.27 Phytochemical Agents

Antioxidants play a key role in immunity of plant against diseases, infections, and mechanical injuries, etc. These antioxidants are also beneficial to mankind through consumption of different plant parts. A wide variety of antioxidants like polyphenols are observed in sorghum (Fig. 1.2), which are generally divided into hydrolysable tannins (gallic acid esters of glucose and other sugars) and phenyl propanoids, such as lignins, flavonoids, and condensed tannins. In sorghum, we can estimate the amounts of polyphenols contents in these genotypes of sorghum, and determine which have a better yield and/or poor yield with regard to human health. The better yielding tannins genotypes are better for cultivation and also better for human welfare. This

Base unit:	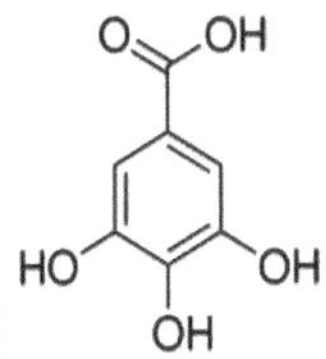 Gallic acid	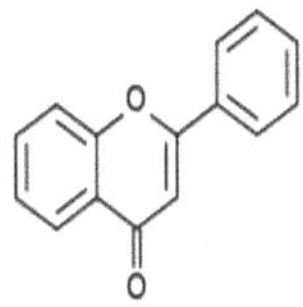 Flavones	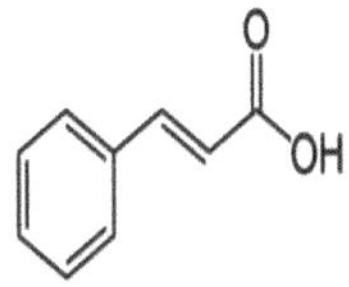 Cinnamic acid
Class/polymer:	Hydrolysable tannins	Flavonoids, condensed tannins	Lignins

FIGURE 1.2 Basic structures of polyphenols.

study to investigate the better genotypes for polyphenols production from the selected varieties of sorghum genotypes.

Experiments were conducted to study the concentrations of total polyphenols, which were estimated by the different methods, such as "Prussian Blue assay" and modified procedure of Hagerman and Butler for the analysis of condensed polyphenols. In this analysis, we can determine the polyphenolic concentrations in different genotypes of sorghum in flour, etc.

1.1.28 Sorghum Phytochemicals and Their Impact on Human Health

Sorghum phenols protect plants against insects and diseases, and they can also act as antioxidants in vitro. Free radicals play a role in diseases such as cancer, atherosclerosis, rheumatoid arthritis, inflammatory bowel disease, and cataracts and phenolic compounds may decrease the risk of these diseases by lowering the amount of free radicals. Other roles of antioxidants include antifungal, antibacterial, and antiviral agents. Sorghum is a rich source of various phytochemicals including tannins, phenolic acids, anthocyanins, phytosterols, and policosanols. These phytochemicals have potential to significantly impact human health. Sorghum fractions possess high antioxidant activity in vitro relative to other cereals or fruits. These fractions may offer similar health benefits commonly associated with fruits. Available epidemiological evidence suggests that sorghum consumption reduces the risk of certain types of cancer in humans compared to other cereals. The high concentration of phytochemicals in sorghum may be partly responsible. Sorghums containing tannins are widely reported to reduce caloric availability and hence weight gain in animals. This property is potentially useful in helping reduce obesity in humans.

Sorghum phytochemicals also promote cardiovascular health in animals. Such properties have not been reported in humans and require investigation, since cardiovascular disease is currently the leading killer in the developed world. The available information reviews on sorghum phytochemicals, how the information relates to current phyto nutrient research, and how it has potential to combat common nutrition-related diseases including cancer, cardiovascular disease, and obesity (Awika and Rooney, 2004).

1.2 COMPARISON WITH OTHER CEREALS

Grain sorghum is mostly used for human food as well as in small quantities for animal feeds. The utilization of sorghum grains in the world is shown in Table 1.5. This table shows that most of the sorghum grains are used as food in Asia and Africa. Developed countries are using very small quantities of sorghum grains as a food, and the remainder is used mostly for animal feeds.

Sorghum grains are rich in fiber and minerals as well as having a sufficient quantity of carbohydrates, proteins, and fat. The comparison of sorghum grain with other cereals is given in following Tables 1.6–1.12.

TABLE 1.5 Sorghum Utilization (Average Million Tones)

Region	Food	Feed	Other uses	Total
Africa	8.0	0.4	2.3	10.7
Asia	15.1	6.3	2.1	23.5
Central America	0.3	8.4	0.2	8.9
South America	-	4.6	0.3	4.9
North America	-	12.6	0.1	12.7
Europe	-	1.4	-	1.4
USSR	-	2.3	0.3	2.6
Oceania	-	0.4	-	0.4
World	**23.4**	**36.4**	**5.3**	**65.1**
Developing countries	23.2	15.6	4.8	43.6
Developed countries	0.2	20.8	0.5	21.5

TABLE 1.6 Average Composition of Cereal Grains

Grains	Moisture (%)	Crude protein (%)	True protein (%)	Oil/ether extract (%)	Crude fiber (%)	CHO (NFE) (%)	Ash (%)
Barley	15.0	9.0	8.5	1.5	4.5	67.4	2.6
Sorghum	11.0	9.6	8.6	3.8	1.9	71.3	2.4
Maize	13.0	9.9	9.4	4.4	2.2	69.2	1.3
Millet	13.0	10.5	9.9	3.9	8.1	60.7	3.8
Oats	13.0	10.4	9.5	4.8	10.3	58.4	3.1
Rice	11.4	8.3	7.2	1.8	8.8	64.7	5.0
Rice polished	13.0	6.7	6.4	0.4	1.5	77.6	0.8
Rye	13.0	11.6	10.7	1.7	1.9	69.8	2.0
Wheat	13.0	12.2	11.0	1.9	1.9	69.3	1.7

TABLE 1.7 Amino Acid Composition of Sorghum, Wheat and Rice Grains (g/16 g Nitrogen)

Amino acid	Sorghum	Wheat	Rice
Lysine	2.60	2.72	3.68
Histidine	2.06	2.08	2.08
Arginine	4.20	4.64	7.68
Aspartic acid	7.21	3.40	4.85
Threonine	3.00	2.88	3.68
Serine	3.66	4.30	4.56
Glutamic acid	20.30	32.50	11.69
Proline	6.25	11.60	5.27
Glycine	3.45	3.20	6.41
Alanine	8.64	2.00	3.23
Cystine	0.97	2.24	1.44
Valine	4.39	4.48	6.08
Methionine	1.39	1.44	2.40
Isoleucine	3.83	3.52	4.80
Leucine	12.27	6.56	8.00
Tyrosine	3.25	2.88	4.64
Phenylalanine	4.55	4.48	4.48
Trptophan	1.12	1.12	1.28
Total	**93.14**	**96.04**	**86.25**
Energy (Kcal)	349	341	345

TABLE 1.8 Mineral Composition of Sorghum, Wheat, and Rice Flour (mg/100 g)

Mineral	Sorghum	Wheat	Rice
Sodium	21.00	19.3	2.2
Potassium	537.00	315.0	7.4
Calcium	25.00	48.0	10.0
Phosphorus	526.00	183.0	160.0
Magnesium	212.00	132.0	90.0
Iron	8.48	4.90	0.70
Zinc	3.91	2.20	1.40
Copper	0.86	0.51	0.14
Manganese	3.50	2.29	0.59

TABLE 1.9 Vitamins Content in Sorghum, Wheat, and Rice Grains

Vitamin	Sorghum	Wheat	Rice
β-Carotene (μg/100 g)	47.00	64.00	-
Thiamine (mg/100 g)	0.37	0.45	0.06
Riboflavin (mg/100 g)	0.13	0.17	0.06
Niacin (mg/100 g)	3.10	4.30	1.90
Total B6 (mg/100 g)	0.21	0.57	-
Folic acid (μg/100 g)	20.00	35.80	8.00

TABLE 1.10 Oxalic Acid, Phytic Phosphorus and Dietary Fiber in Sorghum, Wheat, and Rice Grains

Constituent	Sorghum	Wheat	Rice
Oxalic acid (mg/100 g)	10.00	8.00	3.00
Phytic Phosphorus (mg/100 g)	172.00	238.00	83.00
Phytic P as percent of total P	77.00	80.00	52.00
Total dietary fiber (%)	12.69	11.40	-

TABLE 1.11 Synergistic Effects of Legume (Green Gram) on PER of Sorghum

Source of protein	PER
Casein	2.87
100% Sorghum	0.93
90% Sorghum + 10% green gram	1.11
80% Sorghum + 20% green gram	1.36
70% Sorghum + 30% green gram	1.75
50% Sorghum + 50% green gram	1.56

TABLE 1.12 Fats and Fatty Acids

Amounts per selected serving	
Total Fat	6.3 g
Saturated Fat	0.9 g
12.00	13.4 mg
14.00	17.3 mg
16.00	781 mg
18.00	67.2 mg
Mono unsatureated Fat	1.9 g
16:1 Undifferentiated	55.7 mg
18:1 Undifferentiated	1851 mg
Polyunsaturated Fat	2.6 g
18:2 undifferentiated	2505 mg
18.03	125 mg
Total Omega-3 fatty acids	125 mg
Total Omega-6 fatty acids	2505 mg

1.2.1 Alternative Uses of Sorghum

Alternative uses of sorghum include the utilization of sorghum grain and sweet stalk in nonfood sectors for the production of commercially valued products. Total consumption of sorghum closely follows the global pattern of output, since most of it is consumed in the countries where it is grown. Sorghum is used for two distinct purposes: human

TABLE 1.13 Sorghum Utilization, 1981–85 Average and Growth From 1961–65 to 1981–85

	1981–85 average (million tonnes)				Annual growth from 1961–65 to 1981 (%)			
Region	**Food**	**Feed**	**Other uses**	**Total**	**Food**	**Feed**	**Other uses**	**Total**
Africa	-	-	8.0	0.4	2.3	-	10.7	1.5
Asia	-	-	15.1	6.3	2.1	-	23.5	-
Central America	0.3	8.4	0.2	-	8.9	2.0	13.2	-
South America	-	4.6	0.3	-	4.9	-	8.5	5.7
North America	-	12.6	0.1	-	12.7	-	0.5	-
Europe	-	-	-	1.4	-	-	1.4	-
USSR	-	-	-	2.3	0.3	-	2.6	-
Oceania	-	-	0.4	-	-	0.4	-	3.5
World	-	-	23.4	36.4	5 3	-	65.1	0.5
Developing countries	23.2	15.6	4.8	-	43.6	0.5	10.3	0.1
Developed countries	0.2	20.8	0.5	-	21.5	3.5	1.7	4.7

food and animal feed. Although in the early 1960s a very large part of the sorghum output was used directly as human food, its share has continuously declined since then. In fact, consumption of sorghum as animal feed has more than doubled, from 30 to 60%, since the early 1960s, while the volume of total food use has remained unchanged or has slightly declined (Table 1.13). In North and Central America, South America, and Oceania, most of the sorghum produced is used for animal feed.

1.2.2 Human Food

While total food consumption of all cereals has risen considerably during the past 35 years, world food consumption of sorghum has remained stagnant, mainly because, although nutritionally sorghum compares well with other grains, it is regarded in many countries as an inferior grain. Per capita consumption of sorghum is high in countries or areas where climate does not allow the economic production of other cereals and where per capita incomes are relatively low. These include especially the countries bordering the southern fringes of the Sahara, including Ethiopia and Somalia, where the national average per capita consumption of sorghum can reach up to 100 kg per year. Other countries with significant per capita consumption include Botswana, Lesotho, Yemen, and certain provinces in China and states in India. In most other countries, food consumption of sorghum is relatively small or negligible compared to that of other cereals.

More than 95% of total food use of sorghum occurs in the countries of Africa and Asia (Table 1.5). In Africa, human consumption accounts for almost three-quarters of total utilization, and sorghum represents a large portion of the total calorie intake in many countries. For example, in Burkina Faso about 45% of the total annual calorie intake from cereals comes from sorghum, although its share has declined from 55% in the early 1960s. China and India account for about 90% of total food use in Asia.

Available data from Africa indicate that despite an increase in total food use between the early 1960s and the mid-1980s, the average per capita consumption declined from 20 to 15 kg per year (FAO, 1988). Decreases were concentrated in Kenya, Mozambique, Nigeria, and Somalia but occurred also in Botswana, Ethiopia, Lesotho, and Zimbabwe. In Asia, both total and per capita food use of sorghum declined in 1980's.

This decline in per capita consumption in many countries was due in part to shifts in consumer habits brought about by a number of factors: the rapid rate of urbanization, the time and energy required to prepare food based on sorghum, inadequate domestic structure, poor marketing facilities and processing techniques, unstable supplies, and relative unavailability of sorghum products, including flour, compared with other foodstuffs. Changes in consumption habits were concentrated in urban areas. Per capita food consumption of sorghum in rural producing areas remains considerably higher than in the towns. In addition, national policies in a number of countries had a negative influence on sorghum utilization as food. For instance, large imports of cheap wheat and rice and policies to subsidize production of those crops in some countries had considerable negative impact on the production of sorghum.

1.2.3 Animal Feed

In the US animal feed industry, corn is the most common grain used for energy. Other feed grains include sorghum, wheat, barley, and oats. The energy provided by starch from these cereals varies widely depending on the type of feed grains and among animal species (Black, 2001). According to the National Research Council (NRC), the metabolizable energy (ME) of sorghum and corn grain by poultry is 3212 and 3350 kcal/kg, respectively. For beef cattle, the ME for sorghum is 2960 kcal/kg and for corn is 3250 kcal/kg. As for swine, the ME for sorghum and corn are 3340 and 3420 kcal/kg, respectively. It is worth mentioning that the information regarding the genotypes of sorghum used to obtain the ME values is not given by NRC; thus, it can only be assumed that they represent typical normal genotypes. The NRC data mentioned suggest that poultry and swine are able to utilize sorghum better than cattle. In addition, several studies reported similar findings that poultry and swine consuming sorghum digest more starch and utilize more energy than cattle (Rowe et al., 1999; Black, 2001). This variation might be due to their different digestive systems. Starch digestion and utilization are well reviewed by Adams (2001) for nonruminants and by Ørskov (1986) and Adams (2001) for ruminants. The digestive system of ruminants, like cows and sheep, is designed such that the foods consumed are first exposed to the microbial enzymes in the rumen. The microbial enzymes have the ability to digest cell wall materials as well as starch, protein, and other nutrients before passing them to the small intestine. In the small intestine, the undigested feeds in the rumen are further digested by the animals' digestive enzyme. The fermentative products from the microbial enzymes are absorbed in the rumen and utilized as the main source of energy for ruminants. Thus, the digestion of starch in the rumen is the main focus for improving the overall starch digestibility in ruminant animal. The feed consumed by nonruminants is directly exposed to the host digestive enzymes in the stomach and small intestine.

The major part of the digestion and absorption of starch/glucose and other nutrients takes place in the small intestine. The fermentation processes, similar to that in the rumen of ruminant species, takes place in the hind-gut and provides far less energy and nutrients compared to that from rumen digestion in ruminants. Therefore, animal nutritionists, plant breeders, and feed industries have to understand that the nutritional improvement provided by a selected cereal grain in one animal might not provide the same enhancement in another. It is important to take into account the type of digestive system of the animal when developing and selecting grain varieties for animal feed purposes.

Although different animals utilize energy from sorghum differently, sorghum has consistently shown somewhat poorer performance compared to corn for all livestock animals. Sorghum is compared often to corn in the feed industry due to their similarity in chemical composition. Various in vivo studies showed that sorghum usually has lower feed efficiency compared to corn (Cousins et al., 1981; Douglas et al., 1990, Streeter et al., 1990; Elkin et al., 1996). It is generally accepted in the animal feed industry that the feed efficiency of sorghum, measured by dividing feed intake by weight gain, is about 5 to 10% lower than that of corn. However, sorghum still is a competitive feed grain because the price of sorghum grain is usually lower than that of corn.

The lower feeding value of sorghum has been attributed principally to the inferior starch digestibility. Various physical processing methods prior to feeding have been shown to improve sorghum's feeding value close or similar to that of corn, but the additional processing cost oftentimes reduces the benefits of using sorghum as a corn substitute. Feeding value of sorghum not only differs when compared to other cereals but also varies considerably among sorghum cultivars (Elmalik et al., 1986; Streeter et al., 1990; Wester et al., 1992; Elkin et al., 1996). Waxy and heterowaxy sorghum usually demonstrate higher feed efficiency as compared to nonwaxy sorghum (Sherrod et al., 1969; Elmalik et al., 1986). The different feed efficiency among cultivars is also attributed to variation in starch digestibility. Scientists would like to improve digestibility and overall nutritional value of sorghum for animal feed through plant breeding to eliminate the additional processing costs.

Grain use for animal feed has been a dynamic element in the stimulation of global sorghum consumption. The demand for sorghum for feed purposes has been the main driving force in raising global production and international trade since the early 1960s. The demand is heavily concentrated in the developed countries, where animal feed accounts for about 97% of total use, and in some higher-income developing countries, especially in Latin America, where 80% of all sorghum is utilized as animal feed. The United States, Mexico, and Japan are the main consuming countries, followed by Argentina, the former Soviet Union, and Venezuela. These countries together account for over 80% of world use of sorghum as animal feed.

1.2.4 India

India is the world's second largest producer of sorghum. At present most of the sorghum produced in India is consumed as a human food in the form of *roti* or chapatti (unleavened flat bread). Walker (1990) analyzed the supply and demand prospects for sorghum in India. He found that in the past three decades the average per capita sorghum consumption declined markedly in both rural and urban households. Average rural consumption fell from 1.74 to 1 kg per capita per month. Urban consumption dropped from 0.74 to 0.46 kg per capita per month. It was projected that sorghum consumption would continue to fall at about 0.5% per annum. The declining trend in sorghum consumption is partly due to the decline in per capita consumption of total cereals.

Decreases in consumption of sorghum were found to be proportional to increases in expenditure. Increased income is accompanied by increased consumption of wheat and rice, as products made from these cereals are easy to prepare and have better keeping quality. There is also a tendency to eat a greater variety of foods as income and urbanization increase. The price of sorghum relative to those of wheat and rice has not increased in the major sorghum-consuming regions. Therefore, other factors are probably more influential than direct price considerations in explaining the fall in per capita sorghum consumption. Prospects of technological change could perhaps change the scenario for improved production and utilization of sorghum.

1.2.5 China

China is the fourth largest producer of sorghum in the world, and in Asia it is second only to India in area and production of sorghum. About 30% of the sorghum produced is used for human consumption and 60% for animal feed and the manufacture of alcoholic beverages (Kelley et al., 1992). However, the importance of sorghum as a human food has declined over time. The area under sorghum cultivation has also declined, from about 2.8 million hectares in 1979–81 to 1.9 million hectares in 1990. Production has decreased correspondingly, from 7 million tonnes in 1979–81 to 5.3 million tonnes in 1990. In recent years, more attention has been given to sorghum fodder and to developing suitable cultivars for this purpose.

1.2.6 West Africa

The West African semiarid tropics are defined as those areas where rainfall exceeds potential evapotranspiration for 2–7 months annually. This area encompasses all of Senegal, the Gambia, Burkina Faso, Cape Verde, major southern portions of Mauritania, Mali, and Niger and the northern portions of Côte d'lvoire, Ghana, Togo, Benin, and Nigeria. Cereals occupy nearly 70% of total cultivated area in this region and engage 50 to 80% of total farm-level resources (Matlon, 1990). Millets and sorghum account for 80% of cereal production. During the last 25 years, growth in millet and sorghum production has been slow and the total output has been about 1% lower than the population growth per year. Average yield per unit area of millet and sorghum has declined during this period, and the small production increases have primarily resulted from an expansion of cropped area. Many factors have contributed to the decreased productivity, including demographic pressure and ecological degradation.

Reardon and Matlon (1989) reported the food consumption patterns of the population of two villages, one representing the Sahelian savannah and the other the Sudano-Sahelian zone. Market dependence was considerably lower among households in the Sudano-Sahelian village and was more equally distributed across income strata than in the Sahelian village. The poor were especially vulnerable in the rainy season when they were more dependent on the market. In fact, purchased food products contributed 60 to 70% of all calories consumed by poor and middle-income households during the rainy season. In the Sahelian village, white sorghum accounted for only 4% of the cropping area but provided nearly 25% of the calories consumed outside the harvest season by poor households. Red sorghum and maize accounted for only 10% of the cultivated area but provided up to 60% of the calories consumed by the poor during nonharvest seasons.

1.2.7 Eastern and Southern Africa

Sorghum and millets account for 23% of the cereal production of the South African Development Community (SADC) countries, which include Angola, Botswana, Lesotho, Malawi, Mozambique, Namibia, Swaziland, the United Republic of Tanzania, Zambia, and Zimbabwe. However, they are the dominant grain crops only in Botswana and Namibia, where they account for 86 and 50% of total cereal production, respectively. Sorghum and millets are important in those areas that receive less than 650 mm of annual rainfall. The productivity of these crops is low, and in most SADC countries there is no strategy for the development of sorghum and millet subsectors.

In most SADC countries, formal-sector (government-regulated) markets handle only a very small proportion of total sorghum and millet production. They handle less than 10% of total production in Lesotho, Swaziland, the United Republic of Tanzania, Zambia, and Zimbabwe. Most of the sorghum and millets produced in the SADC region is consumed by producing households or sold in informal markets, primarily for traditional beer production. Maize is cheaper than sorghum in many informal markets across the SADC region, and there may be good potential for expanding the production of sorghum and millets in view of the price differences.

The use of sorghum in traditional foods and its gradual shift to the envisaged use of value added health food is gaining importance. Usually most sorghum produced globally for food is consumed in the following eight basic methods:

Unleavened bread: *roti, tortilla.*
Leavened Bread: *injera, kisra, dosai.*

Thick porridge: *to, tuwo, ugali, bogobe, sankati.*
Thin Porridge: *ogi, ugi, ambali, edi.*
Steam cooked products: *couscous, wowoto, noodles.*
Boiled sorghum: *soru.*
Snack food: *popped sorghum.*
Alcoholic and nonalcoholic beverages: *burkutu, busa, ting, obhshera, abrev.*

1.3 DOUGH AND ROTI MAKING QUALITY OF SORGHUM

Sorghum is a staple cereal in many states of the Indian subcontinent. Though sorghum grains are nutritious, the consumption of this cereal is decreasing due to the popularity and easy availability of (read "highly subsidized") and convenience for preparing foods from fine cereals. The other major reasons are dying food habits related to traditional, especially rural foods; requirement of special skill for preparing sorghum rotis (unleavened pancake); and nonavailability of ready-made flour and *suji* from sorghum in the market. There is a need to popularize sorghum foods. Sorghum with its rich fiber and low starch digestibility makes an ideal food for the diabetic and obese population, especially in the urban areas. Usually sorghums with pearly white or yellow bold grain types are preferred for roti preparation. Pericarp color, endosperm texture has significant effects on roti quality. Corneous grains, in general, exhibit higher density and breaking strength, lower percentages of water absorption, and better dough and roti quality (Murthy and Subramanian, 1981).

For many years in the sorghum eating population, particularly in Rabi growing areas, the roti made from Maldandi (M35-1) were preferred for taste and softness, over other genotypes. A study made on the dough quality and roti quality analysis of the Kharif and Rabi sorghum genotypes was done to assess the dough and roti quality using M 35-1 as control or check. A taste panel consisting of 15 members trained for tasting and scoring rotis was used for the assessment of roti quality. Each time 10 members did the evaluation on a hedonic scale 1 to 4 ranging from very good (4) to not good (1). The results of the dough quality and roti quality of Rabi 2005–06, Kharif – 2006, and Kharif 2008 genotypes are presented below.

1.3.1 Dough Quality

For the preparation of roti, the quality of the dough is very important and the quality is assessed by different parameters including (1) water requirement, (2) kneading quality, (3) time required for baking, (4) rolling quality (diameter of the roti that is expanded with an equal amount of flour), (5) diameter of the roti after baking, (6) percent moisture retained in the roti, (7) spreading quality, and (8) hectoliter weight are given in the below tables.

A total of 36 genotypes which included IVHT and AVHT along with checks were analyzed for dough quality. Except for the baking time, the genotypes in IVHT trial showed significant variation for all the parameters tested (Table 1.14). The requirement of water to make a dough with constant amount of flour varied from 24.3 mL (SPH 1582) to 31 mL (SPV 1712 (DS)). The dough made from a higher amount of water was relatively better in quality. The rolling quality varied from 16.67 cm (SPH 1581) to 24.6 cm (SPH 1582). The kneading quality was scored in a scale of 1–3 (1 = poor, 2 = medium to good and 3 = very good). The diameter after baking quality varied from 16 cm (SPH 1581) to 22.03 cm (SPV 1769). The time of roti baking quality varied from 2 min (SPV 1672 (DS&SS)) to 2.83 (SPH 1583). The %moisture retained in the roti varied from 10.93% (SPH 1582) to 25.72% (SPH 1581). The entries with superior dough quality were: SPV 1712 (DS SPH 1582, SPV 1769, SPH 1580, and SPH 1583 (compared to M 35-1). Thus, dough quality evaluation does not correlate to the preference to M 35-1 by the roti consumers.

Seven advanced genotypes which include one hybrid and six varieties were compared with seven check genotypes (CSH16, SPV 1616, CSH 23, SPV 462, CSV 15, CSV17, and local check).

Water requirement: The water required for dough preparation was estimated in different genotypes. The genotypes were significantly different. The water required for dough making with 100 g flour was measured, and the variation among the genotypes for this parameter was 66% (SPV 1817) to 120.57% (SPV 1786). The other genotypes SPH 1605, SPH 1596, and checks CSV 15, SPV 1616 were comparatively superior among the genotypes tested for the water requirement (Table 1.15).

Kneading quality: The kneading quality of the dough was measured on 1–3 scale (score 1 is best, score 2 is good, and score 3 is poor) among the genotypes, and the variation observed was 1.00 (SPH 1604, SPH 1605, SPH 1610, SPH 1596, SPH 1616, SPV 1786, and SPV 1616) to 2.70 (SPV 1817). The genotypes were significantly different. The test genotypes identified with best kneading quality on a par with check genotype SPV 1616 were SPH 1604, SPH 1605, SPH 1610, SPH 1596, SPH 1616, SPV 1786 SPH 1604, SPH 1605, SPH 1610, SPH 1596, SPH 1616, and SPV 1786.

Water absorption capacity (WAC): The water absorption capacity ranged from 131.03% (SPV 1817) to 167.23% (SPH 1610). Two genotypes SPV 1786 (161.03%), SPH 1611 (156.43%), and check CSV 15 (161.77%) were superior to all the check genotypes. The genotypes were

TABLE 1.14 Dough Quality Analysis of Sorghum Genotypes Rabi 05–06

S. no.	Genotypes	Water requirement (mL)	Rolling quality (cm)	Diameter after baking (cm)	Kneading quality	Time of baking (min)	Moisture (%)
1.	SPH 1583	25.00	23.00	21.17	2	2.83	15.01
2.	SPH 1581	29.33	16.67	16.00	1	2.17	25.72
3.	SPV 1762	30.00	24.17	21.67	3	2.17	18.93
4.	SPV 1758	26.33	22.83	21.17	1	2.17	12.73
5.	SPH 1582	24.33	24.60	21.97	3	2.17	10.93
6.	SPV 1763	29.33	21.00	19.17	1	2.17	13.45
7.	SPH 1579	26.33	20.67	19.17	1	2.17	17.68
8.	SPV 1757	24.83	20.50	18.57	1	2.17	17.80
9.	SPV 1755	25.33	22.83	21.17	1	2.33	17.80
10.	SPH 1580	28.67	23.07	21.00	2	2.33	19.58
11.	SPV 1759	27.33	22.83	21.00	1	2.17	12.17
12.	SPV 1761	26.67	21.83	20.33	1	2.17	14.88
13.	SPV 1766	30.33	24.17	21.17	3	2.17	15.41
14.	SPV 1767	26.67	19.17	17.67	1	2.17	20.17
15.	CSV 216 R	26.33	23.00	21.10	3	2.67	13.24
16.	CSH 15R	30.00	23.83	21.67	2	2.33	13.53
17.	CSV 14R	29.00	22.33	20.33	3	2.17	13.77
18.	SPV 1768	30.00	24.17	20.67	3	2.17	16.60
19.	SPH 1500 (IS)	30.00	23.00	21.33	3	2.17	12.42
20.	SPH 1449 (DS&IS)	30.00	22.33	20.83	2	2.17	12.42
21.	SPV 1709 (SS)	30.00	22.67	20.33	3	2.33	13.35
22.	SPV 1677	26.00	19.50	18.07	1	2.17	10.99
23.	SPV 1626 (SS)	30.00	23.67	21.83	3	2.17	12.94
24.	SPV 1668 (DS)	26.30	22.50	21.17	2	2.17	13.06
25.	SPH 1456	25.50	24.03	22.00	3	2.17	13.28

(Continued)

TABLE 1.14 (Continued)

S. no.	Genotypes	Water requirement (mL)	Rolling quality (cm)	Diameter after baking (cm)	Kneading quality	Time of baking (min)	Moisture (%)
26.	SPV 1672 (DS&SS)	25.00	23.90	22.00	3	2.00	14.30
27.	SPV 1680 (DS)	25.00	24.00	22.00	2	2.33	14.66
28.	SPH 1504 (DS)	26.30	23.83	21.67	3	2.50	15.58
29.	SPV 1712 (DS)	31.00	21.33	19.33	2	2.17	17.17
30.	SPV 1704	29.70	21.33	19.33	1	2.50	12.62
31.	SPV 1769	30.00	24.07	22.03	3	2.67	14.95
32.	CSH 15R	25.70	22.10	20.00	2	2.33	18.19
33.	CSV 14R (DS&SS)	30.00	20.67	19.17	3	2.17	11.18
34.	CSV 216R (DS&SS)	25.70	23.67	21.67	3	2.17	13.69
35.	MAULEE (SS)	25.00	23.50	21.33	2	2.17	12.68
36.	M 35-1	30.70	23.83	21.77	3	2.17	12.67

significantly different. The percentage of WAC is positively correlated to the roti quality. The higher the WAC the superior was the quality of roti.

Spreading quality: The spreading quality of the roti was measured in a 1–3 scale, 1 being the best and 3 being the poor. The range obtained for spreading quality among the genotypes was 1.00 (SPV 1786, SPH 1596, SPH 1610, SPH 1605, and CSV 15) to 3 (SPV 1817). The genotypes were significantly different. The best genotypes for spreading quality identified were SPV 1786, SPH 1596, SPH 1610, SPH 1605, and check genotype CSV 15.

Hectoliter weight: The hectoliter weight of the grain varied from 75.74 kg/hL (SPH 1611) to 79.54 kg/hL (SPV 1786). SPH 1610, SPH 1603, and SPH 1596 genotypes under testing contained similar weight of and starch superior to its respective check CSV 15 which contained weight of 78.06 kg/hL, respectively. Hectoliter weight is significantly different among all genotypes. SPH 1610, SPH 1603, and SPH 1596 genotypes were also superior to their respective check genotype.

TABLE 1.15 Dough Quality of Sorghum Genotypes Kharif 2008

S. no.	Test entry	Water requirement (mL)	Water absorbance (mL)	Kneading quality scale (1–3)	Spreading quality scale (1–3)	Hectolitre weight (Kg/hL)
1.	SPV 1817	66.00	131.03	2.7	3.0	77.49
2.	SPH 1603	105.00	136.03	1.7	2.0	78.71
3.	SPH 1604	107.20	153.80	1.0	1.3	77.42
4.	SPH 1605	115.57	147.67	1.0	1.0	78.36
5.	SPH 1606	107.77	138.03	1.7	1.3	76.23
6.	SPH 1609	97.77	140.77	1.9	1.2	77.40
7.	SPH 1610	108.33	167.23	1.0	1.0	78.45
8.	CSH 16	96.67	143.97	1.3	1.1	77.87
9.	SPV 1616	111.67	157.00	1.0	1.1	77.70
10.	CSH 23	92.77	148.57	1.3	1.3	77.01
11.	SPV 462	100.57	149.47	1.3	1.7	76.21
12.	CSV 15	113.33	161.77	1.1	1.0	78.06
13.	CSV 17	99.43	140.33	1.3	1.3	77.70
14.	Local check	101.67	149.23	1.3	2.0	76.98
15.	SPH 1611	103.90	156.43	1.7	1.7	75.74
16.	SPH 1596	109.43	148.67	1.0	1.0	78.06
17.	SPH 1615	96.67	147.37	1.3	1.3	77.77
18.	SPH 1616	97.10	139.10	1.0	1.7	77.35
19.	SPV 1786	120.57	161.03	1.0	1.0	79.54

1.3.2 Roti Quality

Roti prepared in the traditional method was used for analyzing the taste and sensory properties. The taste and sensory properties were measured on a hedonic scale. The properties used for sensory evaluation were (1) roti color and appearance, (2) roti texture, (3) roti taste, (4) roti flavor, (5) grade, and (6) acceptability of the roti (Table 1.16).

Color and appearance: Color and appearance index of the genotypes varied from 1.40 (SPV 1786) to 7.80 (SPV 1817). The genotypes were significantly different. One genotypes under testing, SPH 1610 (2.67), showed color index slightly lower than the advanced variety SPV 1786 (1.4) and on a par with check genotype CSV 15.

TABLE 1.16 Roti Quality Analysis* of Sorghum Genotypes From Kharif 2006 and Kharif 2008

S. no.	Entry	Color	Texture	Aroma	Taste	Overall acceptability
1.	SPV 1626	7.67	6.97	6.67	6.57	6.83
2.	SPV 1704	7.30	6.57	4.53	6.57	6.77
3.	SPH 1449	7.30	7.03	6.57	6.50	6.80
4.	SPH 1501	6.63	7.17	6.73	7.03	6.90
5.	SPH 1582	7.33	7.10	6.73	6.80	6.97
6.	SPH 1579	7.27	6.77	6.63	6.87	6.73
7.	SPV 1761	7.37	7.47	6.77	7.37	6.93
8.	SPV 1762	7.30	7.43	6.83	7.03	7.03
9.	SPV 1757	7.47	7.37	6.80	6.90	7.03
10.	CSH 15R	7.20	7.37	7.10	7.00	7.10
11.	CSV 216R	7.43	6.80	6.90	6.73	5.00
12.	Maulee	7.33	6.93	6.83	6.83	7.13
13.	M 35-1	7.10	7.50	6.73	6.93	7.07
14.	CSV 18	7.43	7.20	6.90	6.97	7.03
15.	DSH 4	6.70	6.57	6.50	6.47	6.30
16.	DSV 4	7.47	7.00	6.83	7.03	7.03
17.	SPV 1817	7.80	6.07	6.13	5.87	7.13
18.	SPH 1603	4.73	4.47	4.07	4.20	4.60
19.	SPH 1604	4.87	4.67	4.00	4.40	4.13
20.	SPH 1605	3.67	4.20	4.00	4.07	3.80
21.	SPH 1606	4.53	5.13	4.43	4.00	4.27
22.	SPH 1609	3.93	4.60	3.40	4.07	3.73
23.	SPH 1610	2.67	2.27	2.07	2.40	2.13
24.	CSH 16	5.40	4.87	4.93	4.80	4.93
25.	SPV 1616	4.00	3.80	3.80	3.93	3.73
26.	CSH 23	5.40	4.20	4.67	4.67	4.80
27.	SPV 462	5.87	5.53	5.13	5.67	5.60
28.	CSV 15	2.53	3.07	2.67	2.73	2.73
29.	CSV 17	5.67	3.93	4.20	4.07	4.20
30.	SPH 1611	5.80	5.27	5.07	5.40	5.53
31.	SPH 1596	5.20	5.00	4.73	4.67	4.67
32.	SPH 1615	5.00	5.33	4.67	4.87	4.87
33.	SPH 1616	5.13	5.67	3.93	4.47	5.13
34.	SPV 1786	1.40	1.47	1.20	2.00	1.33

**Sensory score: 10 = like extremely, 9 = like very much, 8 = like moderately, 7 = like slightly, 6 = neither like nor dislike, 5 = dislikes lightly, 4 = dislike moderately, 3 = dislike very much, 2 = dislike extremely, 1 = not acceptable at all.*

Flavor/aroma: Flavor/aroma of the roti among the genotypes, measured on a scale from 1–10, varied from 1.2 (SPV 1786) to 7.1 (CSH 15R). The genotypes were significantly different. The test genotype SPH 1610 (2.40) and check CSV 15 (2.73) are with superior aroma equal to that of SPV 1786.

Texture: Texture of the roti was measured on a scale from 1–10 and the variation for the roti texture in different genotypes was 1.47 (SPV 1786) to 7.5 (M 35-1). The genotypes were significantly different. The genotypes with superior roti texture equal to that of SPV 1786 were SPH 1610 (2.27) and CSV 15 (3.07).

Taste: Taste of the roti was measured on a scale of 1–10 and the genotypes varied from 2 (SPV 1786) to 7.37 (SPV 1761). The genotypes were significantly different. The genotypes SPH 1610 (2.07) and CSV 15 (2.67) were identified with excellent taste nearer to that of SPV 1786.

Overall acceptability: Overall acceptability of the roti for the genotypes was measured on a scale of 1–10 and variation observed was 1.33 (SPV 1786) to 7.13 (SPV 1817). The genotypes were significantly different. None of the genotypes were close to the acceptability score of check genotype. However, the genotypes with a better acceptability score identified were SPV 1610 (2.13) and CSV 15 (2.73).

The data on soluble protein, crude protein, free amino acids, total soluble sugars, and starch are presented in Table 1.17.

Soluble proteins: The soluble proteins ranged from 0.86% (SPH 1609) to 1.21% (CSH 16). The genotypes were significantly different. All the seven test entries had soluble protein above 1.0%. However, check genotypes CSV 15 and SPV 462 were superior to the local check (1.03).

Crude protein: The percent crude protein ranged from 7.28% (SPH 1616) to 9.96% (SPV 1786). The genotypes were significantly different. The advanced genotype (SPV 1786) showed highest crude protein. Five genotypes SPH 1610 (9.72%), CSV 15 (9.57%), local check (9.41%), check genotype CSV 17 (9.15%), and SPH 1596 (9.04%) showed protein above 9%.

Free amino acids (FAN): The free amino acids content ranged from 55.12% (CSV15) to 75.32% (SPV1616). The genotypes were significantly different. The three genotypes showed FAN content on a par with check genotype SPV 462 (72.41%). They are SPV1817 (72.79%), SPH 1609 (72.55%), and SPH 1611 (70.83%).

Total soluble sugars (TSS): The total soluble sugars ranged from 0.84% (CSH 23) to 2.15% (SPV 1786). The genotypes were significantly different. The SPV 1786 advanced genotype had the highest percent total soluble sugars representing good amylolytic activity. Two genotypes, SPH 1610 (1.92%) and SPV 1817 (1.92%), were superior.

Starch: The starch content varied from (CSH 23) 62.12% to (SPV 1786) 70.32%. The genotypes were significantly different. Two genotypes,

TABLE 1.17 Dough Quality of Sorghum Genotypes (Kharif 2008)

S. no.	Entry	Soluble protein (%)	Crud protein (%)	Free amino acids (mg/100 g meal)	Total soluble sugars (%)	Starch (%)
1.	SPV 1817	0.92	7.68	72.79	1.92	66.22
2.	SPH 1603	1.06	8.61	67.60	1.18	66.47
3.	SPH 1604	0.99	7.96	68.83	1.04	65.97
4.	SPH 1605	0.92	8.85	68.25	1.52	65.68
5.	SPH 1606	0.99	7.55	68.54	1.19	63.12
6.	SPH 1609	0.86	8.86	72.55	1.24	65.78
7.	SPH 1610	1.07	9.72	61.86	1.92	68.90
8.	CSH 16	1.21	8.52	61.89	1.18	63.09
9.	SPV 1616	1.25	9.29	75.32	2.02	69.86
10.	CSH 23	0.94	8.75	68.13	0.84	62.12
11.	SPV 462	1.08	8.55	72.41	1.00	62.39
12.	CSV 15	1.21	9.57	55.12	2.13	68.56
13.	CSV 17	0.91	9.15	70.57	1.10	65.04
14.	SPH 1611	0.97	8.69	70.83	1.03	62.79
15.	SPH 1596	1.01	9.04	60.17	1.38	64.55
16.	SPH 1615	0.94	8.53	65.07	1.08	68.9
17.	SPH 1616	1.07	7.28	68.71	0.99	69.21
18.	SPV 1786	1.03	9.96	68.10	2.15	70.32

SPH 1616(69.21%) and SPV 1786(70.32%) were on a par with that of check genotype SPV 1616 (69.86%) and that of SPV 1786 (70.32%) advanced genotype.

1.3.3 Traditional Food Products of Sorghum and Their Commercialization

The most preferred form of sorghum used traditionally is roti or bhakri (unleavened pan cake). It is eaten as breakfast, lunch, and supper and is frequently stored overnight. Roti is consumed along with different kinds of dishes depending on the socioeconomic status of the consumer (Subramanian and Jambunathan, 1980). The consumption of sorghum roti is a traditional practice in the southern states of India, especially in Maharashtra, Karnataka, and Andhra Pradesh. In general roti made from grains with a pale yellow white color, with an intermediate endosperm texture, without a subcoat, and with a thin pericarp had acceptable organoleptic quality. Presence of a tough, leathery pericarp produced rotis with inferior texture and flavor. Floury grains produced poor quality dough, while waxy grains produced a sticky dough and gummy rotis.

There is a decline in the consumption pattern of sorghum mainly due to the urbanization and easy availability of fine cereals like rice through the public distribution system. But in recent years starting from 2007, there is a wake-up in the urban population that sorghum is a health food for diabetics. A recent survey at NRCS (unpublished) has indicated that nearly 20 ton/month/center is being consumed in small hotels and tiffin outlets. The variety is mostly a local one, and the roti is being sold along with *subji* for Rs.6/roti. The trend of using sorghum in this direction is slowly increasing in Andhra Pradesh. The other traditional products like Annam, Sankati, and Kanji are preferred by sorghum farmers.

To is a staple food in most parts of Africa, south of the Sahara. *To* is prepared from rice, maize, millet, or sorghum depending on taste, cost, custom, geographical areas, and/or availability of grains. The *to* of Upper Volta is generally made from sorghum flour prepared by hand decortication of the grain to remove the bran (pericarp). The decorticated grain is ground into flour by additional hand pounding in the mortar with a pestle. The flour is cooked in water that is acidified by adding extracts of tamarind or juices of lemons. The pH of the *to* is about 4.5. The flour to water ratio varies but is usually about one part flour to four parts of water. The *to* is allowed to cool about 1 h after cooking before it is consumed with a sauce composed of numerous ingredients including tomatoes, okra, or gumbe, chillies, cow pea leaves, cow peas, and amaranthus. The exact ingredients used vary with the season, availability, and cost. *To* sauce is an important source of protein, minerals, vitamins, and other nutrients required to improve the nutritional value of *to*. *To* with sauce is the major staple food consumed in Upper Volta. Other foods from sorghum, such as couscous and fura, comprise only a small proportion of the total grain consumed. Thus, for sorghum quality, *to* should be the major food use that must be considered (Akingbala and Rooney, 1982).

Sorghum is the staple food of Botswana and is preferred over maize, millet, and wheat by most Botswanans. It is consumed as porridge, *bogobe*, and as *beer, chibuku*, or *bojalwa*. *Bogobe* is made from fermented and nonfermented sorghum meal. Fermented *bogobe*, a soft porridge, is known as *motogowa ting* or *ting*. A firm nonfermented *bogobe* is called mosokwane.

Ting is usually eaten in the evening and morning and mosokwane eaten for lunch. A variation of mosokwane exists in which a small amount of wheat flour and sugar is added to sorghum meal before boiling into porridge. This porridge is called *mageu*. Both fermented and nonfermented bogobe may be eaten with meat and vegetables. This fermented porridge requires a starter that is made by fermenting a small quantity of sorghum meal in water for 48 h. The starter is mixed with dry sorghum meal and lukewarm water is added. It is stirred until it becomes slurry and after covering it is left for fermentation (24 h). The fermented meal is added to boiling water and after frequent stirring it is cooked for 12–15 min. Mosokwane is the firm nonfermented porridge containing a mixture of one part of sorghum meal to four parts water. It is prepared by mixing meal and boiling water, stirring it frequently, and cooking it for 20–30 min (Boling and Nancy Eisener, 1981).

Sorghum and millets are the traditional staple cereals in most countries of Eastern and Southern Africa. The cooked stiff porridge that is prepared from all the cereals are commonly known as *ugali* (Kiswahili language) in Kenya, Uganda, and Tanzania; *sof* or *mafo* in Somalia; *nsima* in Malawi and Zambia, and *sadza* in Zimbabwe. For convenience and ease of presentation, we will refer to all the cooked stiff porridges (neutral pH) prepared from sorghum as *ugali*. Today most *ugali* is made from maize, but sorghum *ugali* still enjoys considerable importance and popularity among the rural poor for whom it is an important dish. A wide range of sorghum grain types are used in the preparation of *ugali*. However, the white seeded and highly corneous grains, where available, are preferred by most people in several countries. In Uganda, Rwanda, and parts of Western Tanzania, the brown-seeded high tannin sorghum grain types are most common and are used for *ugali* preparation as well as for brewing, while the white or red-seeded types without testa are preferred for *ugali*. The traditional preparation methods for *ugali* in most countries are basically similar. A clay pot is used to boil an estimated amount of water, depending on the size of the family and the amount of flour available. When the water starts boiling, a little flour is sprinkled on the surface of the water and heating is continued. As soon as the water begins boiling again, most of the flour is poured in the pot and allowed to cook for about 2 min. After this one-fourth to one-half of the hot slurry is removed and kept in a separate container. The remaining boiling water and flour in the pot are vigorously mixed with a wooden stick. Additional slurry or flour is added as required until *ugali* with the right consistency is obtained. Then the ugali in the pot is allowed to continue cooking on a reduced fire for an additional 4–5 min. The *ugali* is removed from the pot into baskets made for this purpose, as ugali kept on plates becomes watery after some time. The whole process of *ugali* preparation might take up to 20 min.

In northern Uganda, tamarind water or mango or lemon juice is added to boiling water to improve the taste and flavor of *ugali*. In Kenya, lemon juice or milk may be added to boiling water. Ugali is usually served with beans or peas, vegetables soup, meat or fish stew, etc. A good ugali should not contain any lumps. After ugali has cooled, a thin crust forms on the surface, but the ugali inside should be soft but not sticky and should be able to keep in a similar condition for up to 24 h (Mukuru et al., 1981).

In the savanna and semiarid regions of Nigeria, sorghum and millets are very important food commodities. For most areas, they are the staple foods. *Ogi, kamu,* and *akamu* are popular names given to the thin porridge from wet-milled, fermented sorghum. *Ogi* is the most common porridge eaten by Nigerians from the coast to the extreme northern boundaries of the country. Sorghum *ogi* is the most popular in the savannas where sorghum is mostly grown; overall, sorghum *ogi* is the most extensively consumed in Nigeria. In the northern parts of the Nigerian savanna where sorghum production is greatest, the thick porridge *tuwo* is more extensively consumed than *ogi*. *Tuwo* is usually a molded or shaped solid processed from dry-milled nonfermented whole grain flour. It exhibits little variation across different places and forms the major source of food for the main meal. The only other important thick porridge is *dalaki*, which is made from wet-milled, fermented, cracked sorghum grains. The resulting paste (starch) is dried and repounded into flour before cooking. Cracked sorghum grains can occasionally be used to make a related product called *saino*. The flour from dehulled or whole sorghum is mixed with water and cooked into a thick stiff porridge that is eaten with a soup (sauce) composed of vegetables, meat, and other items depending upon the availability of the ingredients. The basic formula consists of four cups of flour and nine cups of water. Most of the water is brought to boiling. The remaining cold water is used to make a paste of the flour. Then the paste is added to the boiling water. This prevents lumps from forming. The mixture is stirred vigorously, and small amounts of flour are added to prevent lumps from forming. The thick porridge is cooled and consumed immediately; any leftovers are stored overnight (Tunde Obilana, 1981).

Injera is a leavened, flat, and round Ethiopian traditional bread made from cereals. The best quality sorghum *injera* is made from the dehulled grain. In the various languages of Ethiopia, *injera* has different names. In Oromigna it is *bidena*, but in Tigrigna it is known as *taita*. In Guragigna it is often called *tabita*, while in Wolaytigna it is *solo*. *Injera* is, however, an Amharic word and by far it is the most widely used and understood designation for leavened breads consumed in Ethiopia. The major source of protein and other noncarbohydrate nutrients for the traditional consumers of *injera* is mainly *wot*. *Wot* can be made from meat, pulses, vegetables, or their combinations. Various types of *wot* are recognized, the major ones being shiro (pea flour), kikh (split lentils, broad bean, or pea), vegetables, *doro* (chicken), egg, beef, mutton, and fish. Though these are the major ingredients of *wot*, most recipes would include red pepper, onion, butter or oil, and spices. Milk and milk products may also be served along with *injera* and *wot*. The normally sour *injera* and the spicy *wot* create a food taste combination that is considered ideal by Ethiopians depending on this staple. To prepare the dough of *injera* sift the flour to remove any foreign materials, add 1 liter of water and knead well by hand. Transfer the dough into a container and leave it for 48 h. Before baking the *injera*, sift the flour and pour the boiling water over the flour, mix well, and let it cool to 55°C. Add this preparation to the dough and mix well with water. Sprinkle ground rapeseed over it and then bake it for 2–3 min. Soft endosperm types with white or red pericarp, regardless of subcoat presence, produced unacceptable *injera*. Among the soft sorghum varieties that appeared the same visually, significant differences in keeping properties of *injera* were observed (Gebrekidan and GebreHiwo, 1981). Organoletic tests of the composite bread showed that 70% of the taste panelists rated the overall quality of 70 wheat: 30 sorghum composite bread as good. Proximate composition of the composite bread showed that it contained much lower protein and higher crude fiber contents than 100% wheat bread. The shelf life of composite bread at room temperature was lower than that of 100% wheat bread. It was found that addition of 2–4% pentosans to composite flour (up to 50% substitution level) improved the quality of the bread. Prefermentation of the sorghum flour up to 20 h was found to have positive effects on the baking quality of the wheat and sorghum composite flour. Good quality snacks were produced from wheat:sorghum composite flour up to the level of 60% substitution. It was also found that a blend of 70% sorghum flour and 30% detoxified cassava starch could produce acceptable bread and cakes.

Experience showed that composite flour required additional quantities of sugar, yeast, and water than 100% wheat, to make the bread tasty. Baking of composite dough required higher temperatures, and the texture of composite bread was less acceptable. One of the major problems facing bakers is the smaller volume of composite bread as compared to 100% wheat bread from similar quantities of composite flours and wheat flours (Idowu, 1989).

Wheat is substituted with local cereals like maize and sorghum in biscuit production. It was possible to produce biscuits of acceptable quality using 50:50 wheat and sorghum composite flour. Substitution of wheat flour up to 25% with sorghum flour in production of short and hard dough biscuits and wafers was cost-effective without affecting quality, breakages, and plant efficiency. It was suggested that improved milling techniques to produce sorghum flour with particle size comparable to that of wheat flour can increase the possibility of higher substitution (Priyolkar, 1989).

Sorghum-soy flour blends recorded the highest peak viscosities compared to other blends. Each flour blend showed a decline in its cooking properties, as the level of soy was increased from 20 to 30%. The biscuits from sorghum-soy blends were brown in color with a slight spread (Ogundipe and Omasevwerha, 1989).

The production of weaning foods utilizing sorghum and cowpea, and sorghum and soybean foods was described, and the products were evaluated. Preparation of sorghum and soybean *ogi* (weaning food in Nigeria) involved steeping sorghum grain for 24 h followed by wet milling and filtration through a vibroscreen ($<211\ \mu$). Soya beans were dried at 110°C for about 1 h followed be dehulling and cooking in excess water. The detoxified and cooked soybeans were wet milled, and the slurry was filtered through a vibroscreen ($<211\ \mu$). The sorghum and soybean slurries were mixed by dry weight proportion of 7:3 and left to ferment naturally, priming with 24 h sorghum step liquor. It was fermented until the required acidity (0.6%) by fermented product, which was then spray dried. The dried product was sieved and packed. During further trials, cost of drying was reduced by dewatering the sorghum and soybean slurries to give cakes. The cakes were blended and reduced in a hammer mill. The dried granules were milled into powder and mixed with other desired ingredients like vitamins and minerals. The sorghum and soybean compared well with popular commercial weaning foods in the market and had 16–21% protein.

Ogi is the most popular traditional weaning food in Nigeria and is prepared from sorghum, fermented maize, millet, or rice mash. Sorghum *ogi* contains 6–10% protein and about 90% starch. The quality of *ogi* can be improved by cofermenting sorghum with soybean in a ratio of 3:1. Ready-to-eat weaning foods can be prepared from sorghum by various methods: drum drying a cooked paste, extrusion cooking, and baking of dough into cookie-like products (Akingbala and Rooney, 1982).

The weaning practices in Nigeria were briefly reviewed. It was recognized that sorghum grains contain poor quality protein leading to nutritional deficiencies which can be rectified by supplementing sorghum diets with legumes. The study found that *tuwo* and *kamu* were the major traditional products prepared from sorghum. Agricultural Extension and Research Liaison Services (AERLS) has recommended a range of sorghum-based weaning foods supplemented with soybean, cowpea, and groundnuts. A variety of soups and snack foods were also formulated (Igbedioh and Dunmade, 1989).

Sorghum *sankati* is a type of thick porridge consumed in South India. It is called by different names in various regional languages, for example, *mudda* (Telugu), *mudde* (Kannada), and kali (Tamil). Sorghum *sankati* is consumed in the Rayalaseema tracts of Andhra Pradesh, the southern tracts of Karnataka, and all over Tamil Nadu. In the state of Tamil Nadu, about 60–70% of sorghum consumed is taken in the form of *sankati*. In Andhra Pradesh, approximately 50% of the sorghum produced is eaten as *sankati* while in Karnataka approximately 30% of the sorghum is consumed as *sankati*. *Sankati* is usually consumed with a range of side dishes such as sauce, dhal from pulses, pickles, chutneys, buttermilk, curd curries. A 1:3 proportion of sorghum grits/flour and water is used for preparing *sankati*. Water is allowed to boil in a vessel, and grits are added to boiling water coupled with stirring. After 10 min fines are added followed by stirring and this is continued for 3 min. The vessel is removed from the fire and poured onto a moist plate and made into balls of 10 cm diameter by hand. *Sankati* is eaten fresh or stored overnight in water or buttermilk.

Kisra, a thin pan cake-like leavened bread made from whole sorghum flour, is a staple diet of most people in the Sudan. It is usually served with *stews* (*mullah* and *tabbikh*) and other dishes, or at the simplest level it is served with a relish or sauce or even with just water and condiments. *Kisra* is served regularly for at least one of the three meals of the day. A thick paste, *ajin*, is prepared by mixing on an air-dry basis, 60% flour and 40% water (Culwick, 1951). The *ajin* is left to stand fermenting for approx. 12–24 h, by which time it develops a sour taste. Just before baking, the *ajin* is further diluted to a thin batter by the addition of more water at the rate of 50 mL water to 100 g paste. The baking pan is heated and then rubbed with a damp oily cloth before spreading the batter. The baking pan is set on stones in the case of an open fire or directly on the burning charcoal neatly piled up on the local burner. Then using the undispensable gargariba (a dry piece of palm leaf) dipped in water, batter is made into a very thin layer by using quick, smooth sideways movements. In about 1 min, the *kisra* is done and ready to be peeled off, and the pan is rubbed again with the oily cloth before spreading the next lot of batter. In general, *kisra* made from grains with cream color and less than 40% of the endosperm corneous had the best texture and keeping properties. Genotypes with high tannin and phenol content were rejected in consumer tests.

Maize is preferred for *tortilla* production in Central American countries; sorghum however is being consumed alone, or in mixtures with maize, to produce *tortillas* when maize supplies are low. In Mexico there is an increased interest in using sorghum for making *tortillas*. This is mainly due to low yields in maize production. Sorghum performs very similarly to maize during lime cooking, both in the rheological properties of the dough and in the quality of the *tortilla*. But the darker color produced by the tannins and phenols is frequently present in sorghum *tortilla*. Because of the possible rejection of the pure sorghum *tortilla*, due to the greenish and light pinkish color, *tortillas* made from mixtures of maize and sorghum are thought to be the most promising form in which sorghum will be accepted in the Mexican diet. Also the mixtures of maize and sorghum are a partial solution to low maize supply in Mexico. Sorghum grains without a subcoat, with intermediate texture, colorless pericarp, and low polyphenol content produced the best *tortillas*. Sorghum kernels with visually similar characteristics have significantly different *tortilla* making potential.

Tortilla preparation: Mix 150 g of sorghum kernel with 300 mL of distilled water and add 1.5 g of $Ca(OH)_2$. Boil this mixture for 30 min in a 600-mL beaker on a crude fiber extraction apparatus in order to diminish excess evaporation and to obtain uniform boiling treatment for different samples. Allow the mixture to cool and rest for 17 h. Make sure the beaker is covered with aluminium foil in order to prevent evaporation and contamination. Separate the kernels from the liquids with a sieve and wash the kernels with tap water accumulating the washed liquid into the original collected liquids. This washing will help remove excess alkali and loose pericarp. The wet kernels are ground very fine into dough (*masa*). Using a round press with a diameter of 15 cm, place about 30 g dough in a round ball shape in the middle of the press between two pieces of nylon plastic and press very gently to form the round tortilla. Place the tortilla on a heated metal plate and allow some time to cook on one side and then turn over to the other side for complete cooking.

The process of cooking maize in a solution of lime water to remove the pericarp is called nixtamalization. Sorghum is used alone or mixed with maize in parts of Central America and Southern Mexico where maize supplies are limited. People prefer maize tortillas, but significant quantities of sorghums are used for tortillas in Central America where sorghum with thick pericarp, intermediate endosperm texture without pigmented testa, and purple plant color are currently used. Sorghum tortillas are produced from whole grain, but laboratory and commercial trials have indicated that decorticated sorghums have excellent potential for nixtamalization. Whole sorghum is cooked in excess water containing 0.5–0.8% lime (based on grain weight). The grain is held near the boiling point for 10–20 min and steeped for several hours or overnight. During washing, the pericarp is removed. The washed nixtamal contains 50% moisture; water added during grinding increases the *masa* moisture to 54%. The *masa* is shaped by hand to form tortillas which are baked at 260–280°C for 30–40 s on one side, 30 s on the other side, and 20 s more on the first side. Table tortillas puff during the last bake. They are usually consumed fresh, so tortillas are made daily. Nixtamalized maize and sorghum can be used to produce various snacks, and the masa is cooked in water to produce thick and thin porridges and beverages. In Mexico, sorghum grain is significantly less expensive than maize, so great potential exists for the use of white sorghums for human food.

In India, particularly in the southern regions, boiled sorghum (rice-like) called *Annam* or *soru* is one of the common products prepared, and it accounts for about 10% of the total sorghum grain produced (Fig. 1.3). A similar product of sorghum is known as *khichuri* in Bangladesh, *lehata-wagen* in Botswana, *kaoliang mi fan* in China, *nufro* in Ethiopia, and *Oka baba* in Nigeria. The freshly prepared product is consumed with dhal, sambar, buttermilk with pickles, or onion and green chillies for lunch or supper. Sometimes, it is stored overnight by adding water and consumed the next morning with buttermilk. Sorghum is traditionally used as boiled sorghum after dehulling the grains. The dehulled grain is cooked in water in the proportion of 1:3. Sometimes the grains are also soaked overnight in water and cooked the following morning. The cooking is preferably done in an earthen pot, which is heated using firewood. The grains are cooked to softness and the excess water is drained off. The cooked product has to be fluffy, uniform yellow, or creamy white in color with a sweet taste.

Couscous is a steamed granulated product made from cereal flour. It is the principal cereal food of North Africa, the Sahara, and the Sahel. In North Africa and in the Niger inland delta, couscous is prepared from wheat, whereas in the Sahel it is prepared from pearl millet or sorghum. Couscous can be prepared directly into a steamed product that is eaten with sauce. It can be steamed, sun-dried, stored indefinitely, and is reconstituted in milk, or again steamed and served with a sauce. Whole or dehulled grain is reduced to flour, which is sifted through a sieve with 1-mm mesh openings. The flour is wetted with cool water and agglomerated into small particles with the fingers. Those flour aggregates are forced through a sieve with 1.5-mm mesh openings. The wet aggregates are steamed in a covered perforated pot, which is placed directly over another pot containing boiling water. The juncture between the two pots is sealed with a damp cloth to force the steam through the perforations and into the flour aggregates. The wet cloth is wiped with okra powder to assure a tight seal. After about 15 min steaming, the aggregates form a large single chunk, which is taken out of the bowl, broken up into aggregates, and again steamed for an additional 15 min. The aggregates are again broken up into singular units and sifted through a 2.5-mm sieve. At that point, the steamed aggregates can be dried and stored for future use. If couscous is to be consumed immediately, the aggregates are sprinkled with cool water and mixed thoroughly with the fingers. Baobab (*Adansonia digitata*) leaf

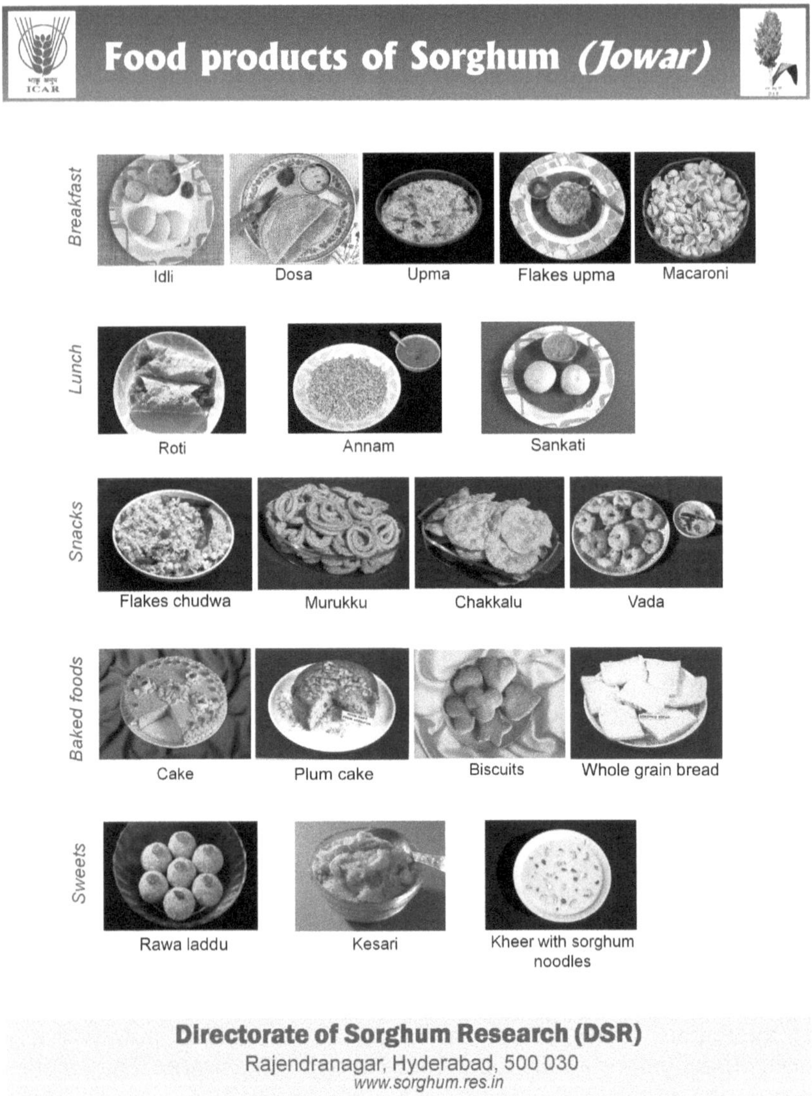

FIGURE 1.3 Food products of Sorghum (Jowar).

powder is mixed with the particles. This powder serves as a lubricant which prevents desiccation and stickiness and improves palatability. Okra powder can be used as a substitute for this. After mixing the aggregates are again placed in the perforated bowl and steamed for about 15 min. The couscous is allowed to cool slowly and then served with sauce.

1.3.4 Noodles

Noodles made from sorghum are not very common and the acceptability of sorghum noodles was not a problem; however, it was not given adequate attention. The technology for noodles from sorghum makes the product cheaper and healthier as sorghum products are known for their high content B vitamin and dietary fiber.

1.3.5 Low Calorie-Low Fat Cookies

Sorghum grain contains 63.4–72.5% starch, 17.8–21.9% amylose, 7.9–11.5% protein, 1.86–3.08% fat, and 1.57–2.41% fiber. The recovery of starch from sorghum grain is 5–8% less than maize but is equally lustrous. The recovery can be increased by increasing grain size and reducing protein and fiber contents. Sorghum, being a cheaper source for the production of starch and starch by-products, could occupy a pivotal position.

1.3.6 Sorghum Health Benefits

- A study published in the Journal of Food Science states that sorghum flour contains compounds which can help to lower cholesterol.
- Since sorghum flour is a gluten-free food, it is a safe alternative to flour for those with celiac disease. This is good news for those who cannot eat wheat or other grains such as rye or barley due to gluten intolerance.
- Sorghum flour is high in antioxidants, which are related to their polyphenolic compound and sorghum wax. These compounds contain policosanols, which are important to good cardiac health and controlling cancers.
- The starch and protein in sorghum take longer than other similar products to digest. This slow digestion results in very low glycimic index is particularly helpful for those with diabetes.
- One of the main cooking benefits of sorghum flour is that it is relatively tasteless and it adapts well to whatever it is cooked with. It can be substituted for regular wheat flour in most baked goods without altering the taste of the finished product.
- Sorghum is rich in potassium and phosphorus. It has good amount of calcium with small amounts of iron and sodium.
- It is very high in iron, fiber, and protein, making it one of the healthier options for those who have gluten intolerance and cannot eat traditional forms of flour.
- Sorghum is a good source of B vitamins, such as thiamin, riboflavin, vitamin B6, biotin, and niacin, but refining produces losses of all

B vitamins. Mineral composition of sorghum is similar to that of millet. The chief minerals present in sorghum grain are potassium and phosphorus, while calcium is low.
- Sorghum food provides nonglutinous flours which is useful for avoiding damage to the lining of the intestine and makes it easy for nutrients to be absorbed.
- It also avoid dermatitis herpetiforms disease.
- Sorghum is a nonacid forming food.
- Malted sorghum flour can be used for anticonstipation drinks.
- Selected sorghum genotypes can provide sufficient amounts of lysine to meet human requirements.
- Sorghum diet is very good to the people suffering from the Jaundice and obesity.
- It also reduces hunger and keeps fuller for longer time.
- It also helps to manage polycystic ovary syndrome in women.

Sorghum plays a very important role in animal feed production in various countries like the USA, Mexico, South America, and Australia. The nutritional value of good quality sorghum has an average feeding value of 96 to 98% of corn. Sorghum can be processed to further improve its food as well as feed value, and techniques such as grinding, crushing, steaming, steam flaking, popping, and extruding have all been used to enhance the grain for food and feeding purposes. The products are then fed to beef or dairy cattle, egg laying hens, poultry, pigs, and are also used in pet foods.

Industrial uses for sorghum include wallboard and biodegradable packaging materials. In addition to the wallboard, grain sorghum can be processed interchangeably with corn for the production of ethanol—many ethanol-producing factories within the United States use grain sorghum as their principal grain due to their favorable geographic location. A very valuable coproduct created from the processing of grain sorghum for ethanol is dry distillers' grain. Grain sorghum has a high protein content so the production of an animal feed supplement from waste products generates additional revenues for the ethanol processing plant.

References

Adams, C.A., 2001. Raw material processing: digestion and absorption of nutrient. Total Nutrition: Feeding Animals for Health and Growth. Chapter 4. Nottingham University Press, United Kingdom, pp. 65–92.

Akingbala, J.O., Rooney, L.W., 1982. Variation in amylose content among sorghums. In: Rooney, L.W., Murty, D.S. (Eds.), Proceedings of the International Symposium on Sorghum Grain Quality, ICRISAT, 28–31 October 1981, Patancheru, India, pp. 269–279.

AOAC, 1995. 16th ed. Official Methods of Analysis of AOAC International, vol. II. AOAC International, Suite, Arlington, USA, Chapter 27.3.6, Chapter 27.4.4.
Awika, J.M., Rooney, L.W., 2004. Sorghum phytochemicals and their potential impact on human health. Phytochemistry. 65, 1199–1221.
Axtell, J.D., Kirleis, A.W., Hassen, M.M., Mason, N.D., Mertz, E.T., Munck, L., 1981. Digestibility of sorghum proteins. Proc. Natl. Acad. Sci. U.S.A. 78, 1333–1335.
Banks, W., Greenwood, C.T., 1966. The fine structure of amylose: the action of pullulanase as evidence of branching. Arch. Biochem. Biophys. 117, 674–675.
Béleon, A., Colonna, P., Planchot, V., Ball, S., 1998. Starch granule: structure and biosynthesis. Int. J. Biol. Macromol. 23, 85–112.
Black, J.L., 2001. Quality feed grains – research highlights and opportunities. In: Proceedings of the 10 Australian Barley Technical Symposium. Available at http://www.regional.org.au/au/abts/2001. Internet (accessed on 02.11.02.).
Blanshard, J.M.V., 1986. The significance of structure and function of the starch granule in baked products. In: Blanshard, J.M.V. (Ed.), Chemistry and Physics of Baking. Royal Society of Chemistry, London, pp. 1–13.
Boling, M.B., Nancy, E., 1981. Bogobe: Sorghum porridge of Botswana. In: Procedings of the International Symposium on Sorghum Grain Quality, 28–31 October, Patancheru, India.
Bramel-Cox, P.J., Hancock, J.D., Kumar, K.A., Andrews, D.J., 1995. Sorghum and millets for forage and feed. In: Dendy, D.A.V. (Ed.), Sorghum and Millets Chemistry and Technology. American Association of Cereal Chemist, St. Paul, MN, USA, pp. 325–364.
Breuer Jr., L.H., Dohm, C.K., 1972. Comparative nutritive value of several sorghum grain varities and hybrids. J. Agri. Food Chem. 20 (1), 83–86.
Butler, L.G., Riedl, D.J., Lebryk, D.G., Blytt, H.J., 1984. Interaction of proteins with sorghum tannin: mechanism, specificity and significance. J. Am. Oil Chem. Soc. 61, 916–920.
Cousins, B.W., Tanksley Jr., T.D., Knabe, D.A., Zebrowska, T., 1981. Nutrient digestibility and performance of pigs fed sorghums varying in tannin concentration. J. Anim. Sci. 53 (6), 1524–1537.
Culwick, G.M., 1951. Diets in the Gezira Irrigated Area, Sudan. Government Printing Press, Khartoum, Sudan.
Cura, J.A., Jansson, P.E., Krisman, C.R., 1995. Amylose is not strictly linear. Starch/Stärke. 47, 207–209.
Davis, A.B., Hoseney, R.C., 1979. Grain sorghum condensed tannins. I. Isolation, estimation, and selective adsorption by starch. Cereal Chem. 56 (4), 310–314.
Douglas, J.H., Sullivan, T.W., Bond, P.L., Struwe, F.J., 1990. Nutrient composition and metabolizable energy values of selected grain sorghum varieties and yellow corn. Poult. Sci. 69, 1147–1155.
Dubois, M., Giller, K.A., Hamilton, J.K., Rebers, P.A., Smith, F., 1956. Colorimetric method for determination of sugars and related substances. Anal. Chem. 28, 350–356.
Earp, C.F., Rooney, L.W., 1982. Scanning electron microscopy of the pericarp and testa of several sorghum varieties. Food Microstruct. 1, 125–134.
Elkin, R.G., Freed, M.B., Hamaker, B.R., Zhang, Y., Parsons, C.M., 1996. Condensed tannins are only partially responsible for variations in nutrient digestibilities of sorghum grain cultivars. J. Agric. Food. Chem. 44, 848–853.
Elmalik, M., Klopfenstein, C.F., Hoseney, R.C., Bates, L.S., 1986. Digestibility and nutritional quality of sorghum grain with contrasting kernel characteristics. Nutr. Rep. Int. 34 (5), 811–820.
FAO, 1988. In: Alexandratos, N. (Ed.), World Agriculture Toward 2000: An FAO Study. Bellhaven Press, London, p. 338.
FAO, 1995. Production and Utilization. Ch. 2 in Sorghum and Millets in Human Nutrition. FAO, Rome, pp. 13–30.

French, D., 1972. Fine structure of starch and its relationship to the organization of starch granules. J. Jpn. Soc. Starch Sci. 19 (1), 8–25.

French, D., 1984. Organization of starch granules. In: Whistler, R.L., BeMiller, J.N., Paschall, E.F. (Eds.), Starch Chemistry and Technology, second ed. Academic Press, Orlando, USA, pp. 183–247.

Gallant, D.J., Bouchet, B., Buléon, A., Pérez, S., 1992. Physical characteristics of starch granules and susceptibility to enzymatic degradation. Eur. J. Clin. Nutr. 46, S3–S16.

Gallant, D.J., Bouchet, B., Baldwin, P.M., 1997. Microscopy of starch: evidence of a new level of granule organization. Carbohydr. Polym. 32, 177–191.

Gebrekidan, B., Gebrehiwot, B., 1981. Sorghum injera preparations and quality parameters. In: Procedings of the International Symposium on Sorghum Grain Quality, 28–31 October, Patancheru, India.

Griffiths, D.W., Mosely, G., 1980. The effect of diets containing field beans of high or low polyphenolic content on the activity of digestive enzymes in the intestines of rats. J. Sci. Food. Agric. 31 (3), 255–259.

Hamaker, B.R., Bugusu, B.A., 2003. Overview: sorghum protein and food quality. In: Paper Presented at Workshop on the Protein of Sorghum and Millets: Enhancing Nutritional and Functional Properties for Africa [CD]. Pretoria, South Africa.

Hamaker, B.R., Kirleis, A.W., Butler, L.G., Axtell, J.D., Mertz, E.T., 1987. Improving the in vitro protein digestibility of sorghum with reducing agents. Proc. Natl. Acad. Sci. U.S.A. 84, 626–628.

Hibberd, C.A., Wagner, D.G., Schemm, R.L., Mitchell Jr., E.D., Weibel, D.E., Hintz, R.L., 1982. Nutritive characteristics of different varieties of sorghum and corn grain. J. Anim. Sci. 55 (3), 665–672.

Hizukuri, S., 1986. Polymodal distribution of the chain lengths of amylopectins, and its significance. Carbohydr. Res. 147, 342–347.

Hood, L.F., 1982. Current concepts of starch structure. In: Lineback, D.R. (Ed.), Food Carbohydrates. The AVI publishing Company, Inc., Westport, CT, pp. 217–236.

Hoseney, R.C., Davis, A.B., Herbers, L.H., 1974. Pericarp and endosperm structure of sorghum grain shown by scanning electron microscopy. Cereal Chem. 51, 552–558.

Hubbard, J.E., Hall, H.H., Earle, F.R., 1950. Composition of the component parts of the sorghum kernel. Cereal Chem. 27, 415–420.

Idowu, A., 1989. Bread from composite flours. In: Proceedings of a Symposium on the Current Status and Potential of Industrial Uses of Sorghum in Nigeria, 4–6 December, Kano, Nigeria, pp. 22.

Igbedioh, S.O., Dunmade, V.B., 1989. Uses of Sorghum in infant weaning practices: the AERLS experience. In: Proceedings of a Symposium on the Current Status and Potential of Industrial Uses of Sorghum in Nigeria, 4–6 December, Kano, Nigeria, pp. 39.

Imberty, A., Buleon, A., Tran, V., Perez, S., 1987. Recent advances in knowledge of starch structure. Starch. 43, 375–384.

Jane, J., Xu, A., Radosavljevic, M., Seib, P.A., 1992. Location of amylose in normal starch granules. I. Susceptibility of amylose and amylopectin to cross-linking reagents. Cereal Chem. 69, 405–409.

Jenkins, P.J., Cameron, R.E., Donald, A.M., 1993. A universal feature in the structure of starch granules from different botanical source. Starch. 45 (12), S.417–420.

Kasemsuwan, T., Jane, J., 1994. Location of amylose in normal starch granules. II. Locations of phosphodiester cross-linking revealed by phosphorus-31 nuclear magnetic resonance. Cereal Chem. 71, 282–287.

Kelley, T.G., Parthasarathy Rao, P., Singh, R.P., 1992. Trends in sorghum production and utilization in Asia. Economics Group, Resource Management Programme, Progress Report 108. Patancheru, Inde, ICRISAT.

King, D., Fan, M.Z., Ejeta, G., Asem, E.K., Adeola, O., 2000. The effects of tannins on nutrient utilization in white Pekin duck. Br. Poult. Sci. 41, 630–639.

Knabe, D.A., 1990. Sorghum as swine feed. In: Ejeta, G. (Ed.), Proceedings of the International Conference on Sorghum Nutritional Quality. Purdue University, West Lafayette, IN, pp. 95–107.

Lichtenwalner, R.E., Ellis, E.B., Rooney, L.W., 1978. Effect of incremental dosages of the waxy gene of sorghum on digestibility. J. Anim. Sci. 46 (4), 1113–1119.

Lineback, D.R., 1984. Properties of certain starches. Bakers Dig. 52, 16–21.

Longstaff, M., McNab, J.M., 1991. The inhibitory effects of hull polysaccharides and tannins of field beans (Viciafaba) on the digestion of amino acids, starch, lipids, and on digestive enzyme activities in young chicks. Br. J. Nutr. 65, 199–219.

MacLean, W.C., Lopez de Romana, G., Placko, R.P., Graham, G.G., 1981. Protein quality and digestibility of sorghum in preschool children: balance studies and plasma free amino acids. J. Nutr. 111, 1928–1936.

Matlon, P.J., 1990. Improving productivity in sorghum and pearl millet in semi-arid Africa. Food Res. Inst. Stud. 22, 1–43.

Maxson, E.D., Rooney, L.W., Lewis, R.W., Clark, L.E., Johnson, J.W., 1973. The relationship between tannin content, enzyme inhibition, rat performance, and characteristics of sorghum grain. Nutr. Rep. Int. 8 (2), 145–152.

Michael, Brouk, 2012. Sorghum in Beef production Feeding Guide. Kansas state University, Manhattan, Kansas, http://sorghumcheckoff.com/wp-content/uploads/2012/06/beefguideforweb.pdf.

Miller, F.R., Lowry, R.S., Monson, W.G., Barton, G.W., Cruzado, H.J., 1972. Estimates of dry matter digestibility difference in grain of some *Sorghum bicolor* (L.) Moench varieties. Crop. Sci. 12, 563–566.

Miller, O.H., Burns, E.E., 1970. Starch characteristics of selected grain sorghums as related to human foods. J. Food. Sci. 35, 666–668.

Mukuru, S.Z., Mushonga, J.N., Murthy, D.S., 1981. Sorghum ugali. In: Proceedings of the International Symposium on Sorghum Grain Quality, 28–31 October, Patancheru, India.

Murthy, D.S., Subramanian, V., 1981. Sorghum roti: I. Traditional methods of consumption and standard procedures for evaluation. In: Proceedings of the International Symposium on Sorghum Grain Quality, ICRISAT, Patancheru, India, pp. 73–78.

Nikuni, Z., 1978. Studies on starch granules. Starch/Stärke. 30, 105–111.

Ogundipe, H.O., Omasevwerha, O., 1989. Processing and acceptability studies of soy-enriched non wheat Biscuits. In: Proceedings of a Symposium on the Current Status and Potential of Industrial Uses of Sorghum in Nigeria, 4–6 December, Kano, Nigeria, pp. 24.

Ong, M.H., Jumel, K., Tokarczuk, P.F., Blanshard, J.M.V., Harding, S.E., 1994. Simultaneous determinations of the molecular weight distributions of amyloses and the fine structures of amylopectins of native starches. Carbohydr. Res. 260, 99–117.

Oostergetel, G.T., van Bruggen, E.F.J., 1989. On the origin of a low angle spacing in starch. Starch. 41, 331–335.

Oria, M.P., Hamaker, B.R., Schull, J.M., 1995a. In vitro protein digestibility of developing and mature sorghum grain in relation to α-, β-, and γ-kafirin disulfide crosslinking. J. Cereal Sci. 22, 85–93.

Oria, M.P., Hamaker, B.R., Schull, J.M., 1995b. Resistance of sorghum α-, β-, and γ-kafirins to pepsin digestion. J. Agric. Food. Chem. 43, 2148–2153.

Oria, M.P., Hamaker, B.R., Axtell, J.D., Huang, C.P., 2000. A Highly digestible sorghum mutant cultivar exhibits a unique folded structure of endosperm protein bodies. Proc. Natl. Acad. Sci. U.S.A. 97 (10), 5065–5070.

Ørskov, E.R., 1986. Starch digestion and utilization in ruminants. J. Anim. Sci. 63, 1624–1633.

Peat, S., Whelan, W.J., Thomas, G.J., 1952. Evidence of multiple branching in waxy maize starch. J. Chem. Soc.4546–4549.
Pour-reza, J., Edriss, M.A., 1997. Effects of dietary sorghum of different tannin concentrations and tallow supplementation on the performance of broiler chicks. Br. Poult. Sci. 38, 512–517.
Price, M.L., Butler, L.G., 1977. Rapid visual estimation and spectrophotometric determination of tannin content of sorghum grain. J. Agric. Food. Chem. 25, 1268–1273.
Priyolkar, V.S., 1989. Use of Sorghum flour in biscuit and wafer production: the Nasco experience. In: Proceedings of a Symposium on the Current Status and Potential of Industrial Uses of Sorghum in Nigeria, 4–6 December, Kano, Nigeria, pp. 22.
Reardon, T., Matlon, P., 1989. Seasonal food insecurity and vulnerability in droughtaffected regions of Burkina Faso. In: Sahn, D.E. (Éd.), Seasonal variability in third world agriculture. The consequene es for food security. Etats-Unis, International Food Policy Research Institute, Johns Hopkins University, Baltimore, Maryland, pp. 118–136.
Ring, S.H., Akingbala, J.O., Rooney, L.W., 1982. Variation in amylose content among sorghums. In: Rooney, L.W., Murty, D.S. (Eds.), Proceedings of the International Symposium on Sorghum Grain Quality, 1981. ICRISAT, Pantacheru, AP, India, pp. 269–279.
Robin, J.P., Mercier, C., Charbonniere, R., Guilbot, A., 1974. Lintnerized starches, gel filtration, and enzymatic studies of insoluble residues from prolonged acid treatment of potato starch. Cereal Chem. 51, 389–406.
Rooney, L.W., Miller, F.R., 1982. Variation in the structure and kernel characteristics of sorghum. In: Rooney, L.W., Murty, D.S. (Eds.), Proceedings of the International Symposium on Sorghum Grain Quality, 1981. ICRISAT, Pantacheru, AP, India, pp. 143–162.
Rooney, L.W., Pflugfelder, R.L., 1986. Factors affecting starch digestibility with special emphasis on sorghum and corn. J. Anim. Sci. 63, 1607–1623.
Rowe, J.B., Choct, M., Pethick, D.W., 1999. Processing cereal grains for animal feeding. Aust. J. Agric. Res. 63, 1607–1623.
Seckinger, H.L., Wolf, M.J., 1973. Sorghum protein ultrastructure as it relates to composition. Cereal Chem. 50, 455–465.
Sherrod, L.B., Albin, R.C., Furr, R.D., 1969. Net energy of regular and waxy sorghum grains for finishing steers. J. Anim. Sci. 29, 997–1000.
Shull, J.M., Watterson, J.J., Kirleis, A.W., 1992. Purification and immune cytochemical localization of kafirins in *Sorghum bicolor* (L. Moench) endosperm. Protoplasma. 171, 64–74.
Southgate, D.A.T., 1976. On Determination of Food and Carbohydrates. Applied Sciences Publishers Ltd., pp. 52–55.
Streeter, M.N., Wagner, D.G., Hibberd, C.A., Owens, F.N., 1990. Comparison of corn with four sorghum grain hybrids: site and extent of digestion in steers. J. Anim. Sci. 68, 3429–3440.
Subramanian, V., Jambunathan, R., 1980. Traditional methods of processing sorghum (*Sorghum bicolor*) and Pearl millet (*Pennisetum americanum*) grains in India. Rep. Intl. Assoc. Cereal Chem. (ICC). 10, 115–188.
Sullins, R.D., Rooney, L.W., 1974. Microscopic evaluation of the digestibility of sorghum lines that differ in endosperm characteristics. Cereal Chem. 51, 134–142.
Takeda, Y., Hizukuri, S., Takeda, C., Suzuki, A., 1987. Structures of branched molecules of amyloses of various origins, and molar fractions of branched and unbranched molecules. Carbohydr. Res. 165, 139–145.
Taylor, J.R.N., Novellie, L., Liebenberg, N.V.D.W., 1984. Sorghum protein body composition and ultrastructure. Cereal Chem. 61, 69–73.
Taylor, J.R.N., Schüssler, L., Liebenberg, N.V.D.W., 1985. Protein body formation in the starchy endosperm of developing *Sorghum bicolor* (L.) Moench seeds. S. Afr. J. Bot. 51, 35–40.
Tunde Obilana, A., 1981. Traditional sorghum foods in Nigeria: their preparation and quality parameters. In: Proceedings of the International Symposium on Sorghum Grain Quality, 28–31 October, Patancheru, India.

Walker, T.S., 1990. Demand and supply prospects far sorghum: implication for crop improvement strategy. Presented at the 20th Annual Sorghum Workshop, Mahatma Phule Agricultural University, Rahuri, Inde, 30 April–2 May.
Weaver, C.A., Hamaker, B.R., Axtell, J.D., 1998. Discovery of grain sorghum germplasm with high uncooked and cooked in vitro protein digestibility. Cereal Chem. 75 (5), 665–670.
Wester, T.J., Gramlich, S.M., Britton, R.A., Stock, R.A., 1992. Effect of grain sorghum hybrid on in vitro rate of starch disappearance and finishing performance of ruminants. J. Anim. Sci. 70, 2866–2876.
Wheeler, E.L., Ferral, R.E., 1971. A method for phytic acid determination in wheat and wheat fractions. Cereal Chem. 48, 312–320.
Williams, V.R., Wu, W.T., Tsai, H.Y., Bates, H.G., 1958. Varietal differences in amylose content of rice starch. J. Agric. Food. Chem. 6, 47–48.
Willis, R.B., Montgomery, M.E., Allen, P.R., 1996. Improved method for manual, colorimeric determination of total Kjeldhal nitrogen using salicylate. J. Agric. Food. Chem. 44, 1804–1807.
Zhang, G., 1997. Sorghum starch hydrolysis and digestibility (M.S. thesis). Purdue University. West Lafayette, IN.
Zobel, H.F., 1988. Molecules to granules: a comprehensive starch review. Starch/Stärke. 40 (2), 44–50.
Zobel, H.F., 1992. Starch granule structure. In: Alexander, R.J., Zobel, H.F. (Eds.), Developments in Carbohydrate Chemistry. Amer. Assoc. Cereal Chemistry, St Paul, MN., pp. 1–36

Further Reading

Da, S., Akingbala, J.O., Rooney, L.W., Scheyring, J.F., Miller, F.R., 1981. Evaluation of to quality in a sorghum breeding program. In: Proceedings of the International Symposium on Sorghum Grain Quality, 28–31 October, Patancheru, India.
FAO, 2002a. FAOSTAT agriculture data; in agricultural production: crops primary. Available at http://apps.fao.org/page/collections?subset = agriculture; Updated 28 May 2002. Internet (accessed 08.08.02.).
FAO, 2002b. FAOSTAT agriculture data; in commodity balances: crops primary equivalent. Available at http://apps.fao.org/page/collections?subset = agriculture; Updated 4 July 2002. Internet (accessed 08.08.02.).
Nutrient Data, 2005. USDA SR-21.

CHAPTER

2

Malting and Brewing of Sorghum

C.V. Ratnavathi[1] and U.D. Chavan[2]

[1]ICAR-Indian Institute of Millets Research, Rajendranagar, Hyderabad, India [2]Mahatma Phule Krishi Vidyapeeth, Rahuri, Maharashtra, India

OUTLINE

Sorghum Biochemistry: An Industrial Perspective.
DOI: http://dx.doi.org/10.1016/B978-0-12-803157-5.00002-2

Abstract

Sorghum is traditionally used for alcoholic and nonalcoholic fermented beverages in African countries. The potential of sorghum for its malt and beer was studied. Alpha amylase and diastatic activity of different cultivars were studied. Sorghum showed highest amylase activities at 96 h germination. Lager beer is produced from 100% sorghum using exogenous enzymes. The lager beer from sorghum malt and adjunct showed superiority in composition, such as increased free alpha amino nitrogen and color units in sorghum malt and adjunct beer. A good adjunct should have low saccharification time and should have the highest percent extraction of grain. The highest amount of alcohol and free alpha amino nitrogen were recovered with adjuncts of CSH 18 and CSH 16 with SPV 824 malt. Sensory analysis indicated that no significant differences existed between the sorghum beer and a commercial malted barley beer with regard to aroma, mouth-feel, after-taste, and clarity.

2.1 INTRODUCTION

Sorghum is traditionally used for alcoholic and nonalcoholic fermented food beverages in African countries. Most of the food beverages are nutritious as they are either malted or fermented. Apart from this, sorghum also has the great potential of being used as a raw material in the brewery sector. The potential of sorghum as malt and adjunct were extensively reviewed by Owuyama (1997). Alcoholic drinks such as *burukutu, tala, pito, sorghum wine,* and *kafir beer* and nonalcoholic beverages such as *kunu-zaki* are common among the people in savanna regions of Africa and Asia (Owuama, 1991). Nasha is a traditional weaning food (infant porridge) prepared by fermentation of sorghum flour (Graham et al., 1986).

Barley is used normally for the preparation of malt. Lager beer is a fermented beverage made from barley malt and unmalted cereal adjunct. Malts are prepared from cereals (barley in particular) after the process of steeping, germination, and kilning the grains under controlled conditions. Barley is traditionally used for malting and brewing lager beer. Sorghum was first used as a substrate for lager beer production during World War II (Hahn, 1966), and its usage has continued either as a substitute to barley or as an adjunct in varied percentages. The use of sorghum for lager beer is known in Mexico and Nigeria (Okafor and Aniche, 1980).

2.2 USE OF SORGHUM AS MALT

In Nigeria, following the ban on the import of barley malt in 1988, the brewing industry has been utilizing sorghum and maize as raw materials for lager beer production. Sorghum can be used as raw grains, grits, or malted material. The bulk of the sorghum grain used by the brewers in Nigeria was from the varieties SK 5912 and farafara. The kernels of sorghum varieties differ in size, structure, texture, hardness, pigmentation, and biochemical characteristics (Novellie, 1962b; Palmer, 1989; Swanston et al., 1994). However, the selection of a genotype for malt depends not only on size of the kernel but also on malt parameters such as malting loss, free amino nitrogen (FAN), beta glucans, and diastatic activity. The malting quality of the cultivars was found to be influenced by seasonal effects. (van Loggerenberg and Pretorius, 2004).

In India, sorghum is used largely as a food grain, while in many African states it is also of importance in commercial brewing. Recent studies have indicated the feasibility of using sorghum malt for brewing lager beer. While the production of lager beer from barley malt along with sorghum as a cereal adjunct poses no problem, lager beer brewing from 100% sorghum is confronted with problems relating to equipment, sorghum malting, mash gelatinization, saccharification, lautering, wort fermentability, body fullness, and acceptability. Development of nonbiological haze caused by polyphones and insoluble proteins present in sorghum malt and the presence of high lipid content are other unfavorable aspects encountered during sorghum brewing. There is scope, however, to identify sorghum cultivars with desirable malting qualities through searching for natural variability or identifying sorghum cultivars with desirable malting qualities so that barley malt with sorghum malt may provide a cheaper indigenous cereal alternative for lager beer brewing in countries having substantial sorghum production.

Alpha- and beta-amylases and other glucanases mediate starch hydrolysis during germination. Among these malting enzymes, alpha-amylase is the most important in sorghum, accounting for about 75% of the saccharifying activity. The beta-amylase content of sorghum malt is very low compared with that of barley malt. Certain South African sorghum cultivars were reported to possess beta-amylase activity that accounted for 18–39% of starch saccharification. Synthesis of all malting enzymes in sorghum takes place during germination. Activities of alpha- and beta-amylases have been found to increase steadily in an almost constant proportion of 3:1 during the course of germination until a peak of amylolytic activity is attained. Whereas the malting quality of African bird-proof cultivars, measured in terms of the extent of modification and production of peptone soluble diastolic activity, was observed to be higher than in nonbird-proof types, the water-extractable diastolic activities in kilts from the former cultivars were much lower than in those from the latter types. This apparent anomaly is attributable to the interaction between diastase enzymes and polyphenols during aqueous extraction resulting in insoluble polyphenol-enzyme complexes. The enzymes in these insoluble complexes are, nevertheless, active during mashing, and since the polyphenols and enzymes are located in different tissues, malting is unaffected. By contrast, the enzymes present in malt from nonbird-proof low-tannin cultivars are largely water-extractable.

The production of amylases and other diastatic enzymes during sorghum malting is influenced by ambient factors such as temperature, moisture, and humidity, as well as by cultivar. Although the temperature and soaking conditions that are optimal for the development of diastolic power in sorghum are established, different durations of steeping and germination have been used for sorghum malting. Genetic variability in sorghum for all important grain characters except malt ability is well understood. While the possibility of identifying specific malting conditions that may enhance the diastolic power in certain genotypes cannot be excluded, investigations aimed at assessing genetic variability for malt ability in sorghum demand malting of diverse genotypes under identical conditions that would preferably promote optimal production of malting enzymes in all of them. A study was undertaken to identify the steeping and germination times that are optimal for the development of alpha-amylase in different sorghum genotypes and to establish the regression between alpha-amylase activity and diastatic activity (Ratnavathi and Bala Ravi, 1991).

2.3 MALTING STUDIES

Four sorghum cultivars, IS 1347, IS 20503, SPV 736, and CSH 9, having different endosperm characteristics were germinated for 24, 48, 72, 96, and 120 h following 24 h steeping to determine the germination

TABLE 2.1 Mean Alpha-Amylase Activities (g Maltose Liberated in 3 min/mL Extract) in Four Sorghum Cultivars After Different Periods of Germination

		Germination time (h)					
S. no.	**Cultivar**	**24**	**48**	**72**	**96**	**120**	**Mean**
1.	IS 1347	120	443	530	845	230	434
2.	IS 20503	218	723	845	980	313	616
3.	SPV 736	150	275	288	408	300	284
4.	CSH 9	90	390	520	620	463	417
5.	Mean	144	458	546	713	326	
6.	C.D. @0.05						
7.	Cultivar mean						55
8.	Germination time mean						61
9.	Interaction mean						123

period at which maximum development of alpha-amylase activity occurred. The mean alpha-amylase activities at different durations of germination are presented in Table 2.1. A steady increase in alpha-amylase activity from 24 to 96 h of germination and a subsequent decline in activity were conspicuous in all cultivars. Analysis of variance of these results (Table 2.1) indicated that production of alpha-amylase was significantly influenced by cultivar, germination time, and their interactions. Cultivars differed significantly in their mean amylase activities. The highest and lowest activities were found in IS 20503 and SPV 736, respectively. Intermediate activities were noted in IS 1347 and CSH 9. All cultivars showed highest amylase activities at 96 h germination, which were significantly different from the mean activities at other durations of germination (Table 2.1). The response of individual cultivars to different times of germination was not alike, however. For instance, amylase activity in IS 20503 at 24 h germination was significantly higher than that of CSH 9. As germination progressed, the amylase activities increased at different rates. Similarly, the decreases in amylase activity at 120 h were significant in all cultivars except SPV 736.

The results of the studies on the effect of steeping time on alpha-amylase production in six cultivars possessing different endosperm characteristics are presented in Table 2.2. Analysis of variance (Table 2.2) showed that the production of alpha-amylase activity was influenced by cultivar, steeping time, and their interactions. A constant malting period of 96 h was used in these experiments. The mean

TABLE 2.2 Mean Alpha-Amylase Activities (g Maltose Liberated in 3 min/mL Extract) in Four Sorghum Cultivars After Different Periods of Steeping

		Period of steeping (h)				
S. no.	**Cultivar**	**12**	**16**	**20**	**24**	**Mean**
1.	IS 1347	680	635	880	845	760
2.	IS 20503	790	720	814	840	791
3.	IS 23860	795	588	830	618	708
4.	IS 23930	365	590	815	550	580
5.	SPV 736	565	805	765	505	660
6.	CSH 9	790	825	940	830	846
7.	Mean	664	694	841	698	
8.	CD @0.05					
9.	Cultivar mean					71
	Steeping time mean					58
10.	Interaction mean					143

alpha-amylase activity increased steadily in all cultivars with steeping times from 12 up to 20 h and decreased thereafter; the highest activity was observed at 20 h steeping and was significantly different from those at other steeping times. Alpha-amylase activities at 24, 16, and 12 h steeping were lower than the peak activity, but the differences between these values were not significant. Differences were also discernible among cultivars in their mean alpha-amylase activities across the range of steeping times (Table 2.2). These results also suggested that while 20 h steeping led to the highest enzyme activities in CSH 9, IS 1347, IS 23860, and IS 23930, the highest activities in SPV 736 and IS 20503 were noted at 16 and 24 h steeping, respectively. However, these highest activities in the latter two cultivars were not significantly different from the activity at 20 h steeping, which was the next highest in both the cultivars.

Having defined the optimum steeping and germination times, alpha-amylase and diastatic activities in 29 cultivars were compared using steeping and germination times at 30°C of 20 h and 96 h, respectively. Cultivar differences in the production of these enzymes were evident in these experiments also. The dextrin activity of aqueous extracts of malt prepared from these cultivars showed a three-fold difference ranging from 427 to 1462 μg maltose released in 3 min/mL extract. In terms of specific activity, this variation ranged from 456 to 1234 μg maltose/mg protein. The diastatic activity of these cultivars varied from 24.2 to 84.3 SDU. Cultivars CSV 8R, SPV 346, SPV 351, SPV462, SPV 475, and SPV 504 had greater diastatic activities than 75 SDU. However, cultivar SPV 504, which showed the highest diastatic activity, did not have the highest alpha-amylase activity. Nonetheless, the correlation coefficient between alpha-amylase activity and diastatic activity in these 29 cultivars ($r = 0.88$) was positive, and significant. Specific activity of alpha-amylase and diastatic activity ($r = 0.55$) were also significantly correlated.

Notwithstanding the fact that pH and other conditions were chosen so as to be optimal for the extraction of alpha-amylase, the possibility that beta-amylase present in the malt was also extracted cannot be ruled out. As this contaminating beta-amylase was not selectively inactivated before alpha-amylase assay, the results for alpha-amylase activity obtained in this study may include a small but unknown fraction of activity contributed by beta-amylase. The presence of beta-amylase can also enhance the saccharifying activity of alpha-amylase. This contamination of beta-amylase may not, however, vitiate the main conclusions from this study, as the activity of all diastatic enzymes is a better criterion of sorghum malting.

The malting processes in use are still empirical. The standardization of conditions for malt preparation was done by Ratnavathi and Ravi (1991). The diastatic activity of sorghum is expressed in Sorghum Diastatic Units (SDU) and varied from 144.5 in IS 14384, 151 in SPV 824 to 200 in WS 1297 though a malting loss of 27–39% was reported in these genotypes, which can be minimized with air rest and minimized water supply. Malting causes a decrease in the caryopsis density (Beta et al., 1995), decreases lysine from 0.25 to 0.18% (Okoh et al., 1989), and reduces milling energy (Swanston et al., 1994). Fine milling (0.2 mm particle size) of malt increases hot water extract, diastatic power, and sugar contents (Lasekan et al., 1995).

Sorghum malts with high diastatic activity have a large albumin-globulin fraction and reduced paste viscosity (Malleshi and Desikachar, 1986a; Beta et al., 1995). Subramanian et al. (1995) reported that the water extract including water-extractable protein of the malt increases with the diastatic activity. However, sorghum grains with diastatic power of 60–80 Kaffir corn diastatic units (KDU/g) are recommended for commercial malting (Novellie, 1962a).

Seeds of 10 cultivars were malted for 24 h at 30°C and amylase activity assayed in water extracts. Three promising cultivars of grain sorghum were used. The grains were steeped in a running tap water for 10, 20, and 30 h followed by germination for 24, 48, and 72 h at 30°C in seed germinator. The experiment was carried out using factorial completely randomized design to 60 meshes, and the flour was analyzed for changes in chemical composition and nutritional quality (Fig. 2.1).

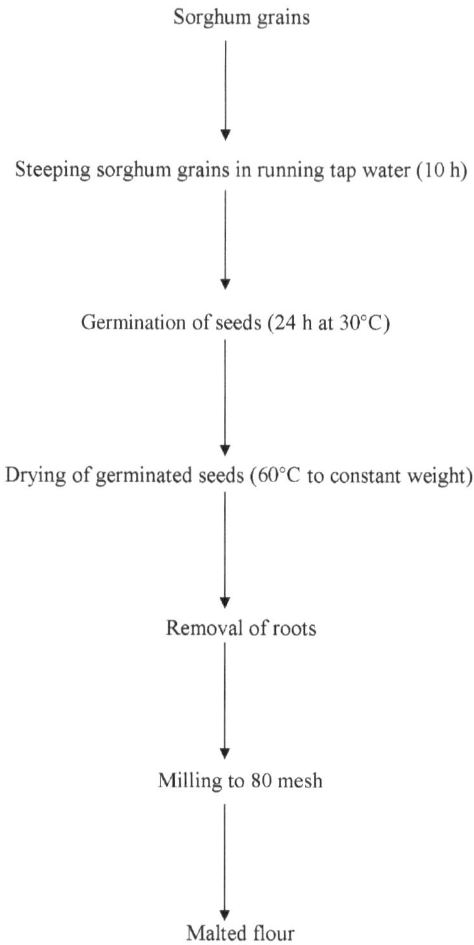

FIGURE 2.1 Flow sheet for preparation of malted sorghum flour.

2.3.1 Amylolytic Activity

The seeds were steeped for 10 h and malted at 30°C for 24 to 72 h. The amylase activity in the unmalted and malted seeds was determined in the water extract. The results are presented in Table 2.3. The amylase activity ranged from 2 to 13.5 units in dry seeds and increased significantly during malting. Among the cultivars studied SPV 86, CSH 6, M 35-1, SPV 351, and SPV 504 exhibited higher amylase activity than other cultivars after malting. The high amylase activity is desirable for rapid starch hydrolysis during malting. The cultivars with increased rate of release of amylase activity may be suitable for brewing purposes.

TABLE 2.3 Effects of Malting on the Amylase Activity of Promising Sorghum Cultivars (Units/g/20 min)

Cultivar	Malting periods (h)			
	0	**24**	**48**	**72**
VARIETIES				
M 35-1	8.5	23.0	29.5	55.0
CSV – 4	5.0	10.0	15.5	22.5
CSV – 504	13.5	18.0	45.0	75.0
CSV – 351	10.0	20.0	38.5	43.0
CSV – 86	3.5	35.5	43.0	46.0
HYBRIDS				
CSH – 1	2.0	15.5	17.5	24.0
CSH – 5	10.0	19.0	26.0	32.0
CSH – 6	10.0	33.5	38.0	40.5
CSH – 9	11.0	15.5	17.5	29.5
CSH – 8R	7.0	8.5	20.5	27.5
Range	2.0–13.5	8.5–35.5	15.5–45.0	22.5–75.0
Mean	8.1	19.8	29.0	39.5

One unit amylase activity is defined as 1 mg maltose released/g meal/20 min. Means of triplicate analysis.

2.3.2 Dry Matter Loss

Steeping of grains followed by germination resulted in significant losses in dry matter (Table 2.4). The dry matter loss due to malting can be kept to a minimum when the grains are steeped for 10 h followed by germination for 24 h.

2.3.3 Proximate Composition

Steeping of grains followed by germination caused significant reduction in the content of crude protein, crude fat, crude fiber, and ash in malted grains (Table 2.5). The losses in nutrients were found to be minimum when the grains were steeped for 10 h followed by germination for 24 h.

2.3.4 Soluble Proteins

The malting of grains resulted in significant increase in the soluble proteins and amino acids in grain sorghum (Table 2.6). High

TABLE 2.4 Effect of Malting on the Malting Loss in Grain Sorghum

Cultivar	Steeping periods (h)	Germination period (h)		
		24	48	72
M 35-1	10	3.2	11.2	19.5
	20	6.2	13.6	20.9
	30	9.5	17.5	25.6
Swati	10	1.9	5.5	19.6
	20	7.5	12.8	20.1
	30	9.8	14.5	25.3
CSH − 5	10	3.8	12.1	18.0
	20	7.3	14.5	25.2
	30	11.5	19.7	27.8

Mean of three replications C.D. at 5% 0.05

TABLE 2.5 Effect of Malting on the Proximate Composition of Grain Sorghum

Cultivar	Steeping period (h)	Germination period (h)		
		24	48	72
I. CRUDE PROTEIN (%)				
M 35-1 (11.56%)	10	11.6	11.3	11.2
	20	10.6	10.5	9.9
	30	10.2	10.2	10.1
Swati (10.73%)	10	10.6	10.7	10.6
	20	10.2	10.2	9.9
	30	9.9	9.9	9.7
CSH − 5 (11.30%)	10	11.3	11.3	11.2
	20	10.6	10.6	10.0
	30	10.5	10.5	9.9
II. CRUDE FAT (%)				
M 35-1 (4.0%)	10	4.0	3.8	3.0
	20	4.2	3.4	2.8
	30	4.4	2.2	2.6

(Continued)

TABLE 2.5 (Continued)

Cultivar	Steeping period (h)	Germination period (h)		
		24	**48**	**72**
Swati (4.0%)	10	4.0	4.0	3.8
	20	4.2	3.8	3.6
	30	4.2	3.4	2.6
CSH – 5 (4.4%)	10	4.0	3.6	3.0
	20	4.2	3.2	2.8
	30	4.2	2.8	2.0
III. CRUDE FIBER (%)				
M 35-1 (2.1%)	10	2.0	1.8	1.7
	20	1.9	1.8	1.8
	30	1.8	1.7	1.6
Swati (1.9%)	10	1.8	1.8	1.7
	20	1.5	1.7	1.6
	30	1.7	1.6	1.5
CSH – 5 (1.8%)	10	1.7	1.6	1.3
	20	1.6	1.6	1.5
	30	1.8	1.5	1.4
IV. ASH (%)				
M 35-1 (1.65%)	10	1.0	1.0	1.0
	20	1.0	0.9	0.9
	30	1.2	1.0	1.0
Swati (1.72%)	10	1.2	1.0	0.8
	20	1.0	1.0	1.0
	30	1.0	1.0	1.0
CSH – 5 (1.60%)	10	1.2	1.2	1.0
	20	1.1	1.0	1.0
	30	1.0	1.0	1.0

Means of three replications. Figures in parenthesis are for dry unmalted seeds. C.D. at 5%: Protein 0.06, Fat 0.08, Ash 0.03, and Fiber 0.11.

TABLE 2.6 Effect of Malting on the Soluble Proteins, Free Amino Acids, and Polyphenols of Grain Sorghum

Cultivar	Steeping period (h)	Germination period (h)		
		24	48	72
I. SOLUBLE PROTEINS (%)				
M 35-1 (1.16%)	10	1.0	1.6	1.9
	20	1.2	1.9	2.4
	30	1.9	2.3	2.8
Swati (0.91%)	10	0.9	1.1	1.9
	20	1.1	1.1	2.2
	30	1.5	2.2	2.4
CSH – 5 (1.08%)	10	0.5	1.6	1.9
	20	1.4	2.0	2.2
	30	1.8	2.3	2.8
II. FREE AMINO ACIDS (MG/G)				
M 35-1 (2.2 mg/g)	10	3.2	5.9	7.1
	20	3.5	5.5	7.1
	30	5.8	7.8	10.2
Swati (1.8 mg/g)	10	3.3	3.8	4.3
	20	3.5	5.3	6.8
	30	4.5	5.5	7.1
CSH – 5 (1.8 mg/g)	10	2.6	5.2	5.6
	20	3.9	6.1	6.6
	30	6.0	7.0	9.5
III. POLYPHENOLS (%)				
M 35-1 (0.16%)	10	0.16	0.22	0.25
	20	0.16	0.20	0.23
	30	0.21	0.27	0.29
Swati (0.14%)	10	0.15	0.17	0.20
	20	0.16	0.20	0.22
	30	0.17	0.22	0.29
CSH – 5 (0.18%)	10	0.16	0.23	0.25
	20	0.18	0.22	0.26
	30	0.25	0.26	0.27

Means of three replications. Figures in parenthesis are for dry unmalted seeds. C.D. at 5%: Soluble proteins 0.04, Free amino acids 2.5, and polyphenols 0.02.

concentration of soluble proteins has been reported to be useful in improving the *bhakari/roti* making quality of the grain sorghum. Hence, malting treatment is useful.

2.4 SUGARS AND STARCH

The malting of grains resulted in significant increases in the content of reducing sugars, with concomitant decreases in the starch content (Table 2.7). Reducing sugars have been correlated with the *bhakari/roti*

TABLE 2.7 Effect of Malting on the Starch and Reducing Sugars of Grain Sorghum

		Germination period (h)		
Cultivar	**Steeping period (h)**	**24**	**48**	**72**
I. STARCH (%)				
M 35-1 (69.50%)	10	68.0	60.6	51.6
	20	64.5	55.1	53.1
	30	62.3	50.5	49.1
Swati (68.20%)	10	67.0	60.0	55.4
	20	64.2	56.7	49.1
	30	65.1	55.1	43.0
CSH – 5 (68.35)	10	66.3	56.7	52.1
	20	64.5	58.6	54.4
	30	63.5	55.1	47.4
II. REDUCING SUGARS (%)				
M 35-1 (0.2%)	10	1.4	2.7	4.8
	20	1.7	3.7	6.4
	30	2.6	4.7	7.3
Swati (0.1%)	10	0.7	1.5	5.4
	20	1.6	4.0	6.4
	30	2.2	3.2	9.0
CSH – 5 (0.1%)	10	1.4	3.6	4.0
	20	1.6	3.8	4.8
	30	2.2	4.1	6.4

Means of three replications. Figures in parenthesis are for dry unmalted grains. C.D. at 5%: Starch 1.3, sugars 0.20.

making quality of grain sorghum. The increase in sugar content due to malting may be useful in improving the organoleptic properties of sorghum *bhakari/roti*.

2.4.1 Starch and Protein Digestibility

The steeping of grains for 10 to 20 h followed by germination for 24 h was found to increase both in vitro protein and starch digestibilities (Table 2.8). The In Vitro Protein Digestibility (IVPD) and IVSD were

TABLE 2.8 Effect of Malting on the In Vitro Protein Digestibility (IVPD) and In Vitro Starch Digestibility (IVSD) of Grain Sorghum

		Germination period (h)		
Cultivar	**Steeping period (h)**	**24**	**48**	**72**
I. IVPD (%)				
M 35-1 (80.8%)	10	82.7	85.3	90.2
	20	79.5	80.9	83.4
	30	78.4	79.4	83.8
Swati (81.6%)	10	84.7	84.6	85.5
	20	82.4	82.4	83.4
	30	83.4	83.8	84.6
CSH – 5 (78.5%)	10	79.2	80.4	85.2
	20	79.5	79.8	83.4
	30	78.9	78.0	83.4
II. IVSD (MG MALTOSE/G/2 H AT 20°C)				
M 35-1 (137.1)	10	204.0	125.7	97.1
	20	136.4	125.7	75.0
	30	121.4	78.5	60.7
Swati (116.4)	10	167.1	164.2	117.8
	20	137.1	92.8	67.1
	30	79.7	78.5	54.2
CSH – 5 (140.5)	10	169.7	109.2	97.1
	20	160.2	101.4	64.3
	30	97.1	91.4	64.3

Means of three replications. Figures in parenthesis are for dry unmalted grains. C.D. at 5%: IVPD 0.13, IVSD 1.28.

however found to be decreased when the malting was carried out for extended periods.

The micromalting technique of sorghum grain (10 g) was developed for rapid screening of diastatic activity of germplasm lines and elite breeding lines in sorghum (Jaya et al., 2001). In contrast to barley, sorghum malt has low levels of β-amylase, β-1,3 and β-1,4-glucanase, and β-D glucans. Sorghum malts show alpha amylase activity of 25 to 183 U/g and β-amylase activity of 11 to 41 U/g (Beta et al., 1995). During the mashing operations, external heat stable enzymes, namely, α-amylase, neutral protease, β-glucanase, cellulose, and amyloglucosidase, are required (Aisien, 1989). The beers from barley and sorghum vary in their texture, nutritional content, and sensory properties (Faparusi et al., 1973; Asiedu, 1992). The differences are mainly due to variations in the preparation process of worts or the addition of adjuncts or spices to worts and organisms associated with fermentation of the wort and postfermentation treatment.

Pilot scale (1000 L) brewing was carried out with a grist comprising of unmalted sorghum (50% of total wet weight of grain) (South African variety) and malted barley (50% of total wet weight of grain). The fermentability of the sorghum worts was lower, and hence, the sorghum beers were lower in total alcohol. Sensory analysis indicated that no significant differences existed between the sorghum beer and both the control beer and a commercial malted barley beer with regard to aroma, mouth-feel, after-taste, and clarity (Goode and Arendt, 2003).

2.5 LAGER BEER FROM 100% SORGHUM

Lager beer is now possible to be produced from 100% sorghum using exogenous enzymes (Arri, 1989). Properties of sorghum malt were in general poor compared to barley malt, but total soluble sugars and cold water (%) extract of the sorghum malt were relatively higher than those of barley malt (Ogundiwin et al., 1989). Production of lager beer from 100% sorghum grain is entirely a new concept. The cultivars that have low gelatinization temperature, low polypeptides, low lipid, high diastatic power, and readily soluble protein are good for lager beer preparation. Beer was prepared using 100% sorghum malt, sorghum adjuncts, and hops in the ratio 96:3:1. Current sorghum brewing techniques have been based on barley brewing technologies with suitable modifications. Sorghum beer had 4% alcohol and higher FAN content than in sole sorghum malt beer. The pentosanase complex of the sorghums, unlike that of barley does not embody a xylosidase, but as in barley, displays arabinosidase activity. Polymeric pentosans of barley worts were found to range from 162 mg/100 mL in the

TABLE 2.9 Analysis of Lager Beer From Sorghum

Component	Malt beer	Adjunct beer
Reducing sugars (%)	0.05 ± 0	0.02 ± 0.02
Total sugars (%)	1.54 ± 0.13	0.95 ± 0.07
Alcohol (%)	4.66 ± 0.12	4.5 ± 0.17
Color (EBC)	5.89 ± 0.03	8.58 ± 0.14
Specific gravity	1.03 ± 0.002	0.98 ± 0.03
Bitterness (BU)	6.56 ± 0.08	7.58 ± 0.15
Proteins (%)	2.68 ± 0.07	1.98 ± 0.07
Free α amino nitrogen (mg/L)	74.61 ± 5.18	119.01 ± 2.63

unmalted grains to 239 mg/100 mL in the 6-day malts while that of the sorghum over the same period was from 41 to 79 mg/mL (Etokakpan, 2004a) The diastatic power of the freshly kilned sorghum malt at 68.1°WK had a 29% drop after 6 months of storage. Freshly kilned sorghum malt displayed high wort turbidity (4.9 EBC) which dropped to 0.95 EBC and 1 EBC after 2 and 6 months of storage, respectively (Etokakpan, 2004b).

The composition of lager beer made from sorghum malt and sorghum adjunct are shown in Table 2.9. The lager beer from sorghum malt and adjunct showed superiority in composition like increased free alpha amino nitrogen and color units in sorghum malt and adjunct beer.

In Nigeria, use of malted sorghum rather than raw grains, as a source of fermentable sugars was found to be advantageous. Grains of sorghum varieties SK 5912, farafara, and HQSV are currently being used for malting. HQSV variety was found to be superior to other varieties for malting purposes. The by-product of malting, namely sprout (dried shoot and root), is currently being used as organic fertilizer and can be a rich source of dhurrin. The increased use of white grain sorghum by breweries might result in competition in the market between grain for food and grain for brewing purposes (Ikediobi, 1989).

2.6 USE OF SORGHUM AS ADJUNCT

Apart from malt, unmalted cereals like maize, rice, and barley either as flakes or broken grains (grits) are used as adjuncts in the brewery sector. Adjuncts are essentially starchy materials with little or no protein content. These adjuncts improve the quality of lager beer. Most of the information available on sorghum malting is from South Africa, and they often referred to the African "bantu" beer (Novellie, 1968).

Earlier Owuyama and Okafor (1990) reported the presence of amylase in the unmalted sorghum and suggested that sorghum can be a favorable brewing adjunct. A study on the amylase in the different cultivars of unmalted sorghum showed that cultivars CSH-5 and SPV 462 with high amylase activity are found suitable for use as adjuncts in the Indian situation (Anonymous and Ratnavathi, 2000). Later a systematic approach for the suitability of sorghum as adjunct in India was carried out (Ratnavathi et al., 2000). Grain samples of 13 sorghum cultivars with diverse chemical compositions were assessed for their suitability as brewing adjuncts based on proximate analysis. Sieving analysis of the grain as well as hot water extractable (HWE), hot water extractable protein (HWEP), and FAN were also determined. Cultivars with high starch and amylose contents together with low protein and fat percentages are better suited as adjuncts depending on their hot water extracts and HWEP yields. Cultivars CSH-5, CSV-11, and CSV-13 among the released cultivars were identified as the best adjuncts which could be used along with barley malt for brewing lager beers. Sorghum beer was prepared using the grain of two cultivars, CSV-15 and CSH-16, as adjunct and sorghum malt prepared from elite breeding line (SPV 824) having high diastatic activity. Of the two cultivars used, hybrid CSH 16 at 15% concentration was found to be suitable as an adjunct. Use of sorghum in preparation of lager beer as malt and as adjunct is very encouraging, and the chemical analysis of the beers prepared using sorghum as malt and as adjunct is presented in Table 2.10.

Another study was conducted by Directorate of Sorghum Research (DSR) in collaboration with Hindustan Breweries Limited; Mumbai to determine the suitability of some nationally released varieties of sorghum as adjunct, in comparison to broken rice (existing commercial adjunct in practice in Indian breweries). The various adjunct parameters that are analyzed in the sorghum are presented in Table 2.11. All sorghum genotypes/cultivars tested are equivalent to broken rice in all the parameters. However, when a comparison was made among sorghum cultivars, hybrids CSH-13 and CSH-14 were having high extractability followed by CSH-17, CSV-15, and SPV-462 with respect to color of adjunct; hybrids CSV-13, CSH-9, and CSH-17 are similar to broken rice (4.5 units). A good adjunct should have low saccharification

TABLE 2.10 Analysis of Sorghum Beer

Component	Sorghum malt and sorghum adjunct beer[a]	Barley malt and sorghum adjunct beer[a]
Reducing sugars	0.05 ± 0	0.02 ± 0.02
Total sugars	1.54 ± 0.13	0.95 ± 0.07
Alcohol	4.66 ± 0.12	4.5 ± 0.17
Color (EBO)	5.89 ± 0.03	8.58 ± 0.14
Specific gravity	1.03 ± 0.002	0.98 ± 0.03
Bitterness (BU)	6.56 ± 0.08	7.58 ± 0.15
Proteins	2.68 ± 0.07	1.98 ± 0.07
Free ∝ amino nitrogen	74.61 ± 5.18	119.01 ± 2.63

[a]*All the data expressed are Mean ± S.D.*
From Final Report, National Agricultural Technology Project, RNPS-24, DSR, 2005

time. Some sorghum genotypes such as CSV-13, SPV-462, and CSH-9 had lower saccharification time (9 min) compared to broken rice (10 min) (Anonymous and Ratnavathi, 2005).

A pilot experiment with sorghum grain at DSR in collaboration with Hindustan Breweries Limited was done. Sorghum grain was blended with broken rice in this experiment. No difference was observed and the beer was absolutely similar in color and taste to the normal beer. There was no problem in marketing this beer made from sorghum. It is determined that all the varieties of grain sorghum are suitable (Table 2.11). The hybrids CSH 13, CSH 17, and CSH 14 recorded the highest percentage extraction of 61, 60, and 61%, respectively (Anonymous and Ratnavathi, 2005).

In a similar study by Goode et al. (2002), an increase of the relative proportion of sorghum in the grist resulted in decrease of wort filtrate, color, viscosity, attenuation limit, FAN, high molecular weight nitrogen, and a corresponding increase in pH ($p < 0.01$). Overall, the addition of small proportions of unmalted sorghum, together with commercial hydrolytic enzymes to malted barley, was found to improve the potential for brewing a high quality lager beer.

Lager beer was prepared in a laboratory fermentor at DSR using the isolated thermotolerant yeast strain. Sorghum grain of SPV 824 variety was malted and the malt was ground in a UDY cyclone sample mill (Tecator, Sweden) and unmalted CSH 9, CSH 16, and CSH 18 sorghum grain flours were used as adjuncts. The results are presented in Table 2.12.

TABLE 2.11 Sorghum Grain Analysis as an Adjunct in Lager Beer Preparation

Parameter	CSH 15	CSV 13	CSH 18	SPV 462	SPV 824	CSH 9	CSH 14	CSH 17	CSH 16	CSH 13	Broken rice (check)
PH	5.75	5.7	5.3	5.6	5.4	5.45	5.6	5.63	5.35	5.5	5.6
Gravity	6.8	6.4	6	6.8	6.4	6.5	7	6.9	6.6	7.1	7.4
Moisture (%)	5	4.7	4.9	4.8	4.5	4.6	4.9	5.2	5.3	5	4.5
Saccharification (min)	10	9	10	9	10	9	12	11	13	12	10
Filtration rate	Good	Good	Good	Good	Good	Good	Good	Good	Good	Good	Good
Clarity	Clear	Clear	Clear	Clear	Clear	Clear	Clear	Clear	Clear	Clear	Clear
Color (EBC Units)	4.4	4.5	4.4	4.4	4.4	4.5	4.4	4.5	4.4	4.4	4.5
Extraction (%)	59	55	52	59	55	56	61	60	57	61	64

TABLE 2.12 Analysis of Beer Produced With Different Sorghum Adjuncts

S. no.	Analysis details	SPV 824 malt without adjunct	SPV 824 malt with CSH 09 as adjunct	SPV 824 malt with CSH 16 as adjunct	SPV 824 malt with CSH 18 as adjunct
1.	Reducing sugars	0.05 ± 0.00	0.03 ± 0.01	0.09 ± 0.01	1.8 ± 0.01
2.	Total sugars	1.44 ± 0.1	0.95 ± 0.07	2.14 ± 0.06	1.47 ± 0.13
3.	Free amino nitrogen	74.61 ± 5.18	119.02 ± 2.6	117.65 ± 0.00	120.33 ± 0.58
4.	Alcohol	1.05 ± 0.12	4.50 ± 0.17	5.29 ± 0.12	7.43 ± 0.23
5.	Color	14.33 ± 0.14	8.58 ± 0.14	18.44 ± 0.07	6.88 ± 0.05
6.	Bitterness	6.40 ± 0.09	7.58 ± 0.15	5.99 ± 0.60	6.97 ± 0.06
7.	Proteins	2.05 ± 0.07	1.98 ± 0.07	2.01 ± 0.00	2.14 ± 0.10
8.	Specific gravity	1.04 ± 0.00	1.00 ± 0.00	1.00 ± 0.01	0.98 ± 0.01

The highest amount of alcohol was recovered with adjuncts of CSH 18 and CSH 16 with SPV 824 malt. The highest amount of free alpha amino nitrogen was recorded in SPV 824 malt with CSH 18, CSH 9, and CSH 16 as adjuncts. Protein content in sole SPV 824 malt is 2.05%, while in SPV 824 with CSH 9, CSH 16, and CSH 18 adjunct are 1.98, 2.01, and 2.14%, respectively (Anonymous and Ratnavathi, 2005).

2.6.1 Malting and Brewing

In Nigeria, following the ban on the import of barley malt in 1988, the brewing industry in Nigeria has been utilizing sorghum and maize as raw materials for lager beer production. Sorghum can be used as raw grains, grits, or malted material. The bulk of the sorghum grain currently used by the brewers in Nigeria is from the varieties SK 5912 and farafara. The malting processes in use are still empirical. In contrast to barley, sorghum malt has low levels of β-amylase, β-1, 3, 1, 4-glucanase, and β-D glucans. During the mashing operations, external heat stable enzymes, namely, α-amylase, neutral protease, β-glucanase, cellulose, and amyloglucosidase, are required.

Lager beer can be produced from barley malt and any cereal adjunct including sorghum. It is now possible to brew some sort of beer from 100% sorghum using exogenous enzymes (Arri, 1989). Good quality

TABLE 2.13 Sorghum Beer Analysis

Component	Malt beer	Adjunct beer
Reducing sugars (%)	0.05 ± 0	0.02 ± 0.02
Total sugars (%)	1.54 ± 0.13	0.95 ± 0.07
Alcohol (%)	4.66 ± 0.12	4.5 ± 0.17
Color (EBC)	5.89 ± 0.03	8.58 ± 0.14
Specific gravity	1.03 ± 0.002	0.98 ± 0.03
Bitterness (BU)	6.56 ± 0.08	7.58 ± 0.15
Proteins (%)	2.68 ± 0.07	1.98 ± 0.07
Free alpha amino Nitrogen (mg/L)	74.61 ± 5.18	119.01 ± 2.63

lager beer similar to that of barley can be prepared from sorghum malt and sorghum adjunct. Rain damaged lower grain mold-infected grain can be conveniently used as a raw material. At DSR lager beer was prepared using Sorghum malt and sorghum as an adjunct in the ratio of 96:3:1. Hops were used up to 1%. Malt was used as ground and gritted sorghum (CSH 9) and was gelatinized to get "wort." Then wort was cooled and fermented with yeast for 7–8 days. Later it was filtered and carbonized with CO_2 and bottled. The chemical composition of beer was analyzed as per the EBC methodology (Table 2.13). Sorghum beer had 4% alcohol and higher free alpha amino nitrogen (FAN) content than sole sorghum malt beer. A higher ratio of adjunct use (up to 8–10%) would achieve the required FAN content (140 mg/L).

Properties of sorghum malt were in general poor compared to barley malt, but total soluble sugars and cold water (%) extract of the sorghum malt were relatively higher than those of barley malt (Ogundiwin et al., 1989).

Use of malted sorghum, rather than raw grains, as a source of fermentable sugars was found to be advantageous. Grains of sorghum varieties SK 5912, farafara, and HQSV are currently being used for malting. HQSV was found to be superior to other varieties for malting purposes. The by-product of malting, namely sprouts (dried shoots and roots), currently being used as organic fertilizer can be a rich source of dhurrin. The increased use of white grain sorghum by breweries might result in competition in the market between grain for food and grain for brewing purpose (Ikediobi, 1989).

In Africa, the great majority of nonalcoholic fermentations are sourings, mainly lactic fermentations. The alcoholic beverages can be classified into beers and wines. A typical modern formula and process used to make millions of hectoliters of sorghum beer is as follows: First fermentation is a lactic acid fermentation called souring, with the medium being a 10% malt slurry, for example, 300 kg of a sorghum malt (270 kg dry mass) together with 2700 L water, inoculated with *Lactobacillus leichmannii*, and is held at 50°C for 12–16 h, after which time pH is 3.3 with 0.8 to 1% lactic acid content. To this sour, 2750 kg as is (2420 kg dry mass) refined maize grits are added plus water to a volume of approx. 15,000 L. The slurry is cooked at 75 kPa for 10 min and then cooled to 60°C and mashing starts; at this temperature 800 kg sorghum malt (720 kg) are added. Volume is adjusted to 24,200 L, with total solids now about 14% and the pH 3.9 to 4.0. This mash is held for between 45 and 90 min until approx. 6% fermentable sugars is reached. This mash is centrifuged through Alfa laval or westfalia decanters to achieve a separate of coarse particle. The liquid called wort consists of about 22,000 L. The wort has a pH of 3.9, 0.16% lactic acid, fermentable sugar as glucose 6%, a specific gravity of 1.037, and total solids 11.8%. The residue called straining contains about 37% solids so that about 24% of the initial dry solids are discarded.

Second fermentation is based on a *Saccharomyces cerevisiae* sp. which grows and ferments well at 25 to 30°C. After 48 h the active fermentation beer is ready for consumption. The beer characteristics are pH 3.6, lactic acid 0.26%, total solids 6.2%, glucose 0.15%, ethanol 2.9 to 3%, and acetic acid 0.03% (Novellie, 1968).

Sorghum displays a unique agricultural adaptability to a world in need of more food. Traditionally, Nigeria has employed sorghum in both malted and unmalted forms in a wide variety of porridges and beverages. The three important local brews made from sorghum and popularly consumed by the rural population are known as pito, burukutu, and otika. The temperature of gelatinization of sorghum starch is higher than that of barley, and malt has lower extract yield and diastatic power than barley malt. The malting of sorghum is similar to that of other grains. It starts with the steeping of precleaned grains in water for a specific number of hours. The steep-ripe grains are transferred to germination beds and kept for 6 days. During malting, the endosperm of the grain is degraded by enzymes mobilized during steeping. By the end of germination, the degradation of the food reserve has progressed sufficiently. The malt is then kiln-dried and stored. Maize and sorghum grits are currently used as adjuncts in virtually all Nigerian breweries. Levels range from 25 to 50%. One of the biggest breweries, Guinness Nigeria Limited, carried out a few production runs with 100% unmalted maize grits in 1987. The beer is called "Merit." Other industries have also tried various substitutions for sorghum malt to brew new brands of beer. "Kings," for example, uses 50% sorghum malt as a substitute, and "Mayor" uses 65% sorghum malt substitution. Other industries have brewed 100% sorghum malt.

Fermented porridges are undoubtedly very important items of the African diet. The raw materials, methods of preparation, and general quality characteristics of fermented porridges have much in common with those of fermented cereal beverages, such as sorghum beer. The ways in which sour porridges can be made vary considerably. There are three operations to be considered: grinding (with or without sifting), souring (with or without inoculation), and boiling (in original medium or in fresh water). The order in which these operations are carried out varies from region to region.

1. Whole grain-wet process: The whole grain is soaked for several days. Microorganisms on the grain surface grow on materials leached from the grain. The microbial metabolic products influence the grain itself. The soaked grain is ground and sifted using more water. This can be left to sediment and sour further. The sediment can be filtered through a cloth and pressed to remove water. This preparation can be dispersed in water and boiled to give porridges of various consistencies according to the concentration chosen.
2. Whole grain-dry process: The whole grain is ground dry and not sifted. It is then mixed with warm water, with or without inoculums of previously soured porridge, and allowed to sour overnight or a little longer. When soured according to the consumer's desires, the mixture is boiled with or without further dilution with water.
3. Fractionated grain: There are three major variations—the dry grain is ground, winnowed, and sifted to give a flour or grits; the grain is dry pearled; the grain is moistened and pounded to dehusk, winnowed, and sifted.
4. Souring of an unsoured porridge: porridge is first made in the normal way and then it is soured. Souring may easily be achieved with the thinner types of porridge by mixing in other soured products (Novellie, 1968).

2.7 NONALCOHOLIC BEVERAGES AND WEANING FOODS

Malt prepared from sorghum grain also is used in the preparation of baby food beverages. Milo is a sorghum drink prepared from sorghum malt. Food Specialities (Nigeria) Ltd. has utilized sorghum as a source of its malt extract and produced many weaning food formulations. Sorghum grains with golden-yellow color and low tannin level are preferred for malt extraction (Solabi, 1989) as the tannins bind the proteins and make them unavailable for its absorption. Fine sorghum flour is used in the preparation of baby foods. The production of weaning foods utilizing sorghum and cowpea, and sorghum and soybean foods were evaluated. Preparation of sorghum and soybean *ogi* (weaning food in Nigeria) involved steeping sorghum grain for 24 h followed by wet milling and filtration through a vibro screen (<211 μ). Soybeans were dried at 110°C for about 1 h followed by dehulling and cooking in excess water. The sorghum and soybean slurries were mixed by dry weight proportion of 7:3 and left to ferment naturally, priming with 24 h sorghum steep liquor. It was fermented till the required acidity (0.6%) by fermented product, which was then spray-dried. The dried product was sieved and packed. During further trials, cost of drying was reduced by dewatering the sorghum and soybean slurries to give cakes. The cakes were blended and reduced in a hammer mill. The dried granules were milled into powder and mixed with other desired ingredients like vitamins and minerals. The sorghum and soybean compared well with popular commercial weaning foods in the market, and had 16–21% protein (Olatunji and Daodu, 1989).

It was recognized that sorghum grains contain poor quality protein leading to nutritional deficiencies, which can be rectified by supplementing sorghum diets with legumes. The study found that *tuwo* and *kamu* were the major traditional products prepared from sorghum. In Africa, the majority of nonalcoholic fermentations are made by souring, mainly lactic fermentations. The alcoholic beverages can be classified into beers and wines. A typical modern formula and process used to make millions of hectoliters of sorghum beer is that first fermentation, that is, a lactic acid fermentation called souring, with the medium being a 10% malt slurry, for example, 300 kg as is a sorghum malt (270 kg dry mass) together with 2700 L water, inoculated with *L. leichmannii*, is held at 50°C for 12–16 h, after which time pH is 3.3 with 0.8 to 1% lactic acid content. To this sour, 2750 kg as is (2420 kg dry mass) refined maize grits are added plus water to a volume of approx. 15,000 L. The slurry is cooked at 75 kPa for 10 min and then cooled to 60°C and mashing starts: at this temperature 800 kg sorghum malt is (720 kg) added. Volume is adjusted to 24,200 L, with total solids now about 14% and the pH 3.9 to 4.0. This mash is held for between 45 and 90 min until approx. 6% fermentable sugars is reached. This mash is centrifuged through decanters to achieve a separate of coarse particle. The liquid called wort consists of about 22,000 L. The wort has a pH of 3.9, 0.16% lactic acid, fermentable sugar as glucose 6%, a specific gravity of 1.037, and total solids 11.8%. The residue called strainings contains about 37% solids so that about 24% of the initial dry solids are discarded.

Second fermentation is based on a *S. cerevisiae* sp. which grows and ferments well at 25 to 30°C. After 48 h, the active fermentation beer is ready for consumption. The beer characteristics are pH 3.6, lactic acid 0.26%, total solids 6.2%, glucose 0.15%, ethanol 2.9 to 3%

and acetic acid 0.03%. "Obushera" is a predominantly lactic acid bacteria-fermented gruel produced from sorghum and/or millet flour, whose shelf life is limited to 1–2 days. A study aimed at exploring the possibilities of extending the shelf life of "obushera" with the least effect on its quality characteristics was conducted. Sorghum grains were germinated, dried, and ground into flour, which was used to produce porridge (Byaruhanga and Ndifuna, 2002).

Germinated sorghum flour had lower fat, protein, and carbohydrate contents but higher ash and fiber than nongerminated sorghum flour. Germinated millet flour had higher moisture, protein, and fiber compared to the nongerminated flour, while the latter had higher ash and carbohydrate contents. Germination resulted in an increase in the concentration of sugars in both sorghum and millet grains (Muyanja et al., 2003). All malting parameters were significantly influenced by season. Seasonal effects on diastatic power were greater than locality and cultivar effects. The malting quality of the total group of GM Class cultivars for each season and three cultivars included in trials over all five seasons were found to be similarly influenced by seasonal effects (van Loggerenberg and Pretorius, 2004).

2.8 FERMENTATION

2.8.1 Prefermentation Processing of Cereals

Prefermentation processing of cereals is largely dependent on the end product desired. In most cases, grains are sun-dried prior to fermentation. Treatments such as washing, steeping, milling, and sieving are prefermentation processing steps applied in the cereals (Fig. 2.2).

2.8.2 Fermented Cereals

Plant tissues or animal tissues subjected to the action of microorganisms and or enzymes to give desired biochemical changes and significant modification of food quality are referred to as fermented foods. Fermentation is the oldest known form of food biotechnology; records of barley conversion to beer date back more than 5000 years. According to Steinkraus, the traditional fermentation of foods serves several functions.

1. Enrichment of the diet through development of a diversity of flavors, aromas, and textures in food substrates.
2. Preservation of substantial amounts of food through lactic acid, alcoholic, acetic acid, and alkaline fermentations.

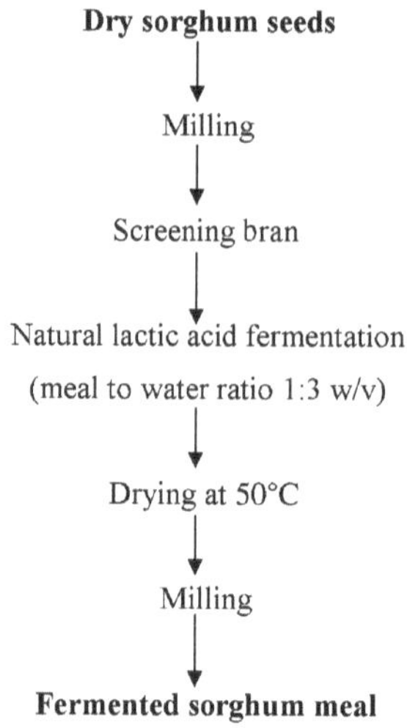

FIGURE 2.2 Flow diagram for preparation of fermented sorghum meal.

3. Enrichment of food substrates biologically with protein, essential amino acid, essential fatty acids, and vitamins.
4. Detoxification during food fermentation processing.
5. A decrease in cooking times and fuel requirements.
6. Easily digestible and available quickly to the body.

Aside from alcoholic fermentations and the production of yogurt and leavened bread, food fermentations continue to be important primarily in developing countries where the lack of resources limits the use of techniques such as vitamin enrichment of foods, and the use of energy- and capital-intensive processes for food preservation. The technology of producing many indigenous fermented foods from cereals remains a household art in these countries. Prospects for applying advanced technologies to indigenous fermented foods and for the production of value-added additive products such as colors; flavors, enzymes, antimicrobials, and heath products during food fermentations.

2.8.3 Indigenous Fermented Cereal Foods

Most bacterial fermentations produce lactic acids, while yeast fermentation results in alcohol production. Many of the indigenous fermentation products of cereals are valued for the taste and aroma active components produced and are used as seasonings and condiments. A summary of flavor compounds formed in such fermented products is compiled by Chavan and Kadam, 1989. A number of fermented products utilize cereals in combination with legumes, thus improving the overall protein quality of the fermented product. Cereals are deficit in lysine, but are rich in cysteine and methionine. Legumes on the other hand are rich in lysine but deficient in sulfur-containing amino acids. Thus, by combining cereals with legumes the overall protein quality is improved. The Chinese concept of *"fan"* (rice) and *"tsai"* (other vegetables) for a balanced and interesting diet is seen throughout the world.

2.8.4 Importance and Advantages/Benefits of Fermented Cereals

Fermented foods contribute to about one-third of the diet worldwide. Cereals are particularly important substrates for fermented foods in all parts of the world and are staples in the Indian subcontinent, in Asia, and in Africa. Fermentation causes changes in food quality indices including texture, flavor, appearance, nutrition, and safety. The advantage of fermentation may induce improvement in palatability and acceptability by developing improved flavors and textures; preservation through formation of acidulants, alcohol, and antibacterial compounds; enrichment of nutritive content by microbial synthesis of essential nutrients and improving digestibility of protein and carbohydrates; removal of antinutrients, natural toxicants, and mycotoxins; and decreased cooking times.

Natural fermentation of cereals increases their relative nutritive value and available lysine. Bacterial fermentations involving proteolytic activity are expected to increase the biological availability of essential amino acids more than yeast fermentation, which mainly degrade carbohydrates. Starch and fiber tend to decrease during fermentation of cereals. Although it would not be expected that fermentation would alter the mineral content of the product, the hydrolysis of chelating agents such as phytic acid during fermentation improves the bioavailability of minerals. Changes in the vitamin content of cereals with fermentation vary according to the fermentation process, and the raw material used in the fermentation. B group vitamins generally show an increase on fermentation. Fermentation of corn meal and soybean-corn meal blend lowers flatus-producing carbohydrates, trypsin inhibitors, and phytates. However, fermentation of cereals with fungi, such as *Rhizopus oligosporus* has been reported to release bound trypsin inhibitor, thus increasing its activity. Fungal and lactic acid fermentations have also been reported to reduce aflatoxin B1, sometimes by opening of the lactone ring which results in complete detoxification.

2.9 FERMENTATION STUDIES

The meals of cultivars M 35-1, Swati, CSH – 5, and M 35-1 + Mung bean (70:30, w/w) were fermented in water (1:3 with 0.1% sorbic acid) for up to 7 days, the slurry dried at 50°C and milled to flour. The flour thus obtained was analyzed for various nutritional quality parameters. The changes in pH and titrable acidity during fermentation were also measured.

2.9.1 Changes in pH and Titrable Acidity

The pH of the fermenting meals decreased sharply (Table 2.14) with concomitant increases in titrable acidity (Table 2.15) within 24 to 48 h fermentation. This indicated that the rapid microbial growth and degradation of reserve nutrients of the meal occur only during first 24 to 48 h of fermentation.

2.9.2 Changes in Total Soluble Solids (TSS) and Dry Weight

The TSS content increased significantly within 24 h of fermentation and thereafter slightly decreased (Table 2.16). However, the dry weight of the

TABLE 2.14 Effect of Fermentation on the pH of Fermenting Sorghum and Sorghum + Mung Bean Meals

Fermentation period (days)	Cultivars/blend					
	M 35-1	**Swati**	**CSH – 5**	**Mung bean**	**M 35-1 + mung bean**	**Mean**
Control	6.3	6.4	6.3	5.9	6.1	6.2
1	4.3	4.4	4.3	4.5	4.1	4.3
2	3.7	3.7	3.7	4.3	3.9	3.9
3	3.6	3.6	3.6	4.3	3.8	3.8
4	3.6	3.5	3.6	4.2	3.8	3.7
5	3.5	3.5	3.5	4.2	3.7	3.7
6	3.5	3.4	3.5	4.2	3.7	3.6
7	3.4	3.3	3.4	4.2	3.7	3.6
Mean	4.0	4.0	4.0	4.5	4.1	4.1

	S.E. ±	C.D. at 5%
Fermentation	0.023	0.065
Cultivars/blend	0.018	0.051
Interaction	0.052	0.015

TABLE 2.15 Effect of Fermentation on the Titrable Acidity of Sorghum and Sorghum + Mung Bean Meals

Fermentation period (days)	Cultivars/blend					
	M 35-1	Swati	CSH – 5	Mung bean	M 35-1 + mung bean	Mean
Control	0.11	0.09	0.11	0.33	0.11	0.15
1	0.44	0.46	0.46	0.43	0.49	0.46
2	1.01	1.05	1.00	0.44	0.62	0.82
3	1.15	1.15	1.15	0.45	0.79	0.94
4	1.27	1.36	1.27	0.46	0.86	1.04
5	1.36	1.41	1.35	0.47	0.98	1.11
6	1.41	1.48	1.45	0.49	1.00	1.17
7	1.48	1.54	1.49	0.50	1.00	1.20
Mean	1.03	1.07	1.03	0.45	0.73	0.86

	S.E. ±	C.D. at 5%
Fermentation	0.003	0.008
Cultivars/blend	0.002	0.007
Interaction	0.007	0.019

TABLE 2.16 Effect of Fermentation on Total Soluble Solids in Sorghum and Sorghum + Mung Bean Blends (Dry Wt. Basis)

Fermentation period (days)	Cultivars/blend					
	M 35-1	Swati	CSH – 5	Mung bean	M 35-1 + mung bean	Mean
Control	9.69	10.76	9.42	22.25	16.07	13.64
1	20.58	20.19	20.73	34.00	23.42	23.78
2	21.36	26.36	17.77	26.69	22.62	22.96
3	20.25	25.83	17.55	28.41	22.41	22.89
4	17.42	25.44	16.71	27.02	22.51	21.82
5	15.76	22.18	15.54	27.35	21.00	20.37
6	10.37	19.04	15.46	28.35	19.38	18.52
7	10.09	14.22	15.36	27.96	17.64	17.05
Mean	15.69	20.50	16.07	27.75	20.63	20.13

	S.E. ±	C.D. at 5%
Fermentation	0.026	0.073
Cultivars/blend	0.020	0.056
Interaction	0.058	0.163

TABLE 2.17 Effect of Fermentation on the Loss in Dry Weight (%) of Sorghum and Sorghum + Mung Bean Blend

	Cultivars/blend					
Fermentation period (days)	**M 35-1**	**Swati**	**CSH − 5**	**Mung bean**	**M 35-1 + mung bean**	**Mean**
1	1.60	4.44	6.64	9.40	9.22	6.26
2	2.30	4.68	7.66	9.60	9.26	6.70
3	3.50	5.08	7.88	9.66	9.30	7.08
4	4.16	5.36	9.42	9.72	9.36	7.60
5	4.30	5.52	15.96	9.78	9.42	9.00
6	4.52	5.98	16.26	9.84	9.50	9.22
7	4.64	6.14	16.90	9.96	9.68	9.46
Mean	3.13	4.65	10.09	8.49	8.22	6.52

	S.E. ±	C.D. at 5%
Fermentation	0.023	0.066
Cultivars/blend	0.019	0.052
Interaction	0.053	0.148

meals continuously and significantly decreased during fermentation (Table 2.17). The minimum loss occurred within 24 to 48 h and among the cultivars; M 35-1 exhibited lesser loss and the CSH − 5 showed higher losses in dry matter during fermentation. The addition of mung bean to sorghum decreased the rate of fall of pH and increased in acidity.

2.9.3 Changes in Composition

Fermentation of meals for 24 to 48 h significantly improved the proteins in all samples except in CSH − 5 (Table 2.18). Addition of mung bean to sorghum exhibited added advantage in improving the protein content. The fermentation for 24 h significantly improved the contents of free amino acids (Table 2.19) and water-soluble proteins (Table 2.20) which is useful for improved protein availability from fermented meal. No definite trends were observed for the changes in crude fat, ash, and crude fiber content due to fermentation. The fat content significantly improved during 24 to 48 h fermentation, while the nitrogen free extract was found to be significantly decreased during fermentation.

TABLE 2.18 Effect of Fermentation on the Protein Content (%) of Sorghum and Sorghum + Mung Bean Blends (Dry Wt. Basis)

	Cultivars/blend					
Fermentation period (days)	**M 35-1**	**Swati**	**CSH − 5**	**Mung bean**	**M 35-1 + mung bean**	**Mean**
Control	11.44	11.66	10.18	24.59	14.78	14.53
1	12.83	12.23	11.80	26.80	17.87	16.31
2	13.12	13.02	12.86	27.02	17.24	16.65
3	13.24	13.82	13.82	27.16	17.62	17.13
4	13.48	13.18	13.37	26.61	17.12	16.75
5	13.62	13.30	13.33	26.48	17.74	16.79
6	13.12	13.02	13.95	26.34	17.74	16.83
7	12.99	13.47	13.54	26.34	16.60	16.59
Mean	12.98	12.96	12.86	26.41	17.03	16.44
		S.E. ±		C.D. at 5%		
Fermentation		0.067		0.189		
Cultivars/blend		0.053		0.149		
Interaction		0.149		0.419		

TABLE 2.19 Effect of Fermentation on the Free Amino Acids (mg/g) of Sorghum and Sorghum + Mung Bean Blends (Dry Wt. Basis)

	Cultivars/blend					
Fermentation period (days)	**M 35-1**	**Swati**	**CSH − 5**	**Mung bean**	**M 35-1 + mung bean**	**Mean**
Control	2.6	1.7	1.7	4.8	2.7	2.7
1	6.8	6.2	6.2	22.3	12.5	10.8
2	7.1	7.1	7.7	23.1	14.8	12.0
3	7.5	7.9	9.3	23.6	14.9	12.6
4	7.9	9.0	10.0	23.9	15.8	13.3
5	8.2	10.0	9.2	25.0	17.8	14.0
6	8.0	10.6	9.7	26.6	18.0	24.6
7	8.4	9.0	8.9	26.8	18.5	14.2
Mean	7.1	7.7	7.8	22.0	14.4	11.8
		S.E. ±		C.D. at 5%		
Fermentation		0.566		1.593		
Cultivars/blend		0.447		1.258		
Interaction		1.265		3.560		

TABLE 2.20 Effect of Fermentation on the Soluble Proteins (%) of Sorghum and Sorghum + Mung Bean Blends (Dry Wt. Basis)

	Cultivars/blend					
Fermentation period (days)	**M 35-1**	**Swati**	**CSH − 5**	**Mung bean**	**M 35-1 + mung bean**	**Mean**
Control	1.40	1.36	1.14	11.67	5.14	4.14
1	2.14	2.06	1.98	15.33	6.24	5.55
2	2.09	2.28	2.12	15.30	6.44	5.65
3	2.14	2.34	2.27	16.22	6.31	5.86
4	2.09	2.28	1.90	15.76	6.48	5.70
5	2.07	2.26	1.82	15.53	6.13	5.56
6	2.05	2.26	1.62	16.15	6.04	5.62
7	2.21	2.25	1.65	15.45	6.26	5.56
Mean	2.02	2.14	1.81	16.43	6.38	5.76

	S.E. ±	C.D. at 5%
Fermentation	0.010	0.028
Cultivars/blend	0.006	0.023
Interaction	0.023	0.065

2.9.4 Changes in Tryptophan and Methionine

The fermentation of different meals for 24 to 48 h significantly improved the content of tryptophan and methionine in proteins of sorghum (Tables 2.21 and 2.22).

2.9.5 Changes in Starch and Sugars

Fermentation treatment continuously and significantly decreased starch content in the fermented meals (Table 2.23). The decrease in starch content was found to be associated with a decrease in the contents of amylase (Table 2.24) and a concomitant increase in the contents of reducing sugars up to 24 to 48 h fermentation (Table 2.25).

2.9.6 Changes in IVPD

The fermentation of different meals significantly improved the in vitro protein digestibility of sorghum and sorghum–mung bean blends within 24 to 48 h (Table 2.26). Fermentation beyond 24 h, however, did not increase the IVPD markedly.

TABLE 2.21 Effect of Fermentation on the Tryptophan (g/16g N) of Sorghum and Sorghum + Mung Bean Blends (Dry Wt. Basis)

Fermentation period (days)	Cultivars/blend					
	M 35-1	**Swati**	**CSH – 5**	**Mung bean**	**M 35-1 + mung bean**	**Mean**
Control	1.04	0.86	1.00	1.13	1.04	1.01
1	1.08	1.85	0.95	1.18	1.16	1.04
2	1.05	1.00	1.00	1.18	1.16	1.08
3	1.00	1.01	0.98	1.18	1.10	1.05
4	1.00	1.01	1.09	1.15	1.09	1.07
5	1.00	1.01	1.09	1.12	1.09	1.06
6	1.06	1.01	1.09	1.17	1.08	1.08
7	1.04	0.95	1.11	1.15	1.08	1.07
Mean	1.03	0.96	1.04	1.16	1.10	1.06

	S.E. ±	C.D. at 5%
Fermentation	0.009	0.024
Cultivars/blend	0.007	0.019
Interaction	0.019	0.054

TABLE 2.22 Effect of Fermentation on the Methionine Content (g/16g N) of Sorghum and Sorghum + Mung Bean Blends (Dry Wt. Basis)

Fermentation period (days)	Cultivars/blend					
	M 35-1	**Swati**	**CSH – 5**	**Mung bean**	**M 35-1 + mung bean**	**Mean**
Control	1.02	1.04	0.99	1.03	1.10	1.04
1	1.32	1.15	1.23	1.03	1.21	1.19
2	1.26	1.09	1.14	1.03	1.19	1.14
3	1.13	1.10	1.14	1.03	1.19	1.12
4	1.13	1.08	1.15	1.14	1.13	1.13
5	1.02	1.01	1.08	1.10	1.08	1.06
6	1.02	0.97	1.09	1.14	1.10	1.06
7	0.98	0.97	1.08	1.40	1.10	1.11
Mean	1.11	1.05	1.11	1.11	1.14	1.10

	S.E. ±	C.D. at 5%
Fermentation	0.022	0.062
Cultivars/blend	0.018	0.051
Interaction	0.050	0.141

TABLE 2.23 Effect of Fermentation on the Starch Content (%) of Sorghum and Sorghum + Mung Bean Blend (Dry Wt. Basis)

	Cultivars/blend					
Fermentation period (days)	**M 35-1**	**Swati**	**CSH − 5**	**Mung bean**	**M 35-1 + mung bean**	**Mean**
Control	76.30	72.14	74.42	62.98	66.76	70.52
1	74.02	69.48	73.16	60.87	64.55	68.42
2	68.12	66.00	69.67	57.43	61.74	64.59
3	66.49	66.00	68.83	55.90	57.75	62.99
4	61.31	64.16	65.26	53.65	54.49	59.77
5	59.94	60.50	62.84	52.83	47.60	56.74
6	58.58	58.39	61.47	50.58	50.04	55.81
7	57.85	57.75	58.74	47.01	47.60	53.79
Mean	65.33	64.30	60.80	55.16	56.32	61.58

	S.E. ±	C.D. at 5%
Fermentation	0.359	1.010
Cultivars/blend	0.283	0.796
Interaction	0.802	2.257

TABLE 2.24 Effect of Fermentation on Amylose (%) Content of Sorghum and Sorghum + Mung Bean Blends (Dry Wt. Basis)

	Cultivars/blend					
Fermentation period (days)	**M 35-1**	**Swati**	**CSH − 5**	**Mung bean**	**M 35-1 + mung bean**	**Mean**
Control	25.08	24.82	23.56	26.50	24.52	24.90
1	23.93	23.02	22.46	25.63	23.02	23.61
2	22.90	21.59	21.15	24.44	21.90	22.40
3	21.82	20.41	18.58	23.10	20.25	20.83
4	20.68	18.56	16.77	21.92	19.19	19.42
5	17.84	17.26	15.59	20.64	18.25	17.92
6	16.91	16.05	14.39	18.87	16.74	16.59
7	14.72	14.33	13.28	17.10	14.43	14.77
Mean	20.48	19.50	18.22	22.27	19.79	20.05

	S.E. ±	C.D. at 5%
Fermentation	0.063	0.177
Cultivars/blend	0.050	0.141
Interaction	0.141	0.397

TABLE 2.25 Effect of Fermentation on the Reducing Sugars Content (%) of Sorghum and Sorghum + Mung Bean Blend (Dry Wt. Basis)

	Cultivars/blend					
Fermentation period (days)	**M 35-1**	**Swati**	**CSH – 5**	**Mung bean**	**M 35-1 + mung bean**	**Mean**
Control	0.29	0.23	0.24	0.25	0.27	0.26
1	4.10	3.11	3.19	0.60	4.90	3.18
2	4.38	3.34	3.14	0.72	5.28	3.37
3	4.56	3.80	3.55	0.87	3.29	3.21
4	1.38	2.07	1.64	0.68	2.05	1.56
5	1.27	1.95	1.06	0.56	1.84	1.34
6	1.10	1.95	0.78	0.52	1.84	1.24
7	1.02	1.95	0.67	0.42	1.85	1.18
Mean	2.26	2.30	1.78	0.58	2.66	1.92

	S.E. ±	C.D. at 5%
Fermentation	0.009	0.025
Cultivars/blend	0.007	0.020
Interaction	0.020	0.056

TABLE 2.26 Effect of Fermentation on In Vitro Protein Digestibility (IVPD) Sorghum and Sorghum + Mung Bean Blend (Dry Wt. Basis)

	Cultivars/blend					
Fermentation period (days)	**M 35-1**	**Swati**	**CSH – 5**	**Mung bean**	**M 35-1 + mung bean**	**Mean**
Control	80.16	79.41	78.60	81.99	83.87	80.81
1	85.76	87.44	84.56	87.01	84.98	85.95
2	85.39	88.84	84.96	87.09	85.29	86.31
3	87.39	89.39	85.07	87.15	86.34	87.07
4	86.91	91.20	88.34	87.86	87.34	88.33
5	85.66	87.34	85.65	88.28	84.12	86.21
6	83.54	88.84	85.22	90.14	84.98	86.54
7	80.85	85.93	79.19	90.19	84.72	84.18
Mean	84.46	87.30	83.95	87.46	85.20	85.67

	S.E. ±	C.D. at 5%
Fermentation	0.238	0.670
Cultivars/blend	0.188	0.529
Interaction	0.532	1.497

2.10 USE OF MALTED AND FERMENTED MEALS IN *BHAKARI/ROTI* MAKING

Several experiments were conducted to investigate the influence of malting and fermentation of sorghum, mung bean, and black gram and their blends in various proportions, on the physical and organoleptic properties of *bhakari/roti*. The combinations included: control sorghum with 20 to 80% malted or fermented sorghum meals. The sorghum–legume combinations included: sorghum + malted or fermented mung bean meal (10 to 30%) and sorghum + malted black gram (20 to 60%). The results are briefly summarized below.

2.10.1 Water Uptake, Dough Rollability, and Moisture Loss

Malting or fermentation of sorghum did not markedly affect the water uptake and dough rollability of the flour. However, these treatments markedly increased the percent moisture loss from *bhakari/roti* during storage at ambient temperature (Table 2.27). Malting of mung

TABLE 2.27 Effect of Malting or Fermentation of Sorghum on Water Uptake, Dough Rollability, and Moisture Loss in *Bhakari* During Storage

Sample[a]	Water uptake (mL)	Dough rollability (cm)	Moisture loss during storage (%, 12 h)
MALTING STUDIESCONTROL:MALTED			
100:00	49.5	19.0	11.7
00:100	50.5	18.6	14.4
80:20	48.0	20.5	16.5
60:40	49.0	21.1	15.9
40:60	48.5	20.7	15.5
20:80	50.0	20.7	15.3
FERMENTATION STUDIESCONTROL:FERMENTED			
100:00	49.5	19.0	11.7
00:100	47.5	19.3	14.8
80:20	48.5	19.4	17.7
60:40	47.5	19.2	15.6
40:60	47.0	19.2	14.2
20:80	47.0	19.1	16.7

[a]*Fifty gram of flour used for experiments. Malting of grains or fermentation of meals for 24 h at 30°C.*

TABLE 2.28 Effects of Blending Malted or Fermented Legumes With Sorghum on Water Uptake, Dough Rollability, and Moisture Loss From *Bhakari* During Storage

Samples	Water uptake (mL)	Dough rollability (cm)	Moisture loss during storage (%, 12 h)
Sorghum control	49.5	19.0	11.7
Sorghum malted	50.5	18.6	14.4
Sorghum fermented	47.5	19.3	14.8
Mung bean control	26.0	21.1	11.8
Mung bean malted	24.0	21.0	17.5
Mung bean fermented	28.0	22.5	13.5
Black gram control	28.0	21.0	10.7
Black gram malted	33.0	22.0	12.3
Sorghum + unmalted mung bean (80:20)	44.0	22.1	18.4
Sorghum + malted mung bean	38.0	21.3	21.0
Sorghum + fermented mung bean (80:20)	45.5	21.2	13.2
Sorghum + unmalted black gram (80:20)	37.0	21.3	19.4
Sorghum + malted black gram (80:20)	43.0	22.2	16.9

bean decreased water uptake and increased moisture loss. However, fermentation of mung bean increased water uptake and decreased moisture loss (Table 2.28). When the malted mung bean was mixed with untreated sorghum flour, it resulted in decreased water uptake and increased moisture loss. However, the reverse trend was observed when fermented mung bean was used (Table 2.28). Use of malted black gram in similar experiments, however, increased water uptake with a decrease in moisture loss.

The results indicated that malting of sorghum and mung bean and their use as a whole or as a supplement to sorghum flour lowers the water uptake and increases moisture losses from *bhakari*. Fermentation of meals and their use in *bhakari* preparation, however, seem to improve these properties. Among the legumes, black gram was found to be relatively superior over the mung bean in the preparation composite *bhakari*.

2.10.2 Organoleptic Evaluation of *Bhakari/Roti*

The fresh *bhakari/roti* prepared from sorghum and sorghum + malted or fermented sorghum meal were evaluated by a panel of judges for organoleptic properties (Table 2.29). The results indicated that malted sorghum meal can be substituted to the extent of 80% and fermented meal to the extent of 60% in the preparation of organoleptically acceptable *bhakari* with improved nutritional quality. Supplementing of malted or fermented sorghum meal however did not improve the sensory properties of *bhakari* over *bhakari* prepared from untreated sorghum meal.

When malted and fermented mung bean or malted black gram meals were mixed with untreated sorghum flour in various proportions to prepare *bhakari*, it was observed that mung bean meal can be mixed, either raw or malted, to the extent of 30% and fermented meal to the extent of 60% with sorghum meal to prepare acceptable product (Table 2.30). Similar experiments using black gram indicated that raw black gram flour can be mixed to the extent of 20% and malted black gram flour to the extent of 30% to prepare acceptable composite *bhakari* (Table 2.31).

TABLE 2.29 Organoleptic Evaluation of *Bhakari/Roti* Prepared From Sorghum Meal Supplemented With Malted or Fermented Sorghum Flours (CSH – 5)[a]

Flour blend	Color and appearance	Flavor	Texture	Sweetness	Mean
CONTROL:MALTED					
100:00	7.0	7.3	7.7	7.5	7.4
80:20	6.5	6.2	7.0	7.7	6.9
60:40	7.0	6.5	7.2	7.7	7.1
40:60	6.5	6.3	7.3	7.7	7.0
20:80	6.8	6.3	7.5	8.0	7.2
C.D. at 5%	0.84	0.98	0.89	0.95	0.92
CONTROL:FERMENTED					
100:00	8.0	8.0	8.5	8.5	8.3
80:20	8.8	8.5	8.0	7.8	8.3
60:40	7.8	7.5	7.3	7.0	7.4
40:60	7.5	7.2	7.1	7.0	7.2
20:80	7.0	6.5	6.8	6.3	6.7
C.D. at 5%	0.57	1.0	1.1	1.2	1.0

[a]*Nine point hedonic scale; ten semitrained judges were used for sensory evaluation. Sensory score: Like extremely (Excellent); 9, Like very much (Very good); 8, Like moderately; 7, Like slightly; 6, Neither like nor dislike; 5, Dislikes lightly; 4, Dislike moderately; 3, Dislike very much; 2, Dislike extremely; 1.*

TABLE 2.30 Organoleptic Evaluation of *Bhakari* Prepared From Composite Flours, Sorghum, and Mung Bean

Flour blend	Color and appearance	Flavor	Texture	Sweetness	Mean
Sorghum CSH – 5	8.3	8.1	8.3	8.5	8.3
SORGHUM:MUNG BEAN (CONTROL)					
90:10	8.3	8.1	7.8	7.5	7.9
80:20	7.9	8.2	7.4	7.4	7.7
70:30	7.4	7.6	7.3	7.1	7.4
C.D. at 5%	1.1	0.72	1.0	0.85	0.92
SORGHUM:MALTED MUNG BEAN					
100:00	8.0	8.0	8.0	8.1	8.0
90:10	8.0	7.8	7.8	8.1	7.9
80:20	7.8	7.4	7.6	8.1	7.7
70:30	7.8	7.8	7.6	8.1	7.8
C.D. at 5%	0.50	1.00	1.00	1.20	0.93
SORGHUM:FERMENTED MUNG BEAN					
100:00	8.5	8.0	8.5	8.0	8.3
80:20	8.5	8.5	8.0	8.3	8.3
60:40	8.3	8.3	8.0	8.0	8.2
40:60	8.0	8.0	7.5	7.9	7.9
C.D. at 5%	0.60	1.20	1.20	1.10	1.00

TABLE 2.31 Organoleptic Evaluation of *Bhakari* Prepared From composite Flours, Sorghum + Black Gram

Flour blend	Color and appearance	Flavor	Texture	Sweetness	Mean
SORGHUM (CSH – 5):BLACK GRAM (CONTROL)					
100:00	8.5	8.2	8.3	8.5	8.4
90:10	8.3	8.0	7.5	7.8	7.9
80:20	7.8	8.0	7.5	7.3	7.7
70:30	7.5	7.5	7.3	6.5	7.2
C.D. at 5%	1.0	0.7	1.1	0.8	0.9
SORGHUM (CSH – 5):MALTED BLACK GRAM					
100:00	7.6	7.6	7.4	7.4	7.5
90:10	7.6	7.8	7.8	7.6	7.7
80:20	7.8	8.0	8.0	8.2	8.0
70:30	8.0	7.6	8.0	8.4	8.0
C.D. at 5%	0.7	1.1	1.6	1.1	1.2

The results thus revealed that both malted and fermented flours of mung bean or malted meal of black gram can be used in the preparation of composite *bhakari* with improved nutritional quality. Both malting and fermentation treatments however do not offer advantage over raw legume flours, except in sweetness in affecting the organoleptic properties of the product. Both malting and fermentation treatments increased the contents of soluble sugars, thus enhancing the sweetness of the products. Further investigations on the removal of the malty flavor of malted meal, the acidic taste of fermented meal, and the biological value of such products with animal feeding experiments are, however, essential.

2.11 SUMMARY

Comparative evaluation of varieties and hybrids of grain sorghum showed that varieties, in general, contained higher protein content, soluble sugars, and soluble proteins and also exhibited higher protein and starch digestibilities, although considerable variations between varieties and hybrids were noticed. The varieties exhibited comparatively superior *bhakari* making quality than the hybrids. The correlations between chemical constituents and *bhakari* making quality were nonsignificant, although the soluble proteins and sugars exhibited positive correlation with *bhakari* quality. Malting studies revealed that the nutritional quality of grain sorghum and legumes can be significantly improved by malting treatment. Malting released amylolytic activity, increased soluble proteins, soluble sugars, protein, and starch digestibility, and decreased the dry weight and starch content significantly. The optimum improvement in the nutritional quality is achieved by steeping the grains for 10 to 20 h followed by germination for 24 h with minimum losses in dry matter.

Fermentation studies revealed similar results as observed during malting experiments. During fermentation, there was a marked decrease in pH (3.7 to 4.3, microbiologically safe) with a concomitant increase in acidity. Both methionine and tryptophan contents were increased due to fermentation treatments. The optimum conditions for natural fermentation with minimum dry matter loss were mixing the meal with water (1:3, w/v) and incubation for 24 to 48 h at 30°C and drying of slurry at 50°C. Both malted and fermented meals of sorghum, mung bean, or malted black gram can be mixed with sorghum flours to the extent of 20 to 30% to prepare acceptable *bhakari* with improved nutritional quality. Among the treatments, fermentation appears to be superior over malting in improving both organoleptic and nutritional quality of sorghum *bhakari*.

References

Aisien, A.O., 1989. Utilization of sorghum in brewing lager beer in Nigeria. In: Proceedings of a Symposium on the Current Status and Potential of Industrial Uses of Sorghum in Nigeria, 4–6 December, Kano, Nigeria, pp. 29.

Anonymous, Ratnavathi, C.V., 2000. Annual Report, National Research Centre for Sorghum, pp. 57.

Anonymous, Ratnavathi, C.V., 2005. Final Report NATP – RNPS – 24, Developing Sorghum as an Efficient Biomass and Bioenergy Crop and Providing Value Addition to the Rain Damaged Kharif Grain for Creating Industrial Demand. DSR, Rajendranagar, Hyderabad – 30.

Arri, B.K., 1989. Problems associated with the use of sorghum for lager beer production. In: Proceedings of a Symposium on the Current Status and Potential of Industrial Uses of Sorghum in Nigeria, 4–6 December, Kano, Nigeria, pp. 29.

Asiedu, J.J., 1992. Sorghum and millet. Processing Tropical Crops. Macmillan Press Ltd, London, pp. 189–223.

Beta, T., Rooney, L.W., Waniska, R.D., 1995. Malting characteristics of sorghum cultivars. Cereal Chem. 72, 533–538.

Byaruhanga, Y.B., Ndifuna, M., 2002. Effect of selected preservation methods on the shelf life and sensory quality of 'Obushera'. Muarik-Bull. 5, 92–100.

Chavan, J.K., Kadam, S.S., 1989. Nutritional improvement of cereals by fermentation. CRC Crit. Rev. Food Sci. Technol. 28 (5), 349.

EtokAkpan, O.U., 2004a. Preliminary study of the enzymolysis of sorghum and barley pentosans and their levels in worts. World J. Microbiol. Biotechnol. 20 (6), 575–578.

EtokAkpan, O.U., 2004b. Changes in sorghum malt during storage. J. Indust. Brew. 110 (3), 189–192.

Faparusi, S.I., Olonboba, M.O., Ekundayo, J.A., 1973. Microbiology of burukutu beer. Zeitschrift for Allgemeine Mikrobiologie. 13, 563–568.

Goode, D.L., Halbert, C., Arendt, E.K., 2002. Mashing Studies with Unmalted Sorghum and Malted Barley. J. Inst. Brew. 108 (4), 465–473.

Goode, D.L., Arendt, E.K., 2003. Pilot scale production of a lager beer from a grist containing 50% unmalted sorghum. J. Inst. Brewing 109, 208–217.

Graham, G.G., Maclean, W.C., Morales, E., Hamaker, B.R., Kirleis, A.W., Mertz, E.T., et al., 1986. Digestability and utilization of protein and energy from nasha, a traditional Sudanese fermented sorghum weaning food. J. Nutr. 116, 978–984.

Hahn, R.R., 1966. Sorghum as a brewing adjunct. Brew. Digest. 41, 70–76.

Ikediobi, C.O., 1989. Industrial production of Sorghum malt in Nigeria. In: Proceedings of a Symposium on the Current Status and Potential of Industrial Uses of Sorghum in Nigeria, 4–6, December, Kano, Nigeria, pp. 32.

Jaya, K., Bala Ravi, S., Ratnavathi, C.V., Suresh, K., Bhalla, J.K., 2001. Micro-scale malting studies in sorghum. In: Poster Presented at the International Conference on New Horizons in Biotechnology Held During April, 18–21 at Trivandrum, India.

Lasekan, O.O., Idowu, M.A., Lasekan, W., 1995. Effect of germination and degree of grind (coarse/fine) on the extract and sugar content of sorghum malts. Food. Chem. 53, 125–128.

van Loggerenberg, M., Pretorius, A.J., 2004. South African Journal of Plant and Soil. 21 (3), 192–195.

Malleshi, N.G., Desikachar, H.S.R., 1986a. Studies on comparative malting characteristics of some tropical cereals and millets. J. Inst. Brew. 92, 174–176.

Muyanja, C.M.B.K., Kikafunda, J.K., Narvhus, J.A., Helgetun, K., Langsrud, T., 2003. Production methods and composition of Bushera: a Ugandan traditional fermented cereal beverage. Afr. J. Food Agric. Nutr. Dev. 3 (1), 10–19.

Novellie, L., 1962a. Kaffir corn malting and brewing studies XII. Effect of malting conditions on malting losses and total amylases activity. J. Sci. Food. Agric. 13, 121–123.
Novellie, L., 1962b. Kaffir corn malting and brewing studies XIII. Variations of diastatic power with variety, season, maturity and age of grains. J. Sci. Food. Agric. 13, 124–126.
Novellie, L., 1968. Kaffir beer brewing. Ancient art and modern industry. Wallerstein Lab. Comm. 31, 17–29.
Ogundiwin, J.O., Ilori, M.O., Okelaye A., 1989. Brewing of clear beer from sorghum grains of SK 5912 variety without addition of external enzymes to achieve Saccharification: a case study. In: Proceedings of a Symposium on the Current Status and Potential of Industrial Uses of Sorghum in Nigeria, 4–6 December, Kano, Nigeria, pp. 31.
Okafor, N., Aniche, G.N., 1980. Brewing a lager beer from Nigerian sorghum. Brew. Distilling Int. 10, 32–35.
Okoh, P.N., Kubiczek, R.P., Njoku, P.C., Iyeghe, G.T., 1989. Some compositional changes in malted sorghum (*Sorghum vulgare*) grain and its value in broiler chicken diet. J. Sci. Food. Agric. 49, 271–280.
Olatunji, O., Daodu A.M., 1989. Potentials of using sorghum in the production of weaninig foods. In: Proceedings of a Symposium on the Current Status and Potential of Industrial Uses of Sorghum in Nigeria, 4–6 December, Kano, Nigeria, pp. 37.
Owuyama, C.I., Okafor, N., 1990. Use of un-malted sorghum as a brewing adjunct. World J. Microbiol. Biotechnol. 6, 318–322.
Owuama, C.I., 1991. Production of burukutu with *Saccharomyces cerevisiae* variants. Appl. Microbiol. Biotechnol. 35, 21–22.
Owuyama, C.I., 1997. Sorghum a cereal with larger beer brewing potential. J. Microbiol. Biotechnol. 13, 253–260.
Palmer, G.H., 1989. Cereals in malting and brewing. In: Palmer, G.H. (Ed.), Cereal Science and Technology. University Press, Aberdeen, pp. 161–242.
Ratnavathi, C.V., Ravi, S.B., 1991. Effect of different durations of steeping and malting on the production of alpha-amylase in sorghum. J. Cereal Sci. 14, 287–296.
Ratnavathi, C.V., Ravi, S.B., Subramanian, V., Rao, N.S., 2000. A study on the suitability of un-malted sorghum as a brewing adjunct. J. Inst. Brew. 106 (6), 383–387.
Solabi, G.A., 1989. Uses of Sorghum malted and fermented foods and drinks. In: Proceedings of a Symposium on the Current Status and Potential of Industrial Uses of Sorghum in Nigeria, 4–6 December, Kano, Nigeria, pp. 37.
Subramanian, V., Rao, N.S., Jambunathan, R., Murthy, D.S., Reddy, B.V.S., 1995. The effect of mashing on the extractability of proteins and its relationship to diastatic activity in sorghum. J. Cereal Sci. 21, 283–289.
Swanston, J.S., Rao, N.S., Subramanian, V., Taylor, K., 1994. The influence of some aspects of grain quality on malting potential in sorghum. J. Cereal Sci. 19, 91–97.

Further Reading

Anonymous, Ratnavathi, C.V., 2003. Annual Report, NATP – RNPS – 24, Developing Sorghum as an Efficient Biomass and Bioenergy Crop and Providing Value Addition to the Rain Damaged Kharif Grain for Creating Industrial Demand. DSR, Rajendranagar, Hyderabad – 30.
Babajide, J.M., Uzochukwu, S.V.A., 2001. Cassava-soy weaning food: biological evaluation and effects on rat organs. Plant Food. Hum. Nutr. 56, 167–173.
Bello, A.B., Rooney, L.W., Waniska, R.D., 1990. Factors affecting quality of sorghum to, a thick porridge. Cereal Chem. 67, 20–25.

El Nour, M.E.M., El-Tigani, S., Dirar, H.A., 1999. A microbiological study of Hussuwa: a traditional Sudanese fermented food from germinated *Sorghum bicolor* cv feterita. World J. Microbiol. Biotechnol. 15, 305–308.

Gaffa, T., Jideani, I.A., Nkama, I., 2002. Traditional production, consumption and storage of Kunu – a non alcoholic cereal beverage. Plant Food. Hum. Nutr. 57, 73–81.

Okafor, N., 1983. Processing of Nigerian indigenous fermented foods a chance for innovation. Niger. Food J. 1, 32–37.

Palmer, G.H., Etokakpan, O.U., Igyor, M.A., 1989. Review: sorghum as brewing material. MIRCEN J. Appl. Microbiol. Biotechnol. 5, 265–275.

Sanni, A.I., Asiedu, M., Ayernor, G.S., 2001. Influence of processing conditions on the nutritive value of *ogi-baba*, Nigerian fermented sorghum gruel. Plant Food. Hum. Nutr. 56 (3), 217–223.

Teniola, O.D., Holzapfel, W.H., Odunfa, S.A., 2005. Comparative assessment of fermentation techniques useful in the processing of *Ogi*. World J. Microbiol. Biotechnol. 21 (1), 39–43.

Thaoge, M.L., Adams, M.R., Sibara, M.M., Watson, T.G., Taylor, J.R.N., Goyvaerts, E.M., 2003. Production of improved infant porridges from pearl millet using a lactic acid fermentation step and addition of sorghum malt to reduce viscosity of porridges with high protein, energy and solids (30%) content. World J. Microbiol. Biotechnol. 19, 305–310.

CHAPTER

3

Mycotoxin Contamination in Sorghum

C.V. Ratnavathi[1], V.V. Komala[1] and U.D. Chavan[2]

[1]ICAR-Indian Institute of Millets Research, Rajendranagar, Hyderabad, India [2]Mahatma Phule Krishi Vidhyapeeth, Rahuri, Maharashtra, India

OUTLINE

Sorghum Biochemistry: An Industrial Perspective.
DOI: http://dx.doi.org/10.1016/B978-0-12-803157-5.00003-4

Abstract

Sorghum is the species cultivated as grain for human consumption and animal feed. Sorghum grain grown in the rainy season (*Kharif*) is becoming severely affected by grain molds; the major fungi involved being *Aspergillus*, *Fusarium*, and *Curvularia*. If the extent of mold is severe, the grain is unsafe for consumption owing to contamination by mycotoxins. A study was conducted in sorghum to evaluate natural contamination of aflatoxin B_1 (AFB_1) in India. A total of 1606 grain sorghum samples were collected during the rainy (*Kharif*) season across four years from seven states of India, representing different geographical regions of the country. AFB_1 contamination during 2007–08 was the highest (13.1%), followed by samples from the year 2004–2005 (2.85%). The samples collected in years 2005–2006 and 2006–2007 showed contamination below 1%. The number of samples (35) showing AFB_1 contamination above the safety limit was also highest during 2007–2008 as compared to samples from the other years. This study, conducted for four years, showed that natural contamination of AFB_1 in sorghum grown in India is within safety limits (20 μg/kg) recommended by the Codex Alimentarius Committee and 73% of samples were positive for toxin. The overall occurrence of toxin from Madhya Pradesh and Rajasthan was below 5 μg/kg.

The inhibitory activity of bioactive polyphenols present in six sorghum genotypes—two red (AON 486 and IS 620), two yellow (LPJ and IS 17779), and two white (SPV 86 and SPV 462) varieties—on *Aspergillus parasiticus* (NRRL 2999) growth and aflatoxin production was evaluated. The production of aflatoxins in the six sorghum genotypes after removal of surface phenolics by acidic methanol treatment was studied and compared with that in untreated grains. Aflatoxin production was found to be fourfold higher in treated grains. The total phenols and bioactive polyphenols extracted by acidic methanol were quantified using the Folin–Denis method and the bovine serum albumin–benzidine conjugate procedure respectively. The effect of extracted sorghum phenolics under in vitro conditions on fungal growth and aflatoxin production was studied at two concentrations (0.01% and 0.1%) of phenolics. Extracted phenolics added to yeast extract sucrose (YES) medium at 0.1% concentration showed an inhibitory effect on aflatoxin production. At 0.01% phenolic concentration, aflatoxin production was minimal on day 3 after infection. At other time points, the aflatoxin content was similar to that in the control. At 9 days after infection the fungal biomass in IS 620 was significantly lower than that in the control. At 0.1% phenolic concentration, aflatoxin production was minimal and the red genotype IS 620 showed maximum resistance. Fungal biomass was lowest at all growth stages in IS 620 as compared with the control.

The potential use of antifungal component eugenol for the reduction of AFB_1 in stored sorghum grain was investigated. Fungal infestation of sorghum results in deterioration of varied biochemical composition of the grain. In this study, three genotypes (M35-1; C-43; LPJ) were inoculated with two highly toxigenic strains of *Aspergillus flavus* with three different eugenol treatments in order to evaluate the AFB_1 production. From this study, it was found that at 8.025 mg/g concentration, eugenol completely inhibited the AFB_1 production. The lowest amount of AFB_1 was observed in genotype M35-1, whereas higher amount AFB_1 was observed in LPJ followed by C-43. In all sorghum genotypes, there was a significant positive correlation existing between protein content and aflatoxin produced, the r values being 0.789 and 0.653, respectively. Starch in three genotypes was found to have a significant negative correlation (r¼ −0.704; −0.609) with aflatoxin produced. The starch content decreased, whereas the protein content in all sorghum varieties increased during infection.

3.1 INTRODUCTION

Sorghum (*Sorghum bicolor* (L.) Moench) is a C_4 monocotyledon belonging to the family of *Poaceae*; subfamily *Panicoideae* is generally referred to as sorghum. *Sorghum* is the species cultivated as grain for human consumption and animal feed. Sorghum is the fifth most important cereal crop in the world, after rice, wheat, maize, and barley. It constitutes the main grain food for over 750 million people who live in the semiarid tropics of Africa, Asia, and Latin America. India is the third largest producer (7.92 MT) of sorghum in the world after USA and Nigeria and has the largest area 7.76 M ha under cultivation of this crop covering the states of Maharashtra, Karnataka, Madhya Pradesh, Andhra Pradesh, Rajasthan, Gujarat, and Tamil Nadu (Codex Alimentarius Commission, 2011). It requires a minimum average temperature of 25°C to give the maximum grain yield. It is grown in rainy (*Kharif*), post rainy (*Rabi*), and summer seasons. Sorghum grain can be used as an alternate source for maize in the production of industrial starch, and sorghum grain is also used as a raw material (brewing adjunct) in the preparation of lager beer (Aisien, 1989). Molded sorghum grain is used for potable ethanol production in distillery industry.

Grain mold (GM) is one of the major biotic constraints of sorghum for feed and food production. The principal GM fungi in India are *Fusarium moniliforme, Curvularia lunata, Phoma sorghina, Alternaria alternata, Exserohilum, Gonatobotrytis sp.*, and *Aspergillus spp*. GM is the result of a complex of fungus–host interactions, which leads to the complete deterioration of sorghum grain. Infestation of sorghum grain by storage fungi results in varying degrees of damage including (1) discoloration of the kernel, (2) reduction in kernel germination, (3) heating, (4) mustiness, and (5) production of mycotoxins. GM causes significant losses in both grain yield and its nutritional quality. Sorghum grain grown in *Kharif* is severely affected by grain molds, and the major fungi involved in this are *Aspergillus, Fusarium*, and *Curvularia*. If the extent of mold is severe, the grain is unsafe for consumption due to the contamination of mycotoxins especially aflatoxin B_1 (AFB_1) and fumonisin B_1 (FB_1) produced from these fungi.

Mycotoxin are a group of chemically diverse secondary metabolites of fungi that have a wide range of toxic effects on humans and animals. It is important to be able to detect and quantify mycotoxins in foods and feeds so that such contaminated materials can be handled so as to protect human and animal health (Thirumala-Devi et al., 2000). Among numerous mycotoxins, aflatoxins (AFs), fumonisins, and ochratoxin A (OTA) are of high priority for control because of their frequent and worldwide distribution in agricultural products.

3.1.1 Natural Occurrence of Mycotoxins in Sorghum

Natural contamination of food grains is greatly influenced by environmental factors such as growing season, type of storage structure, temperature, pH, moisture. As a national center of Sorghum research, a study was carried out to estimate the AFB_1, FB_1, and OTA contamination in sorghum grain samples from the rainy season (*Kharif*) collected from different geographical regions of India and to determine the safety of sorghum as a human food and animal feed. All three toxins were assayed by indirect competitive Enzyme-Linked Immunosorbent Assay (ELISA) method (Devi et al., 2001; Komala et al., 2012; Ratnavathi et al., 2012).

3.1.1.1 *Status of AFB_1 Contamination*

In *Kharif*, the grain is severely damaged by weather and postharvest practices like drying and threshing grain on roads. The most toxic among the AFs is AFB_1, which has been reported to be one of the most potent environmental carcinogens. The safety limit of AFB_1 was set at 20 μg/kg by Codex Committee (Codex Alimentarius Commission, 1989). The natural occurrence of AFB_1 in rain-affected Indian sorghum samples and AFB_1 contamination in Brazilian samples was reported by Da Silva et al. (2000). Sashidhar et al. (1992) reported a systematic study on the mold and mycotoxin contamination in the grain sorghum stored in traditional containers in India. A study was carried out to estimate the AFB_1 contamination in sorghum grain samples collected from different geographical regions of India.

A study for the natural occurrence was conducted. A total of 1606 samples originating from various sources spread across four seasons were evaluated for AFB_1 content. The details of the various sorghum samples collected from different sources are presented in Table 3.1. The levels of AFB_1 in the grains investigated are presented in Table 3.2. The results reveal that the range of toxin in the whole set of samples is 0.01–263.98 μg/kg. The distribution of aflatoxin percent in different states is shown in Fig. 3.1. The distribution of aflatoxin percent in different years is shown in Fig. 3.2.

In the year 2004, 175 grain samples were analyzed for AFB_1 content. Aflatoxin content varied from location to location. The toxin was estimated to be as high as 79.9 μg/kg and as low as 0.30 μg/kg. However, in Gujarat (Surat) location AFB_1 content was highest, whereas in Maharashtra (Akola) the toxin content was recorded lowest.

In the year 2005, a total of 552 samples were analyzed for AFB_1 through ELISA. 32.71% samples were found to be completely free of toxin. Toxin was present at very low concentrations and samples varied significantly for aflatoxin content. However, only one sample collected

TABLE 3.1 Collection of Grain Sorghum Samples from Different Sources during the Years 2005–08

S. no.	Type of sample	2004–05	2005–06	2006–07	2007–08	Total samples
1.	Field	-	27	75	-	102
2.	Research fields	175	319	310	268	1072
3.	Farmer store	-	72	100	-	172
4.	Market	-	120	119	-	239
5.	Poultry	-	6	2	-	8
6.	Brewery	-	2	-	-	2
7.	Distillery	-	6	-	-	6
8.	APMC	-	-	5	-	5
	Total samples	175	552	611	268	1606

TABLE 3.2 Contamination of Sorghum Grain Sample with Aflatoxin B_1 (AFB_1)

S. no.	State	No. of samples analyzed	No. of samples positive for toxin	Range of toxin AFB_1 (μg/kg)	No. of samples above safety limit (20 μg/kg)
YEAR 2004–05					
1.	Andhra Pradesh	31	30	0.40–28.30	2
2.	Maharashtra	37	36	0.30–13.40	-
3.	Rajasthan	37	37	1.10–7.90	-
4.	Madhya Pradesh	6	5	0.60–1.50	-
5.	Tamil Nadu	30	30	1.90–22.10	1
6.	Gujarat	34	13	1.10–79.90	2
	Total	175	151		5 (2.85%)[a]
YEAR 2005–06					
1.	Andhra Pradesh	140	101	0.01–5.26	-
2.	Maharashtra	82	38	0.86–4.95	-
3.	Rajasthan	103	53	0.08–3.26	-
4.	Madhya Pradesh	35	22	0.57–4.46	-

(*Continued*)

TABLE 3.2 (Continued)

S. no.	State	No. of samples analyzed	No. of samples positive for toxin	Range of toxin AFB_1 (μg/kg)	No. of samples above safety limit (20 μg/kg)
5.	Gujarat	84	60	0.10–7.15	-
6.	Karnataka	108	69	0.12–129.36	1
	Total	552	343		1 (0.18%)[a]
YEAR 2006–07					
1.	Andhra Pradesh	188	128	0.01–140.48	3
2.	Maharashtra	114	104	0.15–22.22	1
3.	Rajasthan	88	38	0.06–4.03	-
4.	Tamil Nadu	67	56	0.22–21.76	1
5.	Gujarat	10	10	0.26–55.64	1
6.	Karnataka	144	112	0.07–15.24	-
	Total	611	448		6 (0.98%)[a]
YEAR 2007–08					
1.	Maharashtra	45	37	0.49–139.11	10
2.	Rajasthan	90	77	0.12–15.16	-
3.	Tamil Nadu	88	77	0.01–263.98	24
4.	Karnataka	45	40	0.50–21.20	1
	Total	268	231		35 (13.1%)[a]

[a]*Values in parentheses indicate percent toxicity above safety limit.*

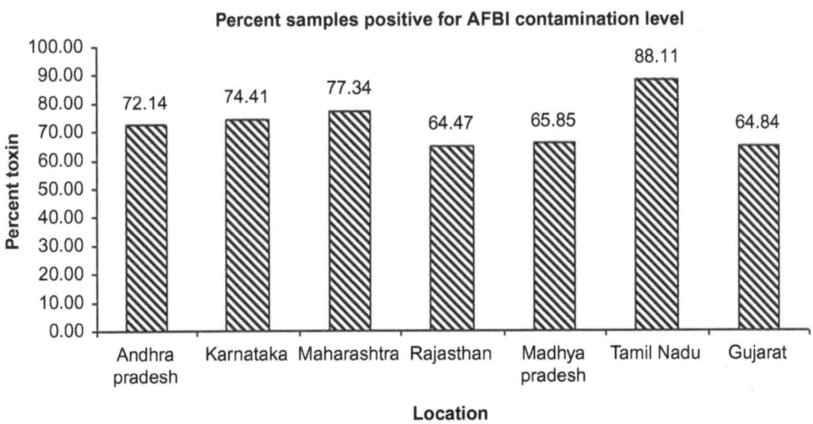

FIGURE 3.1 Percentage of samples positive for aflatoxin B_1 (AFB_1) contamination level.

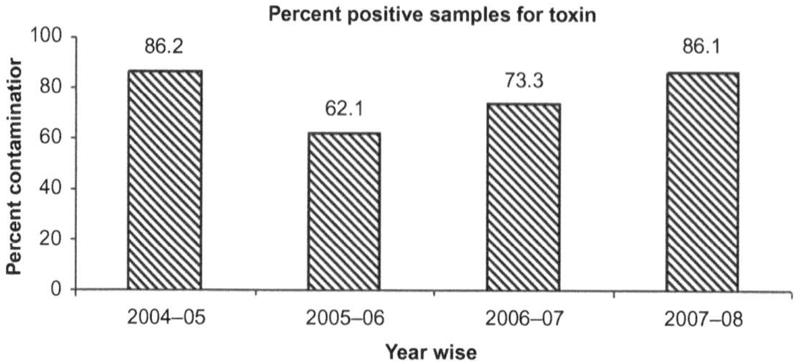

FIGURE 3.2 Percentage of positive samples for aflatoxin B_1 (AFB_1) over four years.

from Chitradurga, Karnataka state recorded high aflatoxin content (129.4 μg/kg) above the safety limit (20 μg/kg as per CODEX Committee). The samples from feed units were free from AFB_1.

During *Kharif* 2006, the percent contamination of Aflatoxin was higher in samples collected from Andhra Pradesh (Palem) compared to Tamil Nadu (Coimbatore). The range of Aflatoxin was 0.01 to 140.48 μg/kg. Out of total 611 samples analyzed, three samples from Palem (SPV 1664 = 140.48 μg/kg, SPH 1576 = 23.84 μg/kg, SPH 1575 = 21.09 μg/kg), one sample from Coimbatore (SPV 1698 = 21.76 μg/kg), and one Sample from Parbhani (SPV 1746 = 22.22 μg/kg) contained aflatoxin above the safety limit (CODEX safety limit = 20 μg/kg).

A total of 258 samples collected in the year of *Kharif* 2007 showed that natural aflatoxin production was lower. The most toxic samples for aflatoxin were present in the grain samples collected from Coimbatore. The range of aflatoxins was 0.00 (SPV 462)–263.98 (SPH 1596) μg/kg. When compared to different states, Rajasthan was free of toxin.

Samples collected from Rajasthan (Udaipur) recorded the lowest levels of aflatoxin as compared to all other locations. AFB_1 contaminations in 1596 samples of sorghum grain from different states in India are shown in Table 3.2. The major portion of the sorghum samples were drawn from experimental fields and farmers' fields. The difference in toxin contamination was statistically significant, and replicated collection of samples also varied significantly in toxin content.

The percentage of samples positive for AFB_1 contamination level was more prevalent in Tamil Nadu (88.11%) when compared to other states. Rajasthan recorded the lowest percentage (64.47%) as shown in Fig. 3.1. AFB_1 contamination was more prevalent in the years 2004 (86.2%) and 2007 (86.1%) than in the years 2005 (62.1%) and 2006 (73.3%), as shown in Fig. 3.2. The important factor influencing AFB_1 production includes annual variations of temperature, relative humidity, pH, and moisture.

This illustrates that the contamination of AFB_1 production is highly influenced by the weather conditions that prevail during the grain development stage, that is, seed set to physiological maturity stage. The aflatoxin content measured in all the samples showed that natural aflatoxin production is lower and 73% of samples were positive for toxin as compared to highly susceptible crops like maize and groundnut. However, only 47 (2.92%) samples were toxic and contained AFB_1 above the safety limit (20 μg/kg) recommended by CODEX (Codex Alimentarius Commission, 1989). The results also show that the locations where the samples were collected had a significant effect on AFB_1 levels. Similar observations were recorded in the studies conducted in wheat in India (Toteja et al., 2006) and also in five different crops surveyed in Nigeria (Odoemelam and Osu, 2009).

3.1.1.2 Status of FB_1 Contamination

FB_1 contamination was studied in *Kharif* grain sorghum samples collected during the years 2006, 2007, and 2008 and are presented in Table 3.3. Most of the samples were positive for the presence of FB_1, and around 7.3% of the samples over the period of 3 years were above the safety limit (200 μg/kg),the natural contamination of FB_1 was lower, and 74.97% (626) of total number of samples were positive for toxin.

The distribution of FB_1 percentage contamination over three years is shown in Fig. 3.3. The highest FB_1 percentage contamination was recorded in Hyderabad, whereas the lowest was in Udaipur. Udaipur recorded the highest number of samples (17.41%) above the safety limit, whereas the lowest number of samples above the safety limit was recorded in Akola (0.71%). None of the samples from Dharwad and Palem recorded FB_1 contamination above the safety limit.

3.1.1.3 Status of OTA Contamination

In the year 2006, a total of 368 grain sorghum samples were collected for assessment of OTA contamination. OTA contamination in sorghum samples collected from different geographical areas of India is shown in Table 3.4. OTA content varied from location to location. The range of OTA content was 0.01 to 29.19 μg/kg. In Hyderabad the percentage of OTA content was highest followed by Dharwad and Akola (Fig. 3.4). In Parbhani the toxin content recorded was the lowest. 55.16% samples were found to be completely free of toxin. Toxin was present at very low concentration in the remaining samples.

From this study, it is observed that contamination of mycotoxins in sorghum is low to medium in *Kharif* produce and grain is mostly safe for consumption. To further reduce the contamination in molded grain, polishing the molded grains, harvesting the crop at physiological maturity, and drying crop artificially through dryers could avoid the

TABLE 3.3 Contamination of Sorghum Grain Samples With Fumonisin B_1 (FB_1)

S. no.	Location	No. of samples analyzed	No. of samples containing toxin	Range of toxin FB_1 (μg/kg)	No. of samples above safety limit	FB_1 levels (μg/kg) (mean ± SD) positive samples		
Year	2006					Mean	±	SD
1.	Palem (AP)[a]	55	42	0.39–154.66	0	27.14	±	33.13
2.	Hyderabad (AP)[a]	60	60	0.33–421.8	8	83.96	±	109.54
3.	Parbhini (MS)[a]	46	45	2.13–370.4	2	46.63	±	64.77
4.	Coimbatore (TN)[a]	52	13	0.82–14.02	0	4.11	±	3.95
5.	Akola (MS)[a]	47	31	0.32–230.95	1	24.46	±	43.21
6.	Dharwad (KA)[a]	51	34	0.43–48.63	0	13.89	±	12.92
7.	Udaipur (RJ)[a]	57	14	0.74–36.51	0	6.05	±	9.19
	Total	368	239		11 (2.98%)[b]			
Year	2007							
1.	Coimbatore (TN)[a]	78	59	0.04–1397.94	15	146.87	±	245.15
2.	Akola (MS)[a]	45	36	0.81–40.11	0	13.77	±	10.79
3.	Dharwad (KA)[a]	45	39	0.96–97.68	0	19.04	±	0.96
4.	Udaipur (RJ)[a]	87	67	0.15–62.38	0	12.30	±	14.36
	Total	255	201		15 (5.88%)[b]			
Year	2008							
1.	Coimbatore (TN)[a]	50	49	0.3–46.6	0	7.90	±	9.21
2.	Akola (MS)[a]	48	31	0.6–48.18	0	8.46	±	10.37
3.	Dharwad (KA)[a]	57	51	0.03–34.23	0	5.33	±	6.16
4.	Udaipur (RJ)[a]	57	54	1.18–1624.68	35	471.34	±	425.07
	Total	212	185		35 (16.5%)[b]			

[a]AP, *Andhra Pradesh;* TN, *Tamil Nadu;* MS, *Maharashtra;* KA, *Karnataka;* RJ, *Rajasthan.*
[b]*Values in parentheses indicate percent toxicity above safety limit 200 (μg/kg).*

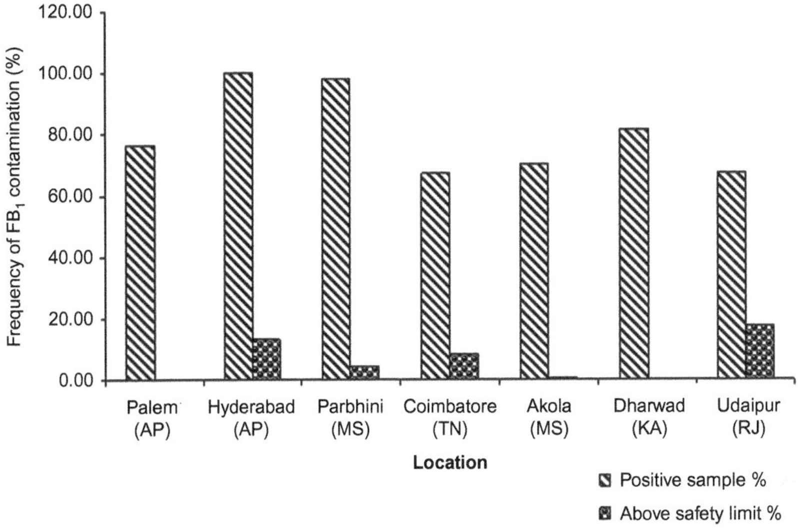

FIGURE 3.3 Frequency of Fumonisin B_1 (FB_1) contamination in different locations of India.

TABLE 3.4 Ochratoxin A (OTA) Contamination in Sorghum Grain Samples

S. no.	Location	No. of samples analyzed	No. of samples containing toxin	Range of toxin OTA (μg/kg)
1.	Palem (AP)	55	13	0.14 3.01
2.	Hyderabad (AP)	60	59	0.37–29.19
3.	Coimbatore (TN)	52	15	0.08–6.55
4.	Akola (MS)	47	19	0.04–5.04
5.	Parbhini (MS)	46	8	0.05–2.8
6.	Dharwad (KA)	51	36	0.01–7.51
7.	Udaipur (RJ)	57	15	0.34–13.72
	Total	368	165	

AP, Andhra Pradesh; *TN*, Tamil Nadu; *MS*, Maharashtra; *KA*, Karnataka; *RJ*, Rajasthan.

contamination of mycotoxins to some extent. Polishing the molded grain before use for poultry and animal feed should be practiced to reduce health hazards. Awareness campaigns among farmers and poultry feed manufacturers and consumers should be organized in *Kharif* sorghum growing areas.

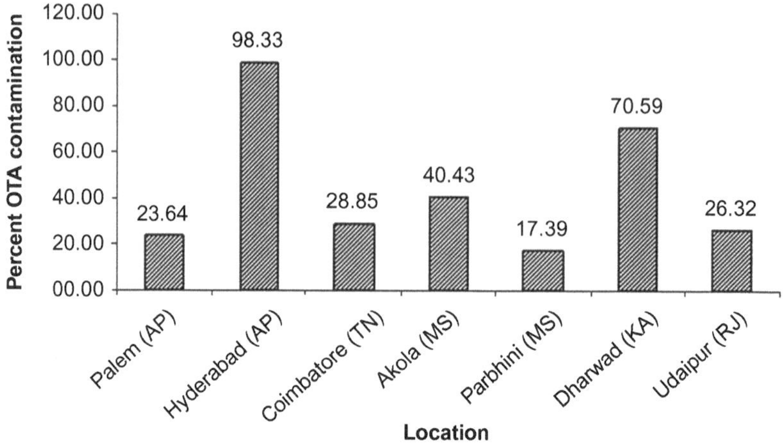

FIGURE 3.4 Percentage of ochratoxin A (OTA) contamination in different geographical areas of India.

3.2 IN VITRO STUDIES ON THE AFLATOXIN ELABORATION IN SORGHUM THROUGH *ASPERGILLUS PARASITICUS*

3.2.1 Substrate Suitability of Sorghum Genotypes to Fungal Infestations

The objective of the study is to investigate the extent of aflatoxin production with *Aspergillus* infection in vitro in different sorghum genotypes with different pericarps, red, yellow, and white, the physical and chemical characteristics of grain during infection, and the changes in grain polyphenols and phytic acid in comparison to maize and groundnut.

A total of 16 sorghum genotypes including six red sorghum genotypes, four yellow sorghum genotypes, and six white sorghum genotypes were used for this study. Groundnut and maize (cv, Madurai) were used for comparison, and they were obtained from local market and the Agriculture Research Institute, Amberpet, Hyderabad, respectively. The fungus strain used in this study was *A. parasiticus* (NRRL 2999), a highly toxigenic strain known to produce copious amounts of AFB_1, AFB_2, AFG_1, and AFG_2. The cultures were maintained on potato dextrose agar (PDA) slants for 6–8 days at 28°C in a BOD incubator (Kalorstat, Mumbai, India). Reference standards of aflatoxins (AFB_1, AFB_2, AFG_1, and AFG_2) were a gift from WHO, Geneva, Switzerland.

3.2.1.1 *Preparation of Samples*

Preparation of samples was done according to the method of Ratnavathi and Sashidhar (2003).

Defatted samples were analyzed for biochemical constituents such as starch, protein, phytic acid, and polyphenols, whereas whole samples were used for aflatoxin and ergosterol analysis. Aflatoxins including AFB_1, AFB_2, AFG_1, AFG_2, and total aflatoxin content were analyzed by the thin-layer chromatography (TLC)-fluorodensitometric method (SLR-TRAFF, Biomed Instruments Inc., Indianapolis, IN) as reported by Egan (1982). The sensitivity of the method is ng/g. Ergosterol was estimated according to the method of Sashidhar et al. (1988). Starch was estimated by enzymatic procedure, as reported by Southgate and protein was estimated by kjeldhal using salicylate (Willis et al., 1996). Fat was estimated by Soxhlet extraction (AOAC, 1995). Phytic acid was quantitated according to the method of Wheeler and Ferrel (1971). Polyphenols were quantitated by precipitation of protein in a microassay using tannic acid as a standard (Ratnavathi and Sashidhar, 1998). Physical characteristics of the sorghum genotypes such as hardness index, 1000 grain weight, and endosperm texture were also analyzed. Hardness index was measured as kg/cm^2, the force required to break the grain using a Kiya hardness tester (Kiya Seisakusho Ltd, Japan). Endosperm texture was classified as per the IBPGR manual (Anonymous, 1993).

3.2.1.2 *Statistical Analysis*

Data were analyzed by two-way analysis of variance (ANOVA) and correlation. Computer software, M. Statistical package, along with Lotus freelance graphics (Ver.2.1), was used in data analysis. The Critical difference (CD) was calculated using the following formula.

$$CD = \text{Standard Error (SE)} \times ''t''$$
$$SE = (2MS_e \times r)^{1/2}$$

MS_e = error means sum of square
r = number of replications
t = critical value of t at 0.01

3.2.1.3 *Status of Aflatoxin Production*

The variations in total aflatoxin produced and also individual aflatoxin (AFB_1, AFB_2, AFG_1, AFG_2) produced at different stages of fungal infection among 16 genotypes of sorghum over a period of 12 days are presented in Tables 3.5–3.8. The level of aflatoxin contamination was in the order red < yellow < white < maize < groundnut. The highest amount of aflatoxin (total) was produced in the genotype CSH 14 on day 6 after fungal infection, that is, 46.17 μg/g.

TABLE 3.5 Aflatoxin Content in Grain of Red Sorghum Genotypes

	Period in infection									
	3 days					6 days				
Genotypes	B_1	B_2	G_1	G_2	Sum	B_1	B_2	G_1	G_2	Sum
IS 14384	1.11 ± 0.11[a]	0.70 ± 0.10	0.51 ± 0.10	0.58 ± 0.02	**2.90 ± 0.33**	2.16 ± 0.20	1.99 ± 020	2.50 ± 0.37	0.62 ± 0.19	**7.27 ± 0.96**
IS 688	2.45 ± 0.50[b]	1.06 ± 0.23	0.30 ± 0.02	1.32 ± 0.15	**5.13 ± 0.90**	1.14 ± 0.33	1.07 ± 0.25	1.07 ± 0.25	1.29 ± 0.05	**6.25 ± 0.70**
AON 486	ND[c]	ND	ND	ND	**ND**	3.30 ± 0.40	1.05 ± 0.01	0.36 ± 0.04	0.63 ± 0.04	**5.34 ± 0.49**
IS 620	1.05 ± 0.09	0.62 ± 0.13	0.25 ± 0.06	0.11 ± 0.01	**2.03 ± 0.29**	2.34 ± 0.47	1.70 ± 0.20	0.72 ± 0.20	0.96 ± 0.18	**5.72 ± 1.05**
IS 18528	ND	ND	ND	ND	**ND**	2.39 ± 0.53	1.47 ± 0.25	1.53 ± 0.47	1.03 ± 0.06	**6.42 ± 1.31**
IS 8014	1.60 ± 0.35	1.50 ± 0.39	0.70 ± 0.20	0.21 ± 0.02	**4.01 ± 0.96**	2.57 ± 0.60	1.77 ± 0.22	1.55 ± 0.45	1.31 ± 0.08	**7.20 ± 1.35**

	Period of infection									
	9 days					12 days				
Genotypes	B_1	B_2	G_1	G_2	Sum	B_1	B_2	G_1	G_2	Sum
IS 14384	7.46 ± 2.10	7.37 ± 2.20	2.83 ± 0.28	1.71 ± 0.30	**19.37 ± 4.88**	3.58 ± 0.44	2.60 ± 0.28	1.38 ± 0.23	0.51 ± 0.03	**8.07 ± 0.98**
IS 688	4.47 ± 0.88	1.77 ± 0.27	0.92 ± 0.06	0.66 ± 0.08	**7.82 ± 1.29**	1.46 ± 0.31	0.64 ± 0.18	0.48 ± 0.16	0.35 ± 0.06	**2.93 ± 0.72**
AON 486	1.71 ± 0.24	0.74 ± 0.13	2.06 ± 0.30	0.62 ± 0.19	**5.13 ± 0.86**	0.61 ± 0.26	0.19 ± 0.04	0.41 ± 0.05	1.35 ± 0.07	**2.55 ± 0.42**
IS 620	2.19 ± 0.25	1.10 ± 0.16	1.03 ± 0.04	1.70 ± 0.06	**6.02 ± 0.51**	1.11 ± 0.15	0.80 ± 0.27	0.67 ± 0.21	0.28 ± 0.08	**2.86 ± 0.71**
IS 18528	5.28 ± 0.60	2.41 ± 0.34	1.55 ± 0.30	0.47 ± 0.12	**9.71 ± 1.36**	2.89 ± 0.84	1.14 ± 0.08	1.22 ± 0.08	0.51 ± 0.13	**5.76 ± 1.13**
IS 8014	3.29 ± 0.48	2.53 ± 0.19	1.05 ± 0.09	0.72 ± 0.01	**7.59 ± 0.77**	3.49 ± 0.04	1.15 ± 0.30	2.07 ± 0.09	1.15 ± 0.23	**7.86 ± 0.66**

[a] *Value represented are mean ± SD of four replications.*
[b] *Aflatoxin values are expressed as μg/g.*
[c] *ND, not detected.*

TABLE 3.6 Aflatoxin Content in Grain of Yellow Sorghum Genotypes

	Period of infection									
	3 days					6 days				
Genotypes	B_1	B_2	G_1	G_2	Sum	B_1	B_2	G_1	G_2	Sum
LPJ	2.35 ± 0.20	3.13 ± 0.44	1.42 ± 0.37	1.40 ± 0.35	**8.30 ± 1.36**	13.65 ± 0.12	3.21 ± 0.37	1.68 ± 0.36	3.02 ± 0.23	**21.56 ± 1.08**
IS 17777	3.34 ± 0.54	1.34 ± 0.54	2.23 ± 0.16	4.30 ± 0.64	**11.21 ± 1.88**	9.31 ± 1.29	8.42 ± 0.12	2.52 ± 0.37	2.17 ± 0.30	**22.42 ± 2.08**
IS 17780	3.25 ± 0.21	0.97 ± 0.03	0.24 ± 0.05	1.26 ± 0.16	**5.72 ± 0.45**	7.11 ± 0.30	7.05 ± 0.20	1.69 ± 0.11	1.30 ± 0.14	**17.15 ± 0.75**
IS 17779	1.64 ± 0.39	0.41 ± 0.04	0.42 ± 0.01	0.09 ± 0.01	**2.56 ± 0.45**	4.08 ± 0.18	1.14 ± 0.16	2.99 ± 0.21	0.97 ± 0.03	**9.18 ± 0.58**

	Period of infection									
	9 days					12 days				
Genotypes	B_1	B_2	G_1	G_2	Sum	B_1	B_2	G_1	G_2	Sum
LPJ	4.08 ± 0.18	6.87 ± 0.47	3.31 ± 0.27	1.54 ± 0.41	**15.80 ± 1.33**	3.02 ± 0.60	0.75 ± 0.27	2.17 ± 0.04	8.13 ± 2.60	**14.07 ± 3.51**
IS 17777	9.48 ± 0.53	3.27 ± 0.41	2.35 ± 0.35	1.35 ± 0.21	**16.45 ± 1.50**	4.92 ± 0.08	2.44 ± 0.17	1.48 ± 0.01	2.98 ± 0.18	**11.82 ± 0.44**
IS 17780	7.99 ± 0.21	9.53 ± 0.11	3.14 ± 0.08	3.17 ± 0.07	**23.83 ± 0.47**	2.06 ± 0.44	0.88 ± 0.14	2.54 ± 0.59	5.50 ± 0.42	**10.98 ± 1.59**
IS 17779	7.26 ± 0.44	4.43 ± 0.60	1.43 ± 0.40	1.26 ± 0.40	**14.38 ± 1.84**	3.73 ± 0.22	0.84 ± 0.18	3.02 ± 0.24	1.52 ± 0.37	**8.81 ± 1.01**

Values represented are mean ± SD of four replications. Aflatoxin values are expressed as μg/g.

TABLE 3.7 Aflatoxin Content in Grain of White Sorghum Genotypes

	Period in infection									
	3 days					6 days				
Genotypes	B_1	B_2	G_1	G_2	Sum	B_1	B_2	G_1	G_2	Sum
CSH 9	2.14 ± 0.03	1.47 ± 0.40	3.08 ± 0.19	1.53 ± 0.66	**8.22 ± 1.28**	14.8 ± 0.28	6.10 ± 0.28	2.88 ± 0.13	4.69 ± 0.36	**28.47 ± 1.02**
CSH 14	3.72 ± 0.17	1.03 ± 0.10	2.72 ± 0.30	1.69 ± 0.24	**9.15 ± 0.81**	25.1 ± 3.00	6.55 ± 0.45	9.42 ± 0.42	5.10 ± 0.19	**46.17 ± 4.06**
SPV 86	5.13 ± 1.60	2.76 ± 0.35	1.01 ± 0.16	1.33 ± 0.17	**10.23 ± 228**	5.37 ± 0.21	1.32 ± 0.05	2.42 ± 0.30	1.05 ± 0.13	**10.46 ± 0.75**
SPV 462	1.31 ± 0.20	1.14 ± 0.20	1.22 ± 0.05	0.58 ± 0.05	**4.25 ± 0.50**	9.97 ± 1.00	6.02 ± 0.40	3.08 ± 0.15	3.00 ± 0.13	**22.07 ± 1.68**
IS 25017	1.55 ± 0.21	1.50 ± 0.42	2.33 ± 0.30	1.54 ± 0.16	**6.92 ± 1.09**	9.80 ± 0.15	2.35 ± 0.10	1.21 ± 0.18	2.69 ± 0.24	**16.05 ± 0.67**
GM 13	6.00 ± 1.40	ND	0.47 ± 0.08	ND	**6.47 ± 1.48**	2.01 ± 0.23	0.84 ± 0.12	1.07 ± 0.12	1.25 ± 0.21	**5.17 ± 0.68**

	Period of infection									
	9 days					12 days				
Genotypes	B_1	B_2	G_1	G_2	Sum	B_1	B_2	G_1	G_2	Sum
CSH 9	7.40 ± 0.85	7.00 ± 0.28	3.07 ± 0.13	3.54 ± 0.42	**21.01 ± 1.68**	5.46 ± 0.38	2.93 ± 0.10	2.00 ± 0.10	2.09 ± 0.16	**12.48.0.74**
CSH 14	5.11 ± 0.03	1.59 ± 0.13	3.84 ± 0.19	1.36 ± 0.06	**11.90 ± 0.41**	3.70 ± 0.10	2.15 ± 0.21	2.00 ± 0.21	0.59 ± 0.10	**8.44 ± 0.62**
SPV 86	3.51 ± 0.30	0.93 ± 0.06	2.96 ± 0.12	1.15 ± 0.04	**8.55 ± 0.52**	3.23 ± 0.11	2.34 ± 0.21	1.52 ± 0.10	1.29 ± 0.05	**8.38 ± 0.47**
SPV 462	5.30 ± 0.54	2.26 ± 0.20	3.00 ± 0.28	2.00 ± 0.04	**12.56 ± 1.06**	2.11 ± 0.10	0.98 ± 0.22	1.65 ± 0.30	4.30 ± 0.10	**9.04 ± 0.72**
IS 25017	3.61 ± 0.45	1.39 ± 0.10	3.25 ± 0.40	1.20 ± 0.11	**9.45 ± 1.06**	0.63 ± 0.18	2.10 ± 0.21	0.44 ± 0.13	1.82 ± 0.30	**4.99 ± 0.82**
GM 13	3.45 ± 0.18	1.32 ± 0.09	2.62 ± 0.14	2.29 ± 0.20	**9.68 ± 0.61**	2.23 ± 0.21	1.49 ± 0.11	1.04 ± 0.00	1.40 ± 0.15	**6.16 ± 0.47**

Values represented are mean ± SD of four replications. Aflatoxin values are expressed as μg/g. *ND*, not detected.

TABLE 3.8 Aflatoxin Content in Grain of Maize and Ground Nut

	Period of infection									
	3 days					6 days				
Genotypes	B_1	B_2	G_1	G_2	Sum	B_1	B_2	G_1	G_2	Sum
Maize (maduri)	2.35 ± 0.08	0.78 ± 0.19	1.70 ± 0.14	0.89 ± 0.05	**5.72 ± 0.46**	7.05 ± 1.19	2.85 ± 0.35	6.78 ± 0.25	2.75 ± 0.21	**19.73 ± 2.00**
Groundnut (commercial)	3.72 ± 0.11	1.95 ± 0.21	3.15 ± 0.08	1.40 ± 0.28	**10.22 ± 0.68**	11.70 ± 0.71	5.10 ± 1.20	10.30 ± 0.71	4.44 ± 0.65	**31.54 ± 3.27**

	Period of infection									
	9 days					12 days				
Genotypes	B_1	B_2	G_1	G_2	Sum	B_1	B_2	G_1	G_2	Sum
Maize (maduri)	5.5 ± 0.42	1.5 ± 0.42	4.6 ± 0.42	1.15 ± 0.07	**12.75 ± 1.33**	3.40 ± 0.28	1.00 ± 0.02	3.00 ± 0.14	0.89 ± 0.16	**8.29 ± 0.60**
Groundnut (commercial)	14.2 ± 0.35	4.1 ± 0.98	12.9 ± 1.10	4.75 ± 1.20	**35.95 ± 3.63**	4.84 ± 1.20	1.97 ± 0.18	4.35 ± 0.64	1.70 ± 0.14	**12.86 ± 2.16**

Value represented are mean ± SD of four replications. Aflatoxin values are expressed as μg/g. *ND*, not detected.

3.2.1.4 *Red Sorghum*

Total aflatoxin production was lower in red genotypes compared to yellow and white genotypes. No aflatoxin was detected in red genotypes AON 486 and IS 18528 on day 3 after infection. In genotype IS 620 on day 3 after infection, the aflatoxin content is 2.03 μg/g. Aflatoxin production peaked among red genotypes 9 days after infection and decreased after that in all red genotypes of sorghum. Aflatoxin production was also different and statistically significant for various time points of infection. In AON 486 and IS 620, aflatoxin produced was least at all stages of infection, that is, 3, 6, 9, and 12 days (0, 5.3, 5.1, and 2.5 μg/g, respectively). In IS 620 and IS 688, aflatoxin produced was lower up to 12 days. The total aflatoxin content in red genotypes ranges from 2.0 μg/g (IS 620, 3 days) to 19.4 μg/g, (IS 14384, 9 days). The two red genotypes IS 14384 and IS 8014 showed high aflatoxin levels (19.37 μg/g in IS 14384 and 7.82 μg/g in IS 688) on day 9 after infection. The ratio of AFB_1, total toxin ranged from 0.32 (AON 486, 12 days) to 0.62 (AON 486, 6 days). There was a great degree of variation with respect to individual aflatoxins (AFB_1, AFB_2, AFG_1, and AFG_2) at different time points (Table 3.5). The variation of AFB_1 was as follows:

Statistically significant among red sorghum genotypes, ranging from 1.05 to 7.46 μg/g at the peak period of production, that is, day 9 after infection, AON 486 showed 1.71 μg/g of AFB_1, which was their lowest content among those red genotypes. The other red genotype, IS 620, resistant to aflatoxin, showed 2.19 μg/g of AFB_1. In IS 14384, significantly higher AFB_1 was produced followed by IS 18528 (5.28 μg/g). However, the amount of AFB_1, at days 3, 6, and 12 after infection was not significantly different. The other three aflatoxin were significantly higher in the red genotype IS 14384 at day 9 after infection. The variation among the other toxins was not significant at all time points in all red sorghum genotypes.

3.2.1.5 *Yellow Sorghum*

The aflatoxin production in yellow genotypes at different periods of infection is presented in Table 3.6. Of the four yellow sorghums tested, IS 17779 showed less aflatoxin production (14.4 μg/g). Peak production of aflatoxin was observed on day 6 after infection in all genotypes except IS 17779, in which it was found to peak at day 9 after infection.

Statistically significant variation for total aflatoxin content as well as for various time points of infection was observed in yellow sorghum genotypes. At day 9 after infection in genotype IS 17780, the total aflatoxin produced was 23.83 μg/g, which was found to be highest among yellow genotypes. In LPJ and IS 17777 toxin production peaked at day 9 after infection. The range of toxin in yellow genotypes was from

2.53 μg/g (IS 17779, 3 days) to 2.9 μg/g (IS 17780, 6 days). In yellow sorghum genotypes, the AFB_1 content was peaking at day 6 after infection, and it was found to be maximum in LPJ, 13.65 μg/g, a local yellow genotype (Table 3.6). The range of AFB_1 contents at peak production, that is, day 6 after infection, was 4.08 and LPJ genotype was 13.65 μg/g. The AFB_1 content in LPJ genotype was statistically significant and higher than significant difference of AFB_1 was observed among the yellow genotypes day 3 and 12 after infection. AFB_2 was significantly higher in IS 17777 and IS 17780 days 6 and 9 after infection (Table 3.6). The highest level of AFG_2 was observed in LPJ and IS 17780 at day 12 after infection. There was no significant variation of the AFG_1 and AFG_2 toxins among the other yellow genotypes.

3.2.1.6 *White Sorghum*

Aflatoxin production in white genotypes is depicted in Table 3.7. Total aflatoxin production in all sorghum genotypes, maize, and groundnut at different periods of infection differed significantly.

Among the six genotypes of white sorghum, four lines, CSH 9, CSH 14, SPV 86, and SPV 462, are released through the All India Coordinated Sorghum Improvement Project. IS 25017 and GM 13 were germplasm accessions having GM resistance. White genotypes showed higher toxin production at day 6 of infection than yellow genotypes (Tables 3.6 and 3.7). The variability for total toxin production in varieties was statistically significant. The individual toxins AFB_1, AFB_2, AFG_1, and AFG_2 were also significantly different (Table 3.7). Aflatoxin production was highest at day 6 after fungal infection in white sorghums except in GM 13, that is, 28.5 and 46 μg/g (Table 3.7). The temporal trend in aflatoxin production SPV 86, SPV 462, and GM 13 was maximum aflatoxin production was observed at day 9 after infection (9.67 μg/g). GM 13 and SPV 86 produced fewer aflatoxin at all stages of infection. At day 3 after fungal infection, no AFB_2 and AFG_2 toxins could be detected in GM 13.

3.2.1.7 *Maize and Groundnut*

Aflatoxin in maize and groundnut at all stages of infection differed significantly. Contents of total and individual aflatoxin are given in Table 3.8. The amount was also high in maize and groundnut compared to red sorghum genotypes (maize, 19.42 μg/g: groundnut, 31.5 μg/g). However, aflatoxin production in two white genotypes (CSH 9 and CSH 14) was found to be higher as compared to maize and groundnut. In both groundnut and maize, the aflatoxin production peaked at days 6 and 9 after infection, respectively.

3.2.1.8 Ergosterol Production

Ergosterol contents were different and statistically significant in all genotypes at all stages of fungal growth ($p < 0.01$) (Ratnavathi and Sashidhar, 2003). The pie chart of ANOVA of Ergosterol also showed that variation exists between genotypes as well as period of infection. The ranking order for the Ergosterol content was different from that of total toxin (yellow > white > red > maize >groundnut). Red genotypes were found to have low amounts of Ergosterol. Varieties are significantly different for the Ergosterol content. As the fungal growth increases, Ergosterol also increased significantly in all of the genotypes up to day 9 after fungal infection, and it decreased or showed no change on day 12 after fungal infection. The range of Ergosterol in red genotypes observed was from 17.0 μg/g (IS 620, 3days) to 228 μg/g (AON 486, 12 days). The lowest contest of Ergosterol (87.5 at day 12) was observed in IS 14384 among the red genotypes at all stages of infection. LPJ, a local cultivar, showed a low content of Ergosterol (132 μg/g). The range of Ergosterol observed in yellow sorghums was 18.0–248 μg/g. In white sorghum, SPV 86, a *Rabi* based cultivar, showed a low amount (13–85 μg/g) of Ergosterol, whereas IS 25017, a germplasm line, had a high amount of Ergosterol (230 μg/g). In IS 25017, the range of Ergosterol observed was 13.3–230 μg/g. In maize and groundnut, Ergosterol contents were highest at day 9 after fungal infection, that is, 193 and 185 μg/g, respectively.

3.3 PHYSICAL AND CHEMICAL CHARACTERISTICS OF DETERIORATED SORGHUM GRAIN

3.3.1 Physical Characteristics

The physical characteristics of grain in sorghum genotypes are presented in Table 3.9. They represent 1000 grain weight, color of the grain endosperm character, and hardness index. Grain size of red genotypes was small, whereas yellow and white sorghum grain are of medium size. The 1000 grain weight ranged from 15.8 to 42 g. The grain weight in red sorghum genotypes varied from 15.8 g (IS 8014) to 22 g (IS 18528) (Table 3.5). In yellow sorghums, it varies from 29.8 g (LPJ) to 32.0 g in 1779. In white sorghums, the grain weight varied from 16.6 g. (GM 13) to 42.0 (SPV 86, post rainy season cultivar). Among the red genotypes, IS 688, AON 486, and IS 8014 were found to have a hard corneous endosperm. IS 14384 and IS 620 have a chalky or flouncy endosperm. The hardness index of red genotypes varied from 7.6 to 9.5 kg/cm^2, the highest being for IS 8014 (Table 3.5). All yellow and white genotypes had corneous endosperm. The hardness index varied from 9.0 to 10.1 kg/cm^2. The hardness index in white sorghums ranged from 8.98 to 12.5 kg/cm^2. IS 25017 was found to have the highest hardness index.

TABLE 3.9 Physical Characters of the Grain in Sorghum Genotypes

Genotype	Hardness index (kg/cm^2)	1000 grain wt (g)	Endosperm nature
RED			
IS 14384	8.80 ± 0.45	18.9 ± 2.0	Floury
IS 688	9.00 ± 0.53	16.3 ± 1.5	Corneous
AON 486	9.10 ± 0.91	19.3 ± 1.8	Corneous
IS 620	7.60 ± 1.00	16.8 ± 2.0	Floury (Soft)
IS 18528	8.60 ± 0.30	22.0 ± 0.5	Corneous
IS 8014	9.50 ± 0.50	15.8 ± 0.5	Corneous
YELLOW			
LPJ	10.10 ± 0.62	29.8 ± 1.0	Corneous
IS 17777	9.00 ± 0.16	31.0 ± 1.0	Corneous
IS 17780	9.80 ± 0.24	30.0 ± 1.5	Corneous
IS 17779	9.35 ± 0.20	32.0 ± 2.0	Corneous
WHITE			
CSH 9	9.70 ± 0.30	29.2 ± 1.5	Corneous
CSH 14	9.20 ± 0.20	28.5 ± 1.8	Corneous
SPV 86	10.70 ± 0.90	42.0 ± 2.5	Corneous
SPV 462	8.98 ± 0.53	28.7 ± 1.5	Corneous
IS 25017	12.50 ± 0.22	22.9 ± 2.0	Corneous
GM 13	9.65 ± 0.30	16.6 ± 1.0	Corneous

Values represented are mean ± SD of four replications.

3.3.2 Chemical Characteristics

1. *Starch*: A significant different in starch content was found to exist among the varieties. It also differs significantly with the period of fungal infection. The starch content in red genotypes was low as compared to yellow and white genotypes. The maximum amount of starch was observed in CSH 9 (70%).

   Among the red genotypes, IS 18528 was found to have the lowest amount of starch (28.7%), and the highest starch content was observed in IS 14384 (47.5%). In yellow sorghum, IS 17777 was found to have high percentage of starch (61%). The other three yellow genotypes were found to have lower contents of starch (44%). The percent starch content in white sorghums varied from 36% (IS 25017) to 70% (CSH 9, on day 6). In general, starch content decreases during the course of infection. Furthermore, the correlation between starch content and aflatoxin produced was not statistically significant.

2. *Protein*: The protein content in all sorghum genotypes tended to increase as the fungal infection increased. Varieties differ significantly for the protein content. Aflatoxin production was significantly and positively correlated to the grain protein content in red sorghum genotypes ($r = 0.413$). Red sorghum genotypes contain slightly higher amounts of protein compared to yellow and white sorghums. The overall protein content ranged from 6.88 to 29.7%. As the protein increased, the toxin production also increased. However, there was no significant correlation between protein and aflatoxin contents in yellow and white sorghums (Table 3.9).

   In red genotypes, protein content ranged from 14.1% (IS 688, 0 day) to 26.9% (AON 486, 12 days), and in yellow sorghums it varied from 8.75% (LPJ, 0 day) to 29.1% (IS 17780, 9 days), whereas in white cultivars protein content ranged from 10.9% (SPV 86) to 21.3% (GM 13). All of the genotypes were significantly different in their protein contents ($p < 0.05$). Protein content during various periods of infection was also significantly different.

3. *Fat*: Among the 16 genotypes of sorghum percent fat content varied from 0.6 to 4.5. In red genotypes, percent fat content ranged from 0.90 (IS 14384) to 4.5 (IS 8014). The highest fat content was observed in IS 8014 and the lowest in IS 688 (12 days, 0.7%). In yellow genotypes it ranged from 1.0% (LPJ, 3 days) to 4.0% (17780, 0 day). In general, a decrease in fat content was observed over the period of infection. In white genotypes, fat content was in the range from 2.6% (CSH 14, 0 day) to 4.0% (GM 13). However, the decrease in fat content during the period of fungal infection was statistically significant ($p < 0.01$). The correlation coefficient of percent fat to aflatoxin content in white sorghum was -0.526 ($p < 0.01$). However, there is no correlation in red and yellow genotypes between fat and aflatoxin elaborated (Table 3.10).

4. *Polyphenols*: The pie chart of two-way ANOVA of polyphenols showed that 16 sorghum genotypes differed significantly in their polyphenol contents. Red sorghums had a high content of polyphenols as compared to yellow and white genotypes. Polyphenol content was increased in response to infection. Polyphenols in red sorghums varied from 1.58% (AON 486) to 8.64% (IS 8014). The amount of polyphenols in yellow sorghums was marginally higher than that of white sorghums, ranging from 0.23 to 6.46 μg/g. In white sorghums, polyphenols were not detected in SPV 462, whereas GM 13 had a low level of polyphenols. At all stages of infection no polyphenols were detected in these genotypes. In white sorghums, polyphenol content increased as the period of infection

TABLE 3.10 Correlation Coefficients of Total Aflatoxin

Sample	Parameter	Total aflatoxin correlation value
1	Polyphenol (red sorghum)	
	Day 3 of infection	0.589
	Day 6 of infection	0.513
2	Polyphenols (white sorghum)	
	All time points	0.505
3	Protein (red sorghum)	0.413
4	Starch	−0.008[a]
5	Fat (white sorghum)	−0.526
6	Phytic acid (yellow sorghum)	−0.569

Level of significance: $p < 0.01$. [a](−) negative correlation.

and toxin production increased. There was a positive, significant correlation between polyphenol content and aflatoxin production in white sorghums ($r = 0.505$) (Table 3.10). In red sorghum genotypes, there was a significant, positive correlation between polyphenol and aflatoxin contents at day 3 and 6 after infection, their values being 0.589 and 0.513, respectively (Table 3.10). Correlation in yellow sorghums, however, was not significant.

5. *Phytic Acid*: The phytic acid in different genotypes showed that genotypes were not significantly different. The decrease of phytic acid during infection was also not significant. Phytic acid content in red sorghum genotypes was slightly lower compared to yellow and white genotypes. A negative significant correlation was found between phytic acid and toxin production in yellow sorghums, the r value being −0.569 (Table 3.10). The correlation in red and white sorghums was also negative, but it was not significant.

3.4 ENZYMATIC CHANGES IN SORGHUM GENOTYPES DURING A. *PARASITICUS* (NRRL 2999) INFESTATION

Sorghum genotypes, a total of six, which include two genotypes of red (pigmented sorghum), two genotypes of yellow sorghum, and two genotypes of white sorghum, were identified as resistant lines to aflatoxin production based on the data obtained from the previous section, that is, "Substrate suitability of Sorghum Genotypes to Fungal

Infestations." The genotypes of white sorghum SPV 86 and SPV 462 were identified for further investigation considering their usage in breeding research as well as in farmer's fields. They were multiplied at the farms of the National Research Centre for Sorghum, Rajendranagar, Hyderabad. The genotypes used in the study included AON 486 and IS 620 (red), LPJ and IS 17779 (yellow), and SPV 86 and SPV462 (white). Grains were inoculated with spores of *A. parasiticus* (NRRL 2999) as detailed previously. The pure cultures were maintained on potato dextrose agar (PDA) for 6–8 days at 28°C in BOD incubator (Kalorstat, Dwarka Equipment Pvt., India) and were used for inoculation. Chemicals bovine serum albumin (BSA) and polyvinyl pyrrolidone (PVP) were obtained from Sigma Aldrich Chemical Company, St. Louis, USA. All the other chemicals used were of analytical grade.

3.4.1 Preparation of the Sample

Preparation of samples was done for enzymatic changes in sorghum genotypes during *A. Parasiticus* (NRRL 2999) infestation according to the method of Ratnavathi and Sashidhar (2000).

3.4.1.1 α- and β-Amylase Activity

α-Amylase and β-amylase were assayed according to the procedure reported by Bernfeld (1955). Protein was estimated in the supernatant by modified Lowry Procedure (Lowry et al., 1951).

3.4.1.2 α-Amylase

The amylase activity in the six genotypes as well as the activity at various time points of infection is different and statistically significant. Inherent α-amylase activity among the cultivars on 0 day ranged from 0.25 units (IS 620) to 1.45 units (IS 1777). α-Amylase activity ranged in all the genotypes from 0.25 units (IS 620, 0 day) to 24.69 units (IS 17779, 9 days) in case of healthy grains, whereas in infected grains activity ranged from 1.7 (AON 486, 3 days) to 8.1 units (LPJ, 9 days). α-Amylase activity was minimum at 0 day time point in all the six genotypes. α-Amylase activity was highest at 9 days after germination in healthy condition, except in the case of red genotypes, AON 486, where peak activity was found on day 6.

In general, the peak level of activity for α amylase was at 9 days after infection in infected grains. However, in AON 486 and SPV 462, it was maximum at 12 and 6 days after infection, respectively. At 12 days of fungal growth, AON 486 was having α-amylase activity (ie, 5.5 units), which was marginally higher than that of healthy grain after 12 days (4.5 units). α-Amylase activity in the infected grain was significantly

TABLE 3.11 Correlation Coefficients of Aflatoxin Versus Various Enzyme Activities

Enzymes	*r* value	Aflatoxin significance
α Amylase	0.406	Significant*
β Amylase	0.436	Significant*
Protease	0.283	Not Significant
Lipase	−0.349	Not Significant

Note: Aflatoxin data from Chapter IV was used for correlation. *, $p < 0.05$.

less in all the six genotypes compared to the respective healthy samples except in SPV 462. SPV 462 showed maximum α-amylase activity on day 6 after infection.

A paired t-test of α-amylase was performed between the activities of healthy and infected grains at all growth stages. The activity in healthy seedlings was different and statistically significant from the activity under infected condition ($p < 0.05$). The amylase activity under infected condition was found to be correlated to the toxin production. The correlation was positively significant with an r value 0.406. The enzyme activity was found to increase as the toxin elaboration increases. Increased enzyme activity resulted in increased toxin elaboration (Table 3.11).

3.4.1.3 β-Amylase

The β-amylase activity in different genotypes (both healthy and infected) is given in Fig. 5.2a–c. The activity was minimum at 0 day time interval in all the six genotypes, with inherent β-amylase activity among the genotypes ranging from 0.18 units (IS 17779) to 0.91 units (AON 486) under healthy condition. In red and white genotypes, the inherent β-amylase activity at 0 day time point was found to be maximum, and in yellow genotypes it was observed to be minimum. β-amylase activity was observed to be maximum at 3 days after infection in IS 620, SPV86, and SPV 462, whereas it was maximum at 6 days after infection in LPJ, IS 17779. However, AON 486 showed maximum activity at 9 days after infection.

Variability for β-amylase activity present in six genotypes at different periods of infection under both healthy and infected condition was statistically significant. Of all the four genotypes, SPV 462 was having maximum activity (9.15 units). In AON 486, activity was high at 12 days in healthy seeds as well as infected seeds. A paired t-test was performed between healthy and infected grain at all the growth stages. The activity in healthy grains was different and statistically significant from that of infected grain ($p < 0.01$). In general, β-amylase activity in healthy grain was always more than that of infected grain except in AON 486 at 6 days where it was greater in the infected grain. The β-amylase activity was correlated to the total aflatoxin produced. The correlation was positively significant which indicated that as the toxin production increased, β-amylase activity also increased ($r = 0.436$) (Table 3.11).

3.4.1.4 *Protease Activity*

The assay was done as reported by Kunitz (1947).

3.4.1.4.1 Status of Protease Activity

The protease activity was assayed at two temperatures 37°C and 50°C and under two situations of healthy and infected grain sorghum. Genotypes as well as various time points of infection are different and statistically significant for protease activity. The inherent protease activity of the genotypes varied from 60 units (AON 486, 0 day) to 179 units (SPV 86, 0 day). Protease activity was found to be highest in SPV 462 at 9 days after germination. The trend observed for protease activity at 37°C was significantly different from the protease activity at 50°C. However, the activity at 50°C was more than that of 37°C. Except in SPV 462, for all the other five genotypes the activity was peaking up at 6 days after fungal growth and in SPV 462 it peaked at 9 days. The paired t-test was performed between the healthy and infected protease activities at different growth stages. The paired t-tests were performed individually at two temperatures. The activity at 37°C in healthy and infected grains was different and statistically significant. Activity in healthy grains was lower compared to the activity in the infected grains ($p < 0.001$). However, the activity at 50°C in healthy grains was not significantly different from that of infected grains. Protease activity was more in infected grains as compared to healthy grains in four of the genotype at all stages of infection. In other two genotypes SPV 462 and IS 17779 at 9 days after fungal growth, the activity was more in healthy grains (1755 and 636 units) than the infected ones (655 and 204 units). Maximum activity of protease at 50°C was observed in SPV 86 at 12 days after infection. The inherent protease activity at both temperatures was equal in AON 486 and activity at 50°C was high in SPV 86. In IS 620, IS 17779, LPJ and SPV 462 activity was high at 37°C at all the stages of infection. No protease activity was detected at both temperatures in AON 486 at 12 days after fungal growth. Protease activity at 50°C was maximum at 12 days after fungal growth. Protease activity at 50°C was maximum at 6 days after fungal infection. At 3 days high protease activity was observed in SPV 86 and LPJ at 50°C under healthy condition. Maximum activity in infected grains was found at 9 days in genotypes AON 486, LPJ, and SPV 462, while it was high at 6 days after infection in IS 620 and IS 17779 at 50°C. No statistically significant correlation was found between protease activity and aflatoxin production at various stages of fungal growth ($r = 0.283$) (Table 3.11).

3.4.1.5 Lipase

Lipase was assayed according to the method reported by Bier (1962).

3.4.1.5.1 Statistical Analysis

The software M. Stat package and Lotus freelance graphics (Ver.2.1) was used to analyze the data. All the statistical tests, namely paired t-test, correlation, and two way ANOVA, were based on the methods reported by Snedecor and Cochran (1968).

3.4.1.6 Status of Lipase Activity

The variability in lipase activity for different genotypes was highest in SPV 462 and lowest in IS 17779 among the genotypes. Genotypes and different time periods of infection for lipase activity were not significantly different. Inherent lipase activity was highest (at 0 day) in AON 486, SPV 462, and SPV 86, as compared to the activity in infected grains. Lipase activity was always less in infected grain in all the genotypes. Genotypes LPJ and IS 17779 were showing maximum lipase activity at 6 days and 9 days, respectively, in healthy grains. In genotypes AON 486, SPV 86 and SPV 462 at 6 days after fungal growth, lipase was not detectable. A paired t-test was performed for the lipase activity of healthy and infected grains. The activity in healthy grains was different and statistically significant and higher to that in infected grains ($p < 0.01$). The correlation between lipase activity and toxin produced was negative ($r = -0.349$). However, it was not significant.

3.5 INHIBITORY EFFECT OF PHENOLICS EXTRACTED FROM SORGHUM GENOTYPES ON THE GROWTH OF A. *PARASITICUS* (NRRL 2999) AND AFLATOXIN PRODUCTION

Fungal damage to kernels or grains may be limited to the pericarp or may involve extensive internal invasion (Forbes et al., 1992). Kernel resistance to fungus can be either due to physical or structural factors such as pericarp resistance to splitting or to resistance that can arise from chemical effects on the fungus from various parts of the kernel (Norton, 1997). Earlier studies have shown that various phenols, tannins, and related pigments have also been reported to be present in groundnut testae and appear to be involved in defense mechanisms against *A. flavus* invasion (Azaizeh et al., 1990). Lansden (1982) examined the fungi static fungi properties of various tannin fractions from groundnut seed coats and determined their effect on aflatoxin production by *A. parasiticus*. The inhibitory effects of different groundnut parts tested on *A. flavus* and *Trichoderma viride* were studied by Lindsey and Turner (1975). Four compounds inhibitory to *A. flavus* growth were extracted from the groundnut cotyledons with acetone. Three of these compounds demonstrated the properties of phenolics. Turner et al. (1975) identified tannin like inhibitor (5, 7-dimethoxy isoflavone) from cotyledons of groundnut, which showed inhibition against *A. flavus*.

Red sorghum genotypes containing high phenolics were found to be poor substrates for the growth of *A. parasiticus* (NRRL 2999) and aflatoxin elaboration (Ratnavathi and Sashidhar, 2003). Among the six red genotypes studied, two genotypes were found to be more resistant for the toxin elaboration. The role of phenolics present in the pericarp of the seed in fungal resistance was reported earlier by several researches (Harris and Burns, 1973; Bandyopadhyay and Mughogho, 1988; Jambunathan et al., 1986). In order to assess and confirm the role of polyphenols in inhibiting the fungal growth and aflatoxin production, detailed experiments were designed using the polyphenol extracted from sorghum genotypes. In the first experiment, intact grains, devoid of surface phenolics, were tested for their resistance to aflatoxin elaboration. In the second experiment, the effect of extracted sorghum phenolics was tested under in vitro condition, in relation to growth and aflatoxin production by *A. parasiticus* (NRRL 2999). Further, the enzyme polyphenol oxidase in the fungus, *A. parasiticus* (NRRL 2999) was also assayed in order to confirm whether the phenolics are metabolized by the fungus.

3.5.1 Materials

Sorghum genotypes, a total of six, which included two genotypes of red (pigmented sorghum), two genotypes of yellow sorghum, and two genotypes of white sorghum, were identified as resistant lines to aflatoxin production (Ratnavathi and Sashidhar, 2003). They were multiplied at the farms of National Research Centre for Sorghum, Rajendranagar, Hyderabad. The genotypes used in the study included AON 486, IS 620 (red), LPJ and IS 17779 (yellow), and SPV 86 and SPV 462 (white). The low polyphenolic sorghum genotypes (yellow and white) were treated as controls in the study.

3.5.2 Chemicals

Bovine serum albumin (BSA) and tyrosinase were obtained from Sigma-Aldrich Chemical Company, St. Louis, and USA. The reference standards aflatoxins (AFB_1, AFB_2, AFG_1, and AFG_2) were obtained as a gift sample from WHO, Geneva under the International Check Sample Programme. L-3, 4-dihydroxy phenylalanine (DOPA) was obtained from Loba Chemical Ltd, India. All other chemicals used were of analytical grade.

3.5.3 Fungal Strain

The fungal strain was obtained from United States Department of Agriculture at Peoria in Illinois, USA. *A. parasiticus* (NRRL 2999 *A. flavus* var. *parasiticus*), which is a known toxigenic strain, is considered to be identical with *A. flavus* (CMI91019b) Betina (1984). This strain is known to produce AFB_1, AFB_2, AFG_1, and AFG_2 in abundance.

3.5.4 Preparation of Samples

Preparation of samples, acidic methanol-treated sorghum grains, phenolic extract, yeast extract sucrose (YES) liquid culture medium, and Inoculation of Liquid culture were done according to the methods of Ratnavathi and Sashidhar (2006). YES is known to support very high aflatoxin production as compared to other semisynthetic media (El-Bazza et al., 1983).

3.5.5 Estimation of Polyphenols and Total Phenols

The total polyphenols were estimated by Folin–Denis method (Ratnavathi and Sashidhar, 1998), and bioactive (protein precipitable) polyphenols were quantified by the BSA benzidine conjugate in the six genotypes (Lerch, 1987). An aliquot (100 μL) of the culture medium at all time points of fungal infection (3, 6, 9, and 12 days) was also analyzed for polyphenols.

3.5.6 Polyphenol Oxidase Assay

Polyphenol oxidase (EC 1.14.18.1) activity was assayed according to the method reported by Lerch (1987). The enzyme was extracted from the fungal biomass of *A. parasiticus* (NRRL 2999) grown on YES medium with or without added phenolic extracts of sorghum. The culture media was also assayed for the presence of enzyme activity.

3.5.7 Statistical Analysis

The software package M. Stat and Lotus Freelance graphics (Ver.2.1) was used to analyze the data. All the statistical tests, namely paired t-test, correlation, and two-way ANOVA, were done based on the methods reported in Statistical method by Martinelli and Kinghorn (1994). The critical difference (CD) was calculated using the following formula

$$CD = \text{Standard Error(SE)} \times \text{"}t\text{"}$$
$$SE = (2MS_e \times r)^{1/2}$$

MS_e = error means sum of square
r = number of replications
t = critical value of t at 0.01

3.5.8 Aflatoxin Elaboration in Acidic Methanol Treated Grains

Aflatoxin production in all the six genotypes at 9 and 12 days after infection was higher compared to the untreated condition. A paired t-test was performed between acidic methanol treated grains and untreated grains. There was a statistically significant difference between treated and untreated grains for aflatoxin elaboration in red sorghum genotypes. There was no significant difference between treated and untreated grains for aflatoxin production in case of yellow and white genotypes.

Total aflatoxin content was threefold higher at 9 days after infection and fourfold higher at 12 days after infection in high tannin genotypes as compared to the respective untreated grain samples. Total aflatoxin content was decreased at 12 days after fungal infection. AON 486 was having less toxin compared to its other red genotype IS 620. IS 17779, a yellow genotype, produced significantly more toxin (21. 4 μg/g, treated). Statistically significant difference was presented in only red genotypes with respect to individual toxins (AFB_1, AFB_2, AFG_1, and AFG_2) at 9 and 12 days after infection ($p < 0.01$). The difference in AFB_2 was also observed to be statistically significant in SPV 462 at 9 days after infection ($p < 0.01$). For all the genotypes, G_1 and G_2 were significantly lower AFB_2 in control at 12 days after infection. No significant difference was observed for G_1 and G_2 toxins at 12 days after infection. The ration of AFB_1 to total toxin ranged from 0.22 to 0.36.

3.5.9 Effect of Addition of Extracted Phenolics (Extracted From Sorghum Genotypes) on the Growth of A. *parasiticus* (NRRL 2999) and Aflatoxin Production

The individual aflatoxins were produced at different stages of fungal growth and at different concentrations of phenolics (0.01% and 0.1%). At 1% level, the added phenol extract was found to be precipitated after 2 days of incubation. Hence, the experiment was discontinued. Genotypes are significantly different for aflatoxin production. A paired t-test was performed between control and each experimental sample for toxin production. The genotypes also differ for individual toxins significantly from the control at 0.1% level of phenolics at all the time periods of infection studied.

3.5.10 0.01% Level of Phenolics

At 3 days after the infection all six genotypes produced less aflatoxin than the control. IS 620, a red genotype produced lower toxin compared to AON 486. The toxin level was peaking at 9 days after fungal growth. At 6 days after fungal growth toxin production in experimental samples was slightly less than the control. All the individual toxins differed significantly from the control at 3 days after infection in the genotypes. The lowest concentration of AFB_1 was observed in IS 17779 (1.62 μg/g) as compared to white genotypes (2.4 μg/g). The toxins AFB_2, AFG_1, and AFG_2 significantly differ from the control in genotypes IS 620, LPJ, and IS 17779 at 3 days after infection. At 9 days after infection, aflatoxin produced was lower compared to the control in two genotypes (SPV 86 and IS 17779). At 12 days, toxin production decreased (LPJ, AON 486, IS 17779, and SPV 462) and remained the same in some (IS 620). Except for AON 486, LPJ, IS 17779, and SPV 462, the toxin content in other genotypes is equal to that of control. Fungal biomass ranged from 0.71 g to 1.48 g in all the six genotypes under study. In the control, the fungal biomass ranged from 0.9 g to 1.5 g. In all six genotypes, fungal biomass are almost equivalent to that of control except in AON 486 and SPV 462 (12 days) at all time points studied. At 3 days of fungal growth in IS 620 fungal biomass was lower and statistically significant compared to the control value ($p < 0.01$). At 9 days of fungal growth in IS 620, LPJ, IS 17779, and SPV 86, fungal biomass was slightly less than the control. Polyphenol oxidase activity was not detected in the fungal biomass of *A. parasiticus* in the control or in the experimental samples and medium.

3.5.11 0.1% Phenolics Level

The overall toxin content decreased when compared to the control at this 0.1% level of phenolics in all the six genotypes tested. No variation was observed among genotypes for toxin production. No toxin was detected at 3 days of fungal growth in IS 620, LPJ, and SPV 462. Traces of toxin were found in SPV 86 at 3 days after infection. Even after 6 days of inoculation no toxin could be detected in LPJ and SPV 462, yellow and white cultivars, respectively. The lowest amount of toxin was observed in IS 620 (0.294 μg/mL) at 6 days after fungal inoculation followed by yellow genotypes IS 17779 (0.389 μg/mL). Maximum amount of toxin was found in SPV 86 at 9 days after infection (2.36 μg/mL). The lowest amount of toxin was found in IS 620 at 9 and 12 days after infection (0.202 μg/mL and 0.143 μg/mL respectively). Aflatoxin B_2 was not detected in AON 486 at 6 days after infection, and aflatoxin G_2 was not detected in IS 17779 at 9 days after infection. All the individual toxins differ significantly from the control in all the six genotypes. At 3 days after inoculation, all the six genotypes were showing significantly reduced fungal biomass compared to control. Except for SPV 86, all the other five genotypes were having less biomass at all stages of fungal growth as compared to the control. At 12 days after inoculation, biomass ranged from 0.79 g (IS 620) to 1.59 g (control). The lowest biomass was found in IS 620 at all four stages of fungal growth. Polyphenol oxidase could not be detected in the fungal biomass of the experimental flask and in the culture medium. No inherent polyphenol oxidase activity was detected in the control fungus or in the medium.

Since the phenolics present in the pericarp are removed by acid methanol treatment, aflatoxin production was found to be increased as compared to the untreated grain samples in the red genotypes. Phenolics externally added to 0.1% concentration showed an inhibitory effect on aflatoxin production. The aflatoxin production was very low in IS 620, LPJ and SPV 462. The total aflatoxin produced was positively correlated with the fungal biomass. The correlation was positively significant ($r = 0.441$ (0.01%) and $r = 0.637$ (0.1%)).

There was an inverse relationship between fungal inhibition and aflatoxin formation. The polyphenol present at 0.1% level was showing a significant inhibition in aflatoxin production. The 0.01% level of phenolics was not effective as 0.1%; however, the aflatoxins production was low compared to control at 3 days after infection. Lansden (1982) isolated 3 fractions of tannins from seed coats of the cultivars of groundnut Florunner and observed that the growth of *A. parasiticus* on Potato Dextrose agar and production of aflatoxin in liquid culture were inhibited by these tannins. The phenolics extracts of AON 486 were quite different from the phenolics present in IS 620. It was evident that phenolics present in AON 486 genotypes were not having any inhibitory effect on the growth of *A. parasiticus* or on the production of aflatoxin. The phenolics in IS 620 are most effective in the inhibition of aflatoxin production. The phenolic extracts LPJ, IS 17779 and SPV 462 were also found inhibitory at 0.1% level. In IS 620, LPJ and SPV 462 no aflatoxin was detected after infection. In LPJ and SPV 462, no aflatoxin could be detected even after 6 days after infection. The presence of phenolics in the grain was more effective compared to the effect of isolated fractions in the genotype AON 486. Phenolics from IS 620 were equally effective when they were present in the intact grain or when they were in the liquid culture. The reason for this may be the nature of polyphenols and its complexities and interactions with free phenolics, which may play an important role along with polyphenols in the inhibition of toxin elaboration.

Lindsey and Turner (1975) reported that leachates of green groundnut kernels of different cultivars inhibited *A. flavus* and *T. viride* in culture. Crude acetone or methanol extracts from cotyledons of these cultivars were inhibitory to spore germination and growth of both fungi. Later, the inhibitory compound was isolated from peanut cotyledons and identified as 5, 7-dimethoxyiso flavine (Turner et al., 1975). The variation in the inhibitory effect of phenolics from different genotypes in the present study can be mainly attributed to the nature of phenolics present in that particular genotype.

Experimental evidence from the present investigation suggests that the inhibitory action of phenolics extracts was possibly mediated due to the lack of the enzyme ployphenol oxidase of the fungus to metabolize them.

The basic reason for the phenolic extracts to inhibit toxin production would be that the fungus was not able to metabolize the phenolics added to the medium. This is due to the lack of polyphenol oxidase activity in *A. parasiticus*. Polyphenol oxidase activity could not be detected in *A. parasiticus* (NRRL 2999) and hence it is not induced. It was reported earlier (Martinelli and Kinghorn, 1994) that polyphenol oxidase activity was also not detected in *Aspergillus nidulans*, which is ontogenically close to *A. parasiticus*. Thus, *A. nidulans* was used as a model system to study the developmental regulation of tyrosinase enzyme cloned from the fungal such as *Agaricus spp.* and *Neurospora crassa* (Martinelli and Kinghorn, 1994).

In summary, phenolics have an inhibitory effect on the growth of *A. parasiticus* as well as aflatoxin production. Specific phenolics present in the extract were responsible for inhibition. Polyphenol oxidase activity was not detected in the fungus *A. parasiticus*; thus, the fungus could not metabolize polyphenols present in the extracts. Hence, polyphenols were showing an inhibitory effect on fungal growth and aflatoxin production.

The yellow and white genotypes, which have low polyphenols, still show inhibition to toxin elaboration, which may be due to the presence of phenolics as well as other factors like antifungal proteins that include chitinases, beta-glucanases, and ribosome-inactivating proteins.

3.6 INDUCTION OF CHITINASE IN RESPONSE TO *ASPERGILLUS* INFECTION IN SORGHUM

Endochitinases (EC 3.2.1.14) are expressed in many plant species in response to pathogen infection or to other environmental stresses (Zhang et al., 1996). Chitin is an important component of the cell wall of many fungal pathogens, and chitinase has been shown to inhibit hyphal growth of several fungi in vitro (Collinge et al., 1993). Thus, one of the postulated functions attributed to chitinases in plants is as a defense against fungal infection (Punja and Zhang, 1993). Seetharaman et al. (1996) studied the changes in the levels of various antifungal proteins, such as sormatin, chitinase, and glucanase, during development, inhibition, and germination in sorghum. All three proteins increased in concentration and peaked at physiological maturity (30 days after anthesis). Levels of ribosomal-inactivating proteins (RIPs) were observed to peak at 15 days after anthesis and decrease subsequently in sorghum. In another study by Seetharaman et al. (1997), the three antifungal proteins sormatin, chitinase, and glucanase were purified from sorghum grains at 30 days after anthesis and tested against *F. moniliforme*, *Curvularia lunata*, and *A. flavus* using hyphal rupture, hyphal extension, and spore germination methods. Marked inhibition of spore germination in all three species of fungus was observed at a concentration of 360 ppm of antifungal proteins. In the present study, the induction of chitinase as a defensive response to fungal infestation (*A. parasiticus* (NRRL 2999)) in low-polyphenol sorghum genotypes was determined to investigate the biochemical basis for fungal resistance. Groundnut, a high-risk commodity for aflatoxin, was used as a negative control, as no chitinase activity has been reported in the literature in this oil seed.

3.6.1 Experimental Materials

Six sorghum (*S. bicolor* (L.) Moench) genotypes, two red (pigmented sorghum), two yellow, and two white, were identified as lines resistant to fungal infection. They were multiplied at the farm of the National Research Center for Sorghum (Rajendranagar, Hyderabad, India). The genotypes used in the study were AON 486 and IS 620 (red), LPJ and IS 17 779 (yellow), and SPV 86 and SPV 462 (white). Grains were inoculated with spores of *A. parasiticus* (NRRL 2999), which is a known toxigenic strain and is considered to be identical with *A. flavus* (CMI91019b) (Betina, 1984). The culture was maintained on potato dextrose agar (PDA) for 6–8 days at 28°C in a BOD incubator (Kalorstat, Dwarka Equipment, Mumbai, India) and used for inoculation. The fungal spore inoculum, containing 1×10^6 spores prepared in 0.1 g/L Tween 20 in glass-distilled water, was added to 10 g of grain sample. Groundnut was used as a negative control.

Preparations of samples and extraction procedure for chitinase were done according to method of Ratnavathi and Sashidhar (2004). Chitinase activity was detected by polyacrylamide gel electrophoresis (PAGE) using a Mighty Small-II apparatus (Hoefer, Indiana Street, CA94107, USA) as reported by Trudel and Asselin (1989).

3.6.2 Detection of Chitinase Activity After PAGE Under Native Conditions

Gels were incubated in 150 mM sodium acetate buffer (200 mL per gel) at pH 5.0 for 5 min, then placed on a clean glass plate (80 mm × 170 mm), and covered with a 75 g/L polyacrylamide overlay gel (60 mm × 130 mm × 0.75 mm) containing 0.1 g/L glycol chitin in 100 mM sodium acetate buffer (pH 5.0). Sliding a 12 mm × 75 mm test tube over the surface of the overlay gel eliminated the liquid between the gels and the glass plate. Gels were incubated at 37°C for 1 h in a container under moist conditions. Following incubation, plastic spacers (3 mm thick) were sealed with 10 g/L agarose on the overlay gel. Care was taken to seal not only the spacers to the gel but also the overlay gel to the glass plate. The area between the spacers was filled (about 20 mL) with freshly prepared 0.1 g/L calcofluor white M2R in 500 mM Tris/HCl (pH 8.9). After 5 min the brightener solution was removed and the gels were incubated for about 1 h at room temperature in distilled water. Lytic zones of gels were visualized by placing the gels on a transilluminator (Ultra Violet Products, Upland, CA, USA).

3.6.3 Chitinase Assay

The assay was done according to the method reported by Jeuniaux (1962). Nag was estimated by the method of Reissig et al. (1955). Chitin suspension used as a substrate is hydrolyzed by chitinase along with externally added chitobiase to form *N*-acetyl glucosamine. On heating with alkali, this forms an intermediate compound, glucoxazoline, which in turn reacts with DMAB reagent to form a colored complex. The absorbance of the colored complex is read at 585 nm.

3.6.4 Assay Procedure

Chitin (2 mg) in 1 mL of citrate/phosphate buffer (pH 5.1) was added to 1 mL of citrate/phosphate buffer (pH 5.1) containing 20 μL of chitobiase and 250 μL of enzyme supernatant. The reaction mixture was made up to 4 mL with quartz-distilled water. After 3 h of incubation at 37°C in a water bath (Julabo SW-21, Seelbach Germany), an aliquot of the reaction mixture was transferred to a centrifuge tube containing 1 mL of quartz-distilled water. The mixture was then boiled for 5 min and centrifuged at 3000 rpm for 5 min in a tabletop centrifuge (Remi C-23, Mumbai, India). An aliquot of 500 μL was taken and estimated for released *N*-acetyl glucosamine (Nag) according to the method reported by Reissig et al. (1955). Enzyme activity was expressed as μg Nag/mg protein. Two sets of blanks were maintained, one not containing the substrate chitin and the other not containing the enzyme extract, in order to take care of any background Nag. Groundnut was used as a negative control. AFB_1, AFB_2, AFG_1, AFG_2, and total aflatoxin contents were determined by the thin layer chromatography (TLC)/ fluoro densitometry (SLR-TRFF, Biomed Instruments Inc., USA) as reported by Egan (1982).

Statistical analysis data were analyzed by two-way analysis of variance (ANOVA), paired t-test, and correlation (Snedecor and Cochran, 1968). The M Stat C statistical package along with Lotus freelance graphics (Ver.2.1) and Microsoft Excel was used in data analysis. The critical difference (CD) was calculated using the formula $CD = \text{standard error (SE)} \times t(1)$ where $SE = (2Ms_e/r)^{1/2}$, Ms_e is the error mean sum of squares, r is the number of replications, and t is the critical value of t at level 0.01.

3.6.5 Levels of Chitinase Activity

Chitinase activity was visualized on gels as lytic zones. Two lytic zones corresponding to chitinase activity were visualized at 365 nm. The zones, one large and one small, were observed towards the upper side of the gel for all six genotypes tested. The visualization of chitinase activity by PAGE was used as a preliminary screening method and was very useful in identification.

The crude enzymes containing chitinase enzyme activity were extracted from infected grains of the sorghum genotypes. Chitinase activity was visualized in all six genotypes and at all times of infection studied. It is likely that both constitutive and induced enzymes were detected in this study, since the in vitro fungal infection by *Aspergillus* spp. may induce chitinases in sorghum. A similar situation was observed in carrot. In carrot under stress, chitinase was induced (Zhang et al., 1996). The genotypes were further analyzed for quantification of chitinase by colorimetry. Chitinase activity was present at all times of fungal infection in both healthy and infected grains. Groundnut was used as a negative control. The genotypes were different and statistically significant for the chitinase activity. The activity was also significantly different for various periods of infection in all genotypes. The chitinase activity was found to be maximum after 12 days of growth under healthy conditions in all genotypes and was highest in the yellow genotypes.

3.6.6 Red Sorghum

The two red sorghum genotypes (AON 486 and IS 620) differed in chitinase activity induced during infection. The activity under healthy conditions also differed significantly in these two genotypes. Maximum chitinase activity was observed at 12 days under healthy conditions. The activity peaked at 6 days after infection in both genotypes compared with the control. However, in AON 486 on day 6 after infection the chitinase activity was fivefold higher (56.5 μg Nag/mg protein) in infected grains than in healthy grains (10.4 μg Nag/mg protein). In IS 620 the activity (30.3 μg Nag/mg protein) was only twofold higher than the control activity (14.3 μg Nag/mg protein).

3.6.7 Yellow Sorghum

The chitinase activity in both yellow genotypes, LPJ and IS 17779, was also high at 12 days under healthy conditions. The chitinase activity was high on day 9 after infection compared with healthy grains. The two genotypes differed significantly from each other. LPJ had threefold higher activity under infected conditions (44.2 μg Nag/mg protein) than under healthy conditions (15.8 μg Nag/mg protein) on day 9. In IS 17779 the expression of chitinase was fourfold higher (57.68 μg Nag/mg protein) than the control activity (9 days after infection). Maximum chitinase activity was observed in LPJ (136.9 μg Nag/mg protein); however, this was considered as an outlier in the control.

3.6.8 White Sorghum

Induction of chitinase activity was highest in the white sorghum genotypes. The increase in activity was high on days 6 and 9 after infection compared with healthy grains. The increase in chitinase activity began on day 3 after infection in these varieties, and the activity in both healthy and infected grains was the same. Chitinase activity peaked at different times in the two white varieties (SPV 462 and SPV 86). SPV 462 showed the maximum increase in activity (69.7 μg Nag/mg protein) on day 6 after infection, whereas in SPV 86 the peak level of induction was on day 9 after infection. On day 3 after infection the chitinase activity in infected grains was twice that in healthy grains of SPV 462. The increase in activity on day 6 after infection was four times that of the control activity, whereas on day 9 after infection it was only twice that of the activity of healthy grains. However, in SPV 86 the increase in activity in the infected grains compared with healthy grains at 3 days after infection was not significant. On day 6 after infection the increase was significant and doubled (15.1 and 22 μg Nag/mg protein, respectively). Induction of chitinase was found to be highest on the 9th day and was fivefold higher (65.3 vs 12.5 μg Nag/mg protein) than the activity of healthy grains on day 9 after infection. Both yellow genotypes had shown significantly higher activity than control grains after 12 days (91.5 and 107 μg Nag/mg protein). However, in SPV 86 and SPV 462 at 12 days after infection the activity was marginally lower than that in control grains.

3.6.9 Aflatoxin Levels

All four aflatoxins (AFB_1, AFB_2, AFG_1, and AFG_2) were produced in five genotypes (IS 620, LPJ, IS 17779, SPV 86, and SPV 462) at all stages of infection. However, aflatoxin could not be detected in the red genotype AON 486 on day 3 after infection. The total aflatoxin produced in AON 486 on days 6, 9, and 12 after infection was lower than in the other five genotypes. The total aflatoxin produced was found to be less in the red genotypes than in the yellow and white genotypes. Peak aflatoxin production in the red genotypes was on day 9 of infection (5.13 and 6.02 μg/g). The correlation between chitinase activity and total aflatoxin was significantly positive ($r2$ (Pearson's correlation coefficient) $= 0.600$, $p \leq 0.001$). In AON 486 and IS 620, the total aflatoxin produced was lowest at all stages of infection. The total aflatoxin produced was less in IS 17779 (14.4 μg/g). The total aflatoxin content peaked on the day after infection in LPJ (21.56 μg/g). In IS 17779 it peaked on day 9 after infection. The correlation between chitinase activity and total aflatoxin was not significant ($r2 = 0.225$, $p = 0.005$). The white genotypes showed maximum total aflatoxin production on day 6 after infection (10.46 and 22.07 μg/g). SPV 86 and SPV 462 showed lower amounts of aflatoxins at all stages of infection compared with the other genotypes. The aflatoxins produced in the two white genotypes were comparable to those in the red genotypes. The total aflatoxin content in groundnut was higher than in sorghum. The total aflatoxin contents on days 6 and 9 after infection were 31.54 and 35.95 μg/g, respectively. There was a significant positive correlation between chitinase activity and toxin production in the white cultivars ($r2 = 0.482$, $p \leq 0.001$). One suitable substrate for *Aspergillus* spp. is groundnut.

The increase in chitinase activity in groundnut was significantly low. The induction of chitinase activity in groundnut was significantly lower than that in sorghum in relation to the infection by *A. parasiticus* (NRRL 2999). There was no significant increase in chitinase activity observed in groundnut at any day after infection. On day 9 after infection the activities of chitinase in both healthy and infected grains were identical.

3.7 INHIBITION OF AFB_1 PRODUCTION BY AN ANTIFUNGAL COMPONENT, EUGENOL ON SORGHUM GRAINS

Fungal deterioration of stored grains is a chronic problem in the Indian storage system because of the tropical hot and humid climate. Infection of *Aspergillus* spp. was found on most sorghum grains collected from different sorghum growing areas in India either at *Kharif* season or during storage. Hence, the present study was carried out on the inhibition of AFB_1 production by an antifungal component, eugenol, on sorghum grains at different concentration levels and estimating the AFB_1 (μg/kg of grain sorghum) by indirect competitive ELISA. It was explored in a four step approach in the laboratory conditions which are:

1. Identification of highly toxigenic strains.
2. In vitro screening of sorghum cultivars with identified strains, to study the chemical quality of infested grains, through electron microscope.
3. Inhibition of AFB_1 production by an antifungal component, eugenol.
4. Chemical parameters of infested grain such as starch and protein were studied.

The above four methods were done according to the procedures of Komala et al. (2012).

3.7.1 Isolation of A. *flavus* strains and AFB_1 production

Mycotoxin-producing ability of 18 strains of *A. flavus* isolated from sorghum grain samples from different locations were studied under laboratory conditions. Sabouraud dextrose supplemented with 0.3% β-cyclodextrin was used for the visual detection of aflatoxin production. Presence of fluorescence surrounding fungal colonies under UV light (365 nm) after 3 days of incubation at 28°C indicated positive for aflatoxin production. The plates were then stored at 28°C for 20 days to facilitate saturation in aflatoxin production. The substrate was used for toxin extraction following standard protocol and AFB_1 was estimated using ELISA.

Colony color of most of the isolates was in different shades of green that varied from light to dark green. The isolates greatly varied in mycotoxin production ability that ranged from 0 to 8000 μg/kg of substrate. Fifty percent of the isolates produced AFB_1 higher than the safety limits (20 μg/kg) and were considered highly toxigenic strains (Table 3.12). The strain A112 produced 8000 μg/kg AFB1, which was the most toxic among the isolates (Fig. 3.5). *Aspergillus* isolated from market (A112-8000 μg/kg, A104-5150 μg/kg) and brewery (A42-6770 μg/kg, A74-2870 μg/kg) samples produced more AFB_1 as compared to field sample isolates. This assumes more significance as most of the *Aspergillus* infection in sorghum was observed in stored samples (market samples), and infection frequency is very low at field level. This variation in level of toxin production is purely due to variation among fungal strains as the experiment was conducted in controlled conditions using a synthetic medium. The two methods used for aflatoxin estimation (fluorescence method vs ELISA) showed strong correlations ($r = 0.505$ ($p = 0.05$)), suggesting that the fluorescence method can be used for rapid screening of aflatoxin producing isolates in a cost effective way.

3.7.2 In Vitro Screening of Sorghum Cultivars

The highly toxigenic strains of *A. flavus* strain number A104 and A112 were used for the screening of the released cultivars of sorghum. Grains from 15 released cultivars of sorghum (CSH 9, CSH 14, CSH 15R, CSH 17, CSH 18, CSH 19R, CSV 13, CSV 14R, CSV 15, CSV 18, CSV 19SS, CSV 216 R, SPV 462, SPV 1430, and SPV 1616) were collected from *Rabi* 2005 season and used for the screening. The grains were inoculated with the two toxic strains (A112 and A104) in vitro and allowed to grow at 28°C for 12 days. These grains were dried at 40°C and then processed for toxin extraction. A112 strain had produced more toxin (4590 μg/kg in CSH 18) compared to A104 strain. These toxigenic strains were isolated from grain samples collected from markets of Mahaboob Nagar and Hyderabad (Fig. 3.5).

The effect of infestation from these strains on the chemical quality of the grain was also studied in detail. The infested grains were cleaned and processed for electron microscopy. The electron micrographs of the grains infested with A112 and A104 were compared with the control grains. A104 strain did not affect the embryo and only spread through the outer layers of the grain (Figs. 3.6 and 3.7a and 3.7b). It completely damaged the starch granules present in the outer layers of the grain, that is, the mesocarp. The other strain A112 had a damaged embryo as well as outer layers of grain (Fig. 3.8a and 3.8b). The damage caused to the grain by A112 strain was greater compared to the A104 strain.

TABLE 3.12 Aflatoxin B$_1$ Production by Isolates of *Aspergillus flavus* from Sorghum

S. no.	Name of *Aspergillus* isolate	Source of sorghum sample	Location of collection	Colony color on Sabouraud agar[a]	Intensity of fluorescence (1–5 scale)	AFB$_1$ (μg/kg)[c]
1.	A30	Market	Guntur	Dark green	3.0 (1.4)[b]	1370
2.	A31	Market	Guntur	Yellowish green	2.0 (0.0)	59
3.	A81	Market	Mahaboob Nagar	Dark green	3.0 (1.4)	−1
4.	A112	Market	Mahaboob Nagar	Dark green	3.5 (0.7)	8000
5.	A151	Market	Lingampally, Hyderabad	Yellowish green	3.0 (0.0)	−4
6.	A103	Market	Samshabad, Hyderabad	Olive green	3.0 (0.7)	−1
7.	A104	Market	Samshabad, Hyderabad	Olive green	4.5 (0.7)	5150
8.	A60	Market	Malakpet, Hyderabad	Yellowish green	3.0 (1.4)	655
9.	A64	Market	Malakpet, Hyderabad	Yellowish green	4.0 (0.0)	1900
10.	A125	Market	Erragadda, Hyderabad	Dark green	3.5 (0.7)	2280
11.	A133	Market	Erragadda, Hyderabad	Olive green	3.5 (2.1)	−1
12.	A256	Market	Indore	Yellowish green	3.5 (0.7)	0
13.	A41	Brewery	Mumbai	Dark green	3.0 (0.0)	1
14.	A42	Brewery	Mumbai	Dark green	4.0 (0.0)	6770
15.	A74	Brewery	Indore	Yellowish green	4.5 (0.7)	2870
16.	A254	Field	Indore	Dark green	4.0 (1.4)	−2
17.	A250	Field	Nalgonda	Yellowish green	3.5 (0.7)	8
18.	A252	Field	Rajendranagar	Yellowish green	2.0 (0.0)	2
19.	Control	–	–	–	–	−1

[a]*Sabouraud dextrose amended with 0.3% β-cyd.*
[b]*Figures in the parenthesis are standard deviations.*
[c]*20 ppb is the safety limit of aflatoxin as per the CODEX committee.*

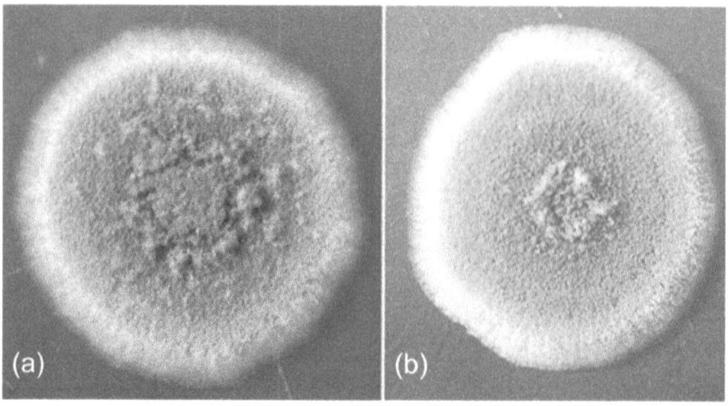

FIGURE 3.5 *Aspergillus flavus* strains of 112 (a) and 104 (b).

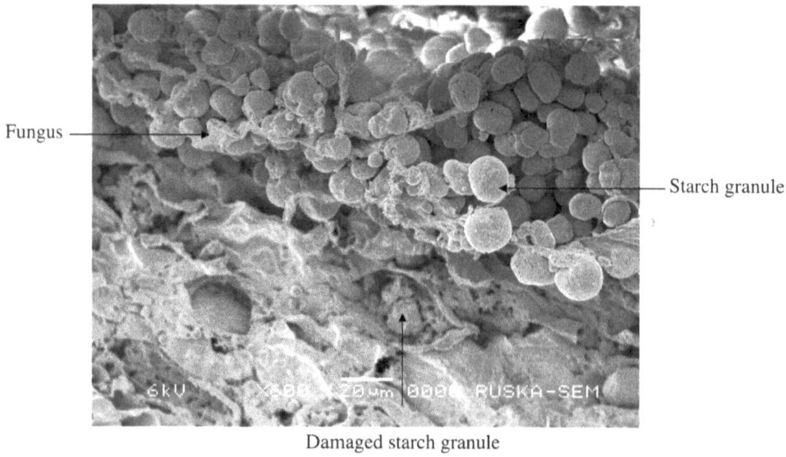

FIGURE 3.6 Sorghum grain longitudinal section (scanning electron micrograph) showing undamaged tissue.

This showed that both strains were virulent; the intensity of virulence for the strain A112 was greater. At the same time, the range of toxin production also varied for a similar set of genotypes (Table 3.13) for these two strains. The range of toxin for A104 was 2110–3170 μg/kg and for A112 was 2540–4590 μg/kg.

3.7.3 Inhibition of AFB1 Production by an Antifungal Component, Eugenol

A total of three sorghum varieties (M35-1, C-43, and LPJ) were collected for eugenol treatment. These varieties were multiplied at the farms of the Directorate of Sorghum Research (formerly National

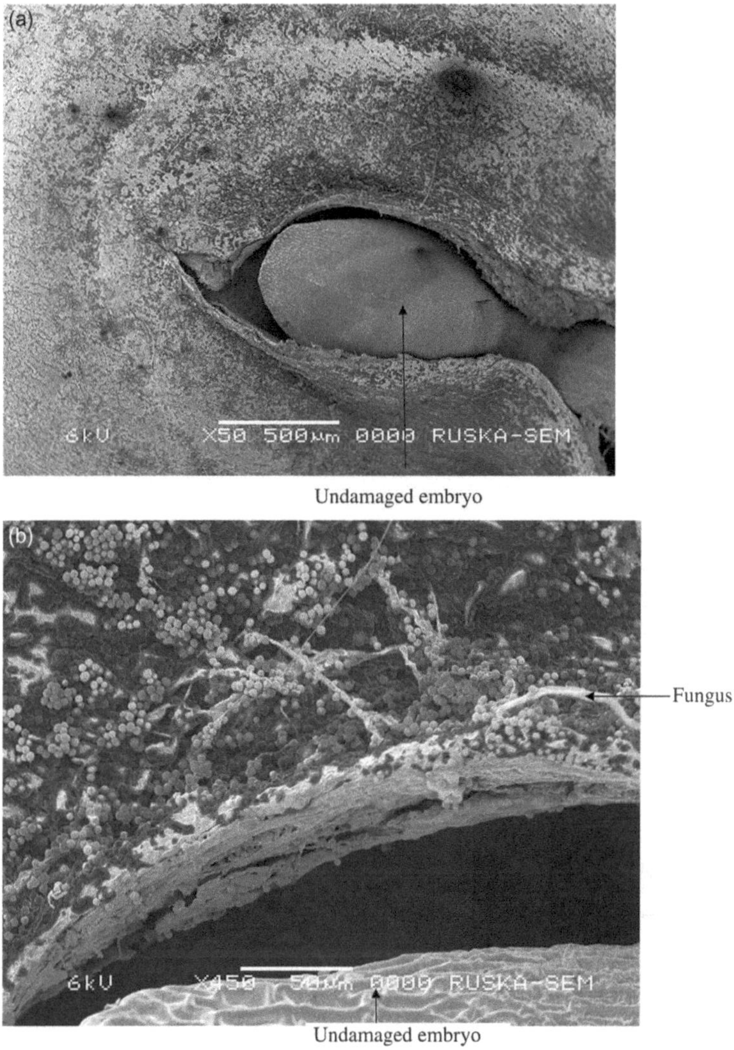

FIGURE 3.7 (a) Sorghum grain 12 days after infection by A104 strain longitudinal section showing embryo undamaged. (b) Sorghum grain 12 days after infection by A104 strain longitudinal section showing embryo undamaged (higher magnification).

Research Centre for Sorghum), Rajendranagar, Hyderabad, and also obtained from a local market of Hyderabad, Andhra Pradesh.

Eugenol treatment was used to inhibit the AFB_1 contamination. The fungal infection among three varieties of sorghum and YES medium in a period of 12 days is shown in Fig. 3.9. The level of aflatoxin contamination

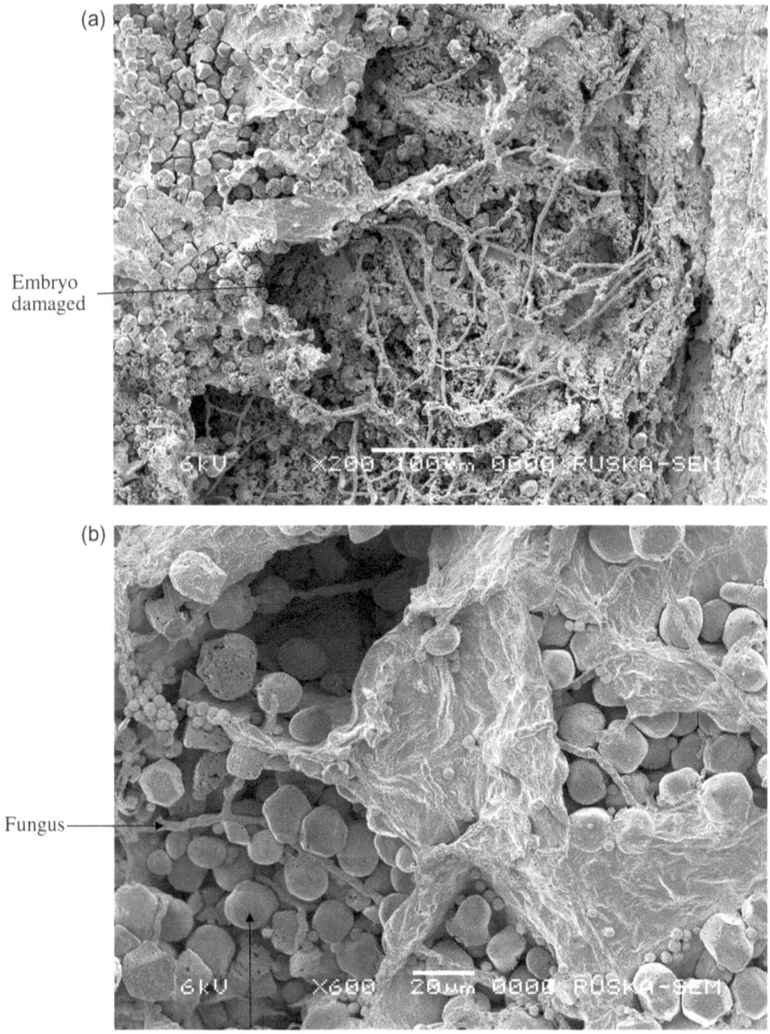

FIGURE 3.8 (a) Sorghum grain 12 days after infection by A112 strain longitudinal section showing embryo damaged. (b) Sorghum grain 12 days after infection by A112 strain longitudinal section showing-damaged tissue and starch granules (higher magnification).

was in the order (strain ± 150 μL Eugenol) < (strain ± 100 μL Eugenol) < (strain ± 50 μL Eugenol) < (only strain) < control. For all three sorghum varieties and YES medium, the strain A112 produced AFB_1 from 1.20 to 4750 μg/kg, whereas A104 produced 16.55 to 1530 μg/kg. This variation in level of toxin production is purely due to the variation of the two fungal

TABLE 3.13 Aflatoxin Production by Toxigenic *Aspergillus* Isolates in Sorghum

	***Asp* 104**	***Asp* 112**
Cultivar	**AFB_1 (μg/kg)**	**AFB_1 (μg/kg)**
CSH 9	3065	3653
CSH 14	2734	3835
CSH 15R	3082	4498
CSH 17	2915	3458
CSH 18	3134	4584
CSH 19R	2629	2873
CSV 13	2915	3578
CSV 14 R	2105	2643
CSV 15	2745	4321
CSV 18	2556	3250
CSV 19 SS	3173	3385
CSV 216R	3025	3837
SPV 462	2886	4416
SPV 1430	2201	2537
SPV 1616	2900	4195
Mean	2804.3	3670.9
Standard error	82.6	170.7
CD 5%	177.2	366.1

strains. This showed that the intensity of virulence for A112 strain was greater. Among the both strains M35-1 showed AFB_1 production ranging from 56 to 1390 μg/kg, which was the lowest content among the three varieties. C-43 showed AFB_1 ranging from 126 to 4430 μg/kg, whereas LPJ showed AFB_1 ranging from 115–4750 μg/kg. YES medium showed AFB_1 ranging from 1.20 to 890 μL/L. No AFB_1 was detected in 150 μL eugenol containing samples.

The antifungal activity of eugenol showed the complete inhibition of AFB_1 production on sorghum grains at a concentration of 8.025 mg/g (150 μL/20 g), whereas in YES medium the fungal growth was inhibited at all the three concentrations because of equal dissemination. The seed germination was also inhibited in all the three concentrations.

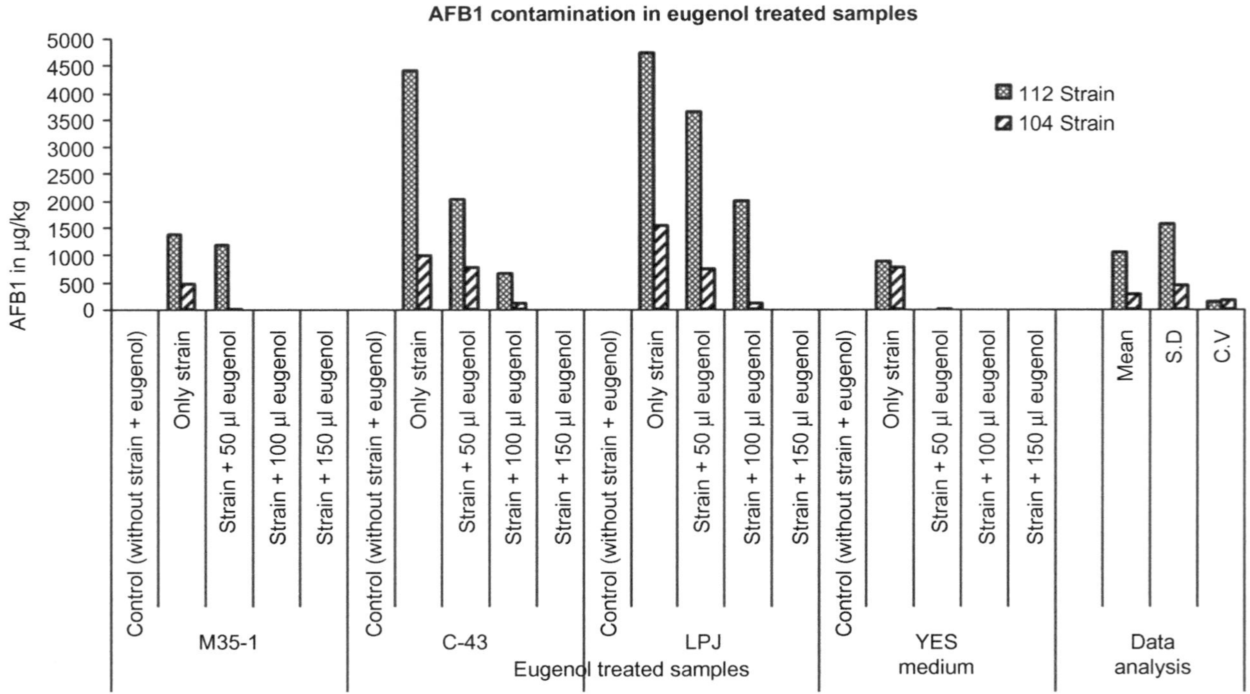

FIGURE 3.9 Inhibition of Aflatoxin B_1 (AFB_1) production by an antifungal component eugenol.

3.7.4 Determination of Starch and Protein

Starch was estimated by the enzymatic procedure reported by Southgate (1976), and protein content was estimated calorimetrically after Kjeldhal digestion using salicylate (Willis et al., 1996).

3.7.4.1 *Protein*

The protein content in all eugenol treated samples tended to increase as the fungal infection increased (Fig. 3.10). The protein content in M35-1 was low as compared to C-43 and LPJ. Maximum amount of protein was observed in LPJ (17.49%). AFB$_1$ production was positively correlated to the protein content. The overall protein content ranged from 8 to 17.49%. The protein content was higher in A112 strain-containing samples compared to A104 strain-containing samples because the intensity of virulence for the strain A112 was greater. As the protein increased, the toxin production also increased.

3.7.4.2 *Starch*

The fungus infecting the grain would draw most of its nourishment from the grain reserves such as starch and protein. Padule and Salunkhe in 1984 reported the decrease in carbohydrate content during

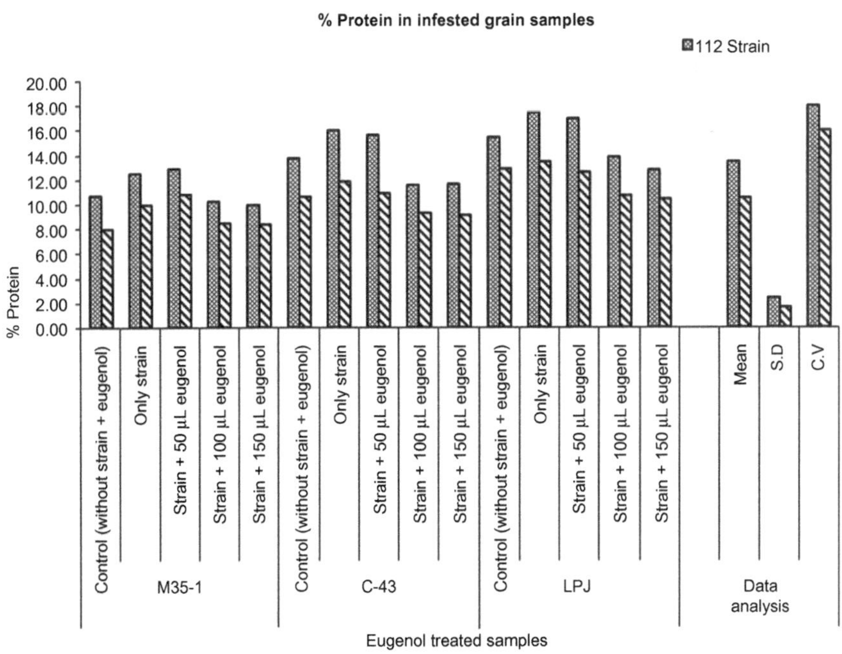

FIGURE 3.10 Percentage of protein content in sorghum infested grain.

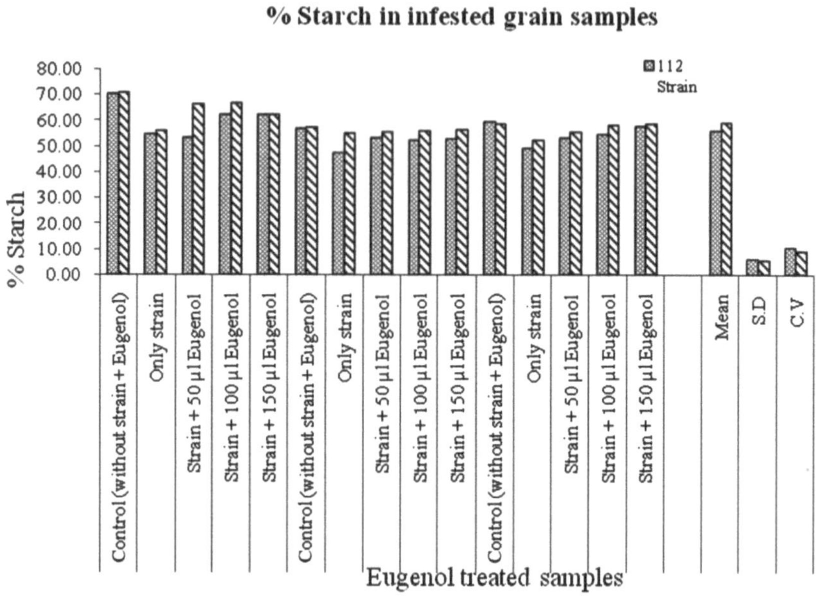

FIGURE 3.11 Percentage of starch content in sorghum infested grain.

fungal infestation of sorghum grain. The starch content in C-43 was low when compared to M35-1 and LPJ s (Fig. 3.11). The maximum amount of starch was observed in M35-1 (70.10%). In general, starch content decreases during the course of infection. Furthermore, the correlation between starch content and AFB_1 produced was not statistically significant. Hence, it was concluded that there is no specific association between starch and toxin production.

3.7.5 Statistical Analysis

Data were analyzed by Microsoft Office Excel 2003. The coefficient variation (CV) and SE were calculated using the following formulas.

$$CV = \text{Standard deviation (SD)}/\text{Arithmetic mean} \times 100$$
$$SE = \text{Standard deviation (SD)}/\sqrt{n}$$

This study showed that eugenol is generally used as a food flavoring agent. In view of its nonmutagenic and noncarcinogenic properties, it is generally regarded as safe. Hence, it is used for inhibition of AFB_1 production in stored sorghum grains. Many research institutes including the Directorate of Sorghum Research, India, have carried out research on mycotoxin contamination and have developed technologies that can significantly reduce contamination, but these technologies were not adopted by farmers due to lack of awareness. Hence this study was aimed at documenting the level of knowledge and extent of adoption of aflatoxin management practices of sorghum. Therefore, there is a need to explore the potential usage of this antifungal component, eugenol. It can be an effective inhibitor of fungal growth and AFB_1 production at the same time and it is economically feasible.

3.8 PEARLING OF BLACK SORGHUM

3.8.1 Pearling of Black Sorghum by Physic-Chemical Methods and Its Utilization

3.8.1.1 Introduction

Sorghum is a major cereal crop cultivated both in *Kharif* (June-sowing) and *Rabi* (Sep-Oct-sowing) seasons in Maharashtra State. The *Kharif* crop is often caught in late rains during grain development stage. Several grain-molds infect the developing grains and cause surface blackening to varying degrees. The infected grains are unsuitable for human and animal consumption due to blackening and presence of aflatoxins. Such produce does not carry any market price. Losses to the tune of 100% are possible if rainy conditions prevail during grain development and subsequent maturation. To offset the economic losses to growers, the State Government has to purchase such produce under a price procurement scheme and bear the loss.

The extent of spoilage is affected by severity of season, type of cultivar, stage of grain development at molding, and the predominant infecting molds. The blackening is limited to the surface of grains initially; however, under severe conditions, the entire grain is spoiled. The severely spoiled grains can be suitably separated, and remaining major lot that has become black only at surface can be pearled by physical or chemical methods to obtain a pearled white product, free of aflatoxin. Hence, the investigations were undertaken to study varietal variations in degree of molding, extent of losses caused in physico-chemical properties of the grains due to blackening, to standardize suitable processing technologies for separation of heavily infested grains, to pearl the black sorghum to obtain a pearly white produce, and to study its shelf-life and its utilization in *bhakari/roti* and bakery products.

3.8.1.1.1 Degree of Moldiness/Blackening in Promising Sorghum Cultivars

Twenty-seven promising sorghum cultivars including varieties and hybrids grown in *Kharif* season of 1997 at Sorghum Research Station, Parbhani were analyzed for degree of blackening/molding.

TABLE 3.14 Mold/blackening Score of Promising Varieties and Hybrids Grown in *Kharif* 1997

S. no.	Percent completely blackened grains	Mold score	Cultivar
1.	0–10	1	CSH – 6
2.	11–20	2	CSH – 5, SPV – 1384
3.	21–30	3	SPV – 1333, 1385, 1403
4.	31–40	4	CSH – 1, SPV – 96, 1381, 1401
5.	41–50	5	SPV – 1293, 1328, 1387
6.	51–60	6	SPV – 1231, 1330, 1408
7.	61–70	7	CSV – 13, SPV – 1022, 1398, 462, 1386
8.	71–80	8	CSH – 9, 14, 15, CSV – 15, PVK – 400
9.	81–90	9	SPV – 1284

The infested grain lots showed a range of blackening levels. In the same lot, some grains were clean white, some showing level of blackening or browning, while others exhibited complete blackening. For recording a mold score, only completely blackened grains were counted in a sample of 200 g in duplicate. The mold score was found to be minimum (less than 2) for CSH – 6, CSH – 5, SPV – 1384, while it was recorded as highest (more than 8) for CSH – 9, CSH – 14, CSH – 15, CSV – 15, PVK – 400, and SPV – 1284 (Table 3.14). A marked variation in the susceptibility to blackening of grain among the various cultivars indicated the possibility to identify/develop a cultivar resistant to the blackening usually associated with *Kharif* season.

3.8.1.1.2 Effect of Blackening on Physico-Chemical Properties

The 1000 grain weight and seed hardness (using a tablet hardness tester) for clean and black grains from the same sample were measured. The blackening of grains to the extent of 100% was found to cause a loss of about 12.5% in mean grain weight (Table 3.15). Among the various cultivars, the lowest weight loss was recorded for SPV-1330 (4.3%), while the highest loss was for SPV-1401 (24.5%). The mold infestation was observed to cause a significant loss especially in seed hardness (40.4%) (Table 3.16) which is an important physical character for pearling. The cultivars SPV – 1293, SPV – 1328, SPV – 1330, SPV – 1333, SPV – 1385, SPV – 1401, and CSH – 5 suffered more than 45% loss in hardness for completely blackening grains. The blackening was found to be limited to surface layers of the grains.

TABLE 3.15 Effect of Blackening on 1000 Grain Weight (g)

S. no.	Cultivar	Clean grains	Black grains	Weight loss (%)
1.	SPV – 96	26.5	23.3	12.1
2.	SPV – 1231	22.1	22.2	4.1
3.	SPV – 1293	26.0	22.9	11.9
4.	SPV – 1328	25.7	21.8	15.2
5.	SPV – 1330	23.4	22.4	4.8
6.	SPV – 1333	31.4	27.7	11.8
7.	SPV – 1384	30.2	26.7	11.6
8.	SPV – 1385	28.7	25.0	12.9
9.	SPV – 1387	26.8	22.3	16.8
10.	SPV – 1401	30.2	22.9	24.2
11.	SPV – 1403	26.2	22.5	14.1
12.	CSV – 13	22.1	19.4	12.2
13.	CSH – 1	34.2	31.0	9.3
14.	CSH – 5	28.4	24.5	13.7
15.	CSH – 9	28.0	25.0	10.7
Range	-	22.1–34.2	19.4–31.0	4.1–24.2
Mean	-	27.3	23.9	12.5
S.D.	-	3.96	4.74	–

The blackening was found to lower the contents of crude proteins by about 14.5% (Table 3.17), with an increase in crude fat by 28.1% (Table 3.18), crude fiber by 13% (Table 3.19), and total ash by 23.2% (Table 3.20). The blackening of grains due to mold infestation was found to cause a mean loss of 25.45% in reducing sugars (Table 3.21), in nonreducing sugars loss to 8.9% (Table 3.22) and by 14.5% in starch (Table 3.23). The proteins and carbohydrates seem to be preferred nutrients for infecting molds.

The blackening due to mold infestation was observed to increase the total phenolics by about 19.3% (Table 3.24). An increase in phenolics contents may be due to secretion of pigments by infecting fungi and migration of glume phenolics to the grain surface due to rains. However, the proportion for the increase in phenolics does not parallel with the degree of blackening. This indicates that the infesting fungi may be secreting the colored compounds that are not fully measured by Folin–Danis reagent.

TABLE 3.16 Effect of Blackening on Grain Hardness

S. no.	Cultivar	Clean grains	Black grains	Loss in hardness (%)
1.	SPV – 96	6.6	4.7	28.8
2.	SPV – 1231	7.1	4.9	31.0
3.	SPV – 1293	6.5	3.5	46.2
4.	SPV – 1328	6.9	3.7	46.4
5.	SPV – 1330	7.1	4.0	43.7
6.	SPV – 1333	6.9	4.5	53.3
7.	SPV – 1384	8.2	5.8	29.3
8.	SPV – 1385	8.0	4.3	46.3
9.	SPV – 1387	7.5	4.7	37.3
10.	SPV – 1401	7.8	4.1	47.4
11.	SPV – 1403	7.9	4.6	41.8
12.	CSV – 13	6.5	4.2	35.4
13.	CSH – 1	6.6	5.1	32.9
14.	CSH – 5	7.2	3.2	55.6
15.	CSH – 9	6.3	4.4	30.2
Range	-	6.3–8.2	3.2–5.8	28.8–55.6
Mean	-	7.2	4.4	40.4
S.D.	-	0.59	0.64	–

TABLE 3.17 Effect of Blackening on Crude Protein Content (%)

S. no.	Cultivar	Clean grains	Black grains	Loss in proteins (%)
1.	SPV – 96	12.3	11.6	5.6
2.	SPV – 1231	8.4	6.9	17.9
3.	SPV – 1293	8.7	8.2	5.7
4.	SPV – 1328	9.2	7.8	15.2
5.	SPV – 1330	8.1	6.9	14.8
6.	SPV – 1333	10.0	9.7	3.0
7.	SPV – 1384	8.2	7.3	11.0
8.	SPV – 1385	9.9	7.1	28.3
9.	SPV – 1387	11.1	9.5	14.4

(Continued)

TABLE 3.17 (Continued)

S. no.	Cultivar	Clean grains	Black grains	Loss in proteins (%)
10.	SPV – 1401	10.0	8.6	14.0
11.	SPV – 1403	10.5	9.1	13.3
12.	CSV – 13	9.3	8.0	9.1
13.	CSH – 1	8.7	6.2	28.7
14.	CSH – 5	9.3	6.2	33.3
15.	CSH – 9	9.2	7.8	15.2
Range	-	8.1–11.1	6.2–9.7	3.0–33.3
Mean	-	9.5	8.0	14.6
S.D.	-	1.14	1.43	–

TABLE 3.18 Effect of Blackening on Crude Fat Content (%)

S. no.	Cultivar	Clean grains	Black grains	Loss in fat (%)
1.	SPV – 96	1.91	1.99	4.20
2.	SPV – 1231	2.75	2.95	7.30
3.	SPV – 1293	2.41	2.78	15.40
4.	SPV – 1328	2.30	3.34	45.20
5.	SPV – 1330	1.31	1.64	25.20
6.	SPV – 1333	2.24	2.89	29.00
7.	SPV – 1384	1.46	1.65	13.00
8.	SPV – 1385	2.35	3.38	43.80
9.	SPV – 1387	2.23	2.58	15.70
10.	SPV – 1401	2.04	3.01	47.50
11.	SPV – 1403	1.67	1.98	18.60
12.	CSV – 13	2.45	2.60	6.10
13.	CSH – 1	1.08	1.56	44.40
14.	CSH – 5	1.37	2.04	48.90
15.	CSH – 9	1.96	3.09	57.70
Range	-	1.08–2.75	1.56–3.38	4.2–57.7
Mean	-	2.00	2.50	28.13
S.D.	-	0.48	0.62	–

TABLE 3.19 Effect of Blackening on Content of Fiber (%)

S. no.	Cultivar	Clean grains	Black grains	Loss in crude fiber (%)
1.	SPV – 96	1.66	1.79	7.83
2.	SPV – 1231	2.05	2.14	4.39
3.	SPV – 1293	1.20	1.87	55.83
4.	SPV – 1328	1.87	2.02	8.02
5.	SPV – 1330	1.93	2.02	4.66
6.	SPV – 1333	0.90	1.03	14.44
7.	SPV – 1384	1.50	1.65	10.00
8.	SPV – 1385	2.03	2.25	10.83
9.	SPV – 1387	1.93	2.17	12.43
10.	SPV – 1401	1.08	1.10	1.85
11.	SPV – 1403	2.13	2.42	13.61
12.	CSV – 13	1.55	2.00	29.03
13.	CSH – 1	2.01	2.10	4.77
14.	CSH – 5	2.16	2.51	16.20
15.	CSH – 9	1.98	2.01	1.51
Range	-	0.90–2.16	1.03–2.51	1.51–55.83
Mean	-	1.73	1.93	13.00
S.D.	-	0.39	0.40	–

TABLE 3.20 Effect of Blackening on Total Ash Content (%)

S. no.	Cultivar	Clean grains	Black grains	Loss in ash (%)
1.	SPV – 96	1.10	1.22	0.90
2.	SPV – 1231	1.00	1.18	18.00
3.	SPV – 1293	1.17	1.22	4.27
4.	SPV – 1328	1.19	1.22	2.50
5.	SPV – 1330	1.01	1.02	1.00
6.	SPV – 1333	1.14	1.33	16.70
7.	SPV – 1384	0.90	1.07	18.90
8.	SPV – 1385	1.16	1.31	12.90
9.	SPV – 1387	0.49	0.96	95.90

(Continued)

TABLE 3.20 (Continued)

S. no.	Cultivar	Clean grains	Black grains	Loss in ash (%)
10.	SPV – 1401	1.16	1.29	11.21
11.	SPV – 1403	0.57	0.94	64.90
12.	CSV – 13	1.14	1.20	5.30
13.	CSH – 1	1.07	1.72	60.70
14.	CSH – 5	1.04	1.28	23.10
15.	CSH – 9	1.16	1.18	1.70
Range	-	0.49–1.19	094–1.72	1.0–95.90
Mean	-	1.02	1.21	23.20
S.D.	-	0.20	0.18	–

TABLE 3.21 Effect of Blackening on Reducing Sugars Content (%)

S. no.	Cultivar	Clean grains	Black grains	Loss in reducing sugars (%)
1.	SPV – 96	0.23	0.18	14.0
2.	SPV – 1231	0.49	0.42	14.3
3.	SPV – 1293	0.30	0.13	56.7
4.	SPV – 1328	0.46	0.41	10.9
5.	SPV – 1330	0.30	0.23	23.3
6.	SPV – 1333	0.27	0.12	55.6
7.	SPV – 1384	0.47	0.40	14.9
8.	SPV – 1385	0.42	0.33	21.4
9.	SPV – 1387	0.41	0.39	4.8
10.	SPV – 1401	0.31	0.14	54.8
11.	SPV – 1403	0.37	0.28	24.3
12.	CSV – 13	0.34	0.30	11.8
13.	CSH – 1	0.31	0.21	32.3
14.	CSH – 5	0.42	0.34	19.1
15.	CSH – 9	0.40	0.31	22.5
Range	-	0.21–0.47	0.12–0.42	4.8–56.7
Mean	-	0.36	0.28	25.4
S.D.	-	0.07	0.09	–

TABLE 3.22 Effect of Blackening on Nonreducing Sugars Content (%)

S. no.	Cultivar	Clean grains	Black grains	Loss in non-reducing sugars (%)
1.	SPV – 96	2.02	1.69	16.3
2.	SPV – 1231	1.11	1.08	2.7
3.	SPV – 1293	1.95	1.87	4.1
4.	SPV – 1328	1.95	1.85	5.0
5.	SPV – 1330	1.38	1.22	11.6
6.	SPV – 1333	1.64	1.40	14.6
7.	SPV – 1384	1.41	1.21	14.2
8.	SPV – 1385	1.53	1.38	9.8
9.	SPV – 1387	1.87	1.64	12.3
10.	SPV – 1401	1.66	1.53	7.8
11.	SPV – 1403	1.41	1.30	7.8
12.	CSV – 13	2.14	1.84	14.0
13.	CSH – 1	2.17	2.04	6.0
14.	CSH – 5	1.90	1.88	1.1
15.	CSH – 9	1.96	1.85	5.6
Range	-	1.11–2.17	1.08–2.04	1.1–16.3
Mean	-	1.76	1.58	8.9
S.D.	-	0.31	0.36	–

TABLE 3.23 Effect of Blackening on Content of Starch (%)

S. no.	Cultivar	Clean grains	Black grains	Loss in starch (%)
1.	SPV – 96	73.80	71.00	3.8
2.	SPV – 1231	74.05	70.00	5.5
3.	SPV – 1293	66.60	56.70	14.9
4.	SPV – 1328	70.00	67.50	3.6
5.	SPV – 1330	82.50	80.05	2.9
6.	SPV – 1333	64.80	58.50	10.3
7.	SPV – 1384	79.00	77.90	1.4
8.	SPV – 1385	82.00	81.04	1.2
9.	SPV – 1387	71.50	66.00	7.7

(Continued)

TABLE 3.23 (Continued)

S. no.	Cultivar	Clean grains	Black grains	Loss in starch (%)
10.	SPV – 1401	60.30	52.20	13.4
11.	SPV – 1403	81.04	78.30	3.3
12.	CSV – 13	72.00	64.00	11.1
13.	CSH – 1	73.80	67.50	8.5
14.	CSH – 5	78.3	71.50	8.7
15.	CSH – 9	70.00	67.02	4.3
Range	-	60.3–82.50	52.2–87.14	1.2–14.9
Mean	-	73.31	65.54	6.7
S.D.	-	6.51	9.53	–

TABLE 3.24 Effect of Blackening on Content of Total Phenolics (mg/100 g)

S. no.	Cultivar	Clean grains	Black grains	Gain in phenolics (%)
1.	SPV – 96	171.36	196.35	14.58
2.	SPV – 1231	142.28	160.65	12.91
3.	SPV – 1293	153.51	189.21	23.25
4.	SPV – 1328	124.95	164.22	31.42
5.	SPV – 1330	199.92	214.20	7.14
6.	SPV – 1333	82.11	103.53	26.08
7.	SPV – 1384	146.37	174.93	19.51
8.	SPV – 1385	210.63	232.05	10.16
9.	SPV – 1387	167.79	192.78	14.89
10.	SPV – 1401	132.09	153.51	16.21
11.	SPV – 1403	149.94	189.21	26.19
12.	CSV – 13	103.53	132.09	27.58
13.	CSH – 1	135.66	167.79	23.68
14.	CSH – 5	146.37	174.93	19.51
15.	CSH – 9	157.08	182.07	15.90
Range	-	82.11–210.63	103.53–232.05	7.14–31.42
Mean	-	148.23	175.16	19.26
S.D.	-	32.71	30.97	–

TABLE 3.25 Relation of Peroxidase (POD) and Polyphenol Oxidase (PPO) Activities of Clean Grains to Mold Infestation and Blackening

S. no.	Cultivar	Blackening (%)	POD (units/g/min)	PPO (units/g/h)
1.	SPV – 96	39.90	148	41
2.	SPV – 400	78.40	144	61
3.	SPV – 462	67.00	144	73
4.	SPV – 1293	49.50	132	32
5.	SPV – 13	62.10	126	59
6.	SPV – 1330	53.70	114	61
7.	SPV – 1384	18.22	154	48
8.	SPV – 1385	26.10	148	53
9.	SPV – 1022	64.20	108	66
10.	SPV – 1403	23.70	150	53
11.	CSH – 14	71.50	142	61
12.	CSH – 1	32.30	166	50
13.	CSH – 5	17.10	156	44
14.	CSH – 6	10.20	212	56

Blackening x POD, $r = -0.715$ at 5%.

3.8.1.1.3 Relationship Between Blackening Intensity and Grain Peroxidase Activity

The clean-white grains of 15 cultivars that have suffered mold infestation and blackening to varying degrees were analyzed for activities of polyphenol oxidizing enzymes, viz., peroxidase (POD) and polyphenoloxidase (PPO). A correlation between degree of blackening and POD activity was found to be significantly negative. The cultivars containing higher grain POD activity were observed to suffer a minimum molding or blackening (Table 3.25). The PPO activity, however, did not exhibit such a relationship. These results indicated that the cultivars with POD activity particularly during grain development may be resistant to blackening. This will be a useful biochemical marker to screen the cultivars for resistance to blackening.

3.8.1.1.4 Standardization of Technique for Separation of Light-Weight Blackened Grains

Initially, a specific gravity of normal healthy grains of 15 cultivars was determined (Table 3.26). The specific gravity ranged between 1.1 to 1.5.

TABLE 3.26 Specific Gravity of Normal Healthy Sorghum Grains of Different 15 Cultivars

S. no.	Cultivar	Specific gravity
1.	SPV – 96	1.3
2.	SPV – 1231	1.1
3.	SPV – 1293	1.3
4.	SPV – 1328	1.3
5.	SPV – 1330	1.2
6.	SPV – 1333	1.3
7.	SPV – 1384	1.5
8.	SPV – 1385	1.4
9.	SPV – 1387	1.3
10.	SPV – 1401	1.5
11.	SPV – 1403	1.3
12.	CSV – 13	1.5
13.	CSH – 1	1.2
14.	CSH – 5	1.2
15.	CSH – 9	1.3
Range	-	1.1–1.5
Mean	-	1.25
S.D.	-	0.14

A specific gravity of 1.2 was then used as a basis to separate light-weight blackened grains from the remaining normal-weight but blackened grains. The grains were suspended in 40% NaCl solution (sp. gr. 1.2) at ambient conditions (1:10 w/v) for about 2 mins and the floaters were removed. The sinkers were collected and dried. The recovery of grains as sinkers was near 70% (Table 3.27) which was similar to that obtained for middle- and heavy-weight fractions combined on the gravity separator (Table 3.28). Hence, this simple technique was used to further evaluate various cultivars and for pearling studies.

3.8.1.1.5 Standardization of Pearling Treatment

Pearling conditions with respect to time and rpm and grain conditioning with added moisture (5%) were standardized using cv. M 35-1 (blackened) to obtain clear pearly white grains with minimum breakage

TABLE 3.27 Percent Sinkers and Floaters Obtained From a Black Sorghum Grains of Different Cultivars

S. no.	Cultivars[a]	Sinkers (%)	Floaters (%)
1.	SPV – 1408	53.8	46.1
2.	CSV – 15	62.1	37.8
3.	SPV – 1293	56.9	43.0
4.	CSV – 13	70.1	29.8
5.	SPV – 1403	81.4	18.5
6.	SPV – 1401	72.8	27.2
7.	CSH – 9	73.4	26.7
8.	SPV – 1384	84.8	15.1
9.	SPV – 1385	73.8	26.2
10.	SPV – 1381	42.5	47.5
11.	SPV – 1398	46.6	53.4
12.	CSH – 6	94.6	5.4
13.	SPV – 1386	76.4	23.6
14.	CSH – 5	96.2	3.7
15.	CSH – 1	80.4	19.5
16.	SPV – 1333	77.7	22.3
17.	CSH – 14	69.1	30.9
18.	SPV – 1022	75.5	24.5
19.	SPV 462	77.0	23.0
20.	SPV – 1387	33.1	66.9
21.	SPV – 96	72.3	27.7
22.	SPV – 1330	74.2	5.8
23.	SPV – 1284	56.9	43.1
24.	SPV – 1231	78.6	21.4
25.	CSH – 16	51.2	48.8
26.	PVK – 400	79.8	20.2
Range	-	33.1–96.2	3.7–66.9
Mean	-	69.7	29.9
S.D.	-	15.3	14.7

[a]*Hundred gram lots were suspended in 40% NaCl solution (sp. gr. 1.24) at 27 ± 2°C for 2 min.*

TABLE 3.28 Gravity Separation of Black Sorghum by Damas Gravity Separator

S. no.	Fraction[a]	Recovery (%)
1.	Light weight	24.65
2.	Middle weight	56.30
3.	Heavy weight	19.00
4.	2 + 3	75.30

[a]*Two kg grain lot of black sorghum (3 CVS pooled) and subjected to gravity separation.*

TABLE 3.29 Standardization of Pearling Conditions (cv. M 35-1, Molded) on Laboratory Rice Polisher

Pearling treatment (min)	Pearled grain (%)	Bran fraction (%)
A. RPM 650		
2 min	86.5	13.1
3 min	80.0	20.0
4 min	76.3	22.9
5 min	73.9	25.7
5 min with conditioning	73.6	26.0
B. RPM 1050		
1 min	74.5	25.1
2 min	63.7	35.9
3 min	61.4	38.2

and complete removal of surface layer. Among the two rpm and varying times tested, a combination of 650 rpm and 5 min pearling on the rice polisher (INDOSAW lab model) was found to produce maximum (74%) clear pearly-white product with about 26% bran or *bhusa* (Table 3.29).

Fifteen sorghum cultivars with varying degree of blackening were evaluated for recovery of pearled product (Table 3.30). The yield of pearled product (at 650 rpm and 5 min pearling) ranged from 56.4 to 90.5 with a mean of 77.2%. Most of the cultivars except SPV – 1293 and CSH – 14 yielded more than 70% pearled produce. In some cultivars, viz., CSH – 5, CSH – 9, SPV – 1403, SPV – 1384, the recovery of pearled product is more than 75%. A considerable variation was found among the cultivars for degree of spoilage and also in the yield of pearled product.

TABLE 3.30 Yield of Polished Grains Including Broken From Sinkers of Different Cultivars

S. no.	Cultivar	Polished grain (%)	Fraction bran (%)
1.	SPV – 96	80.9	19.1
2.	SPV – 462	70.2	29.8
3.	SPV – 1022	81.8	18.2
4.	SPV – 1293	66.6	33.4
5.	SPV – 1330	77.9	22.0
6.	SPV – 1333	71.7	28.9
7.	SPV – 1384	88.1	11.8
8.	SPV – 1385	71.9	28.1
9.	SPV – 1403	88.9	11.0
10.	CSV – 13	75.7	24.3
11.	CSH – 1	76.8	23.2
12.	CSH – 5	81.3	18.7
13.	CSH – 9	90.5	9.5
14.	CSH – 14	56.5	43.5
15.	PVK – 400	79.3	20.7
Range	-	56.4–90.5	9.4–43.5
Mean	-	77.2	22.8
S.D.	-	9.1	9.5

A pearled product of 16 cultivars was separated into whole and broken pearled grains (Table 3.31). A considerable variation among the cultivars was noticed. The yield of whole pearled grains ranged from 23.5 to 95.1 with a mean of 63.45. The cultivars CSH – 9, SPV – 1403, CSH – 5, SPV – 96, and SPV – 1384 yielded more than 70% whole pearled grains.

3.8.1.1.6 Effect of Pearling on Nutrient Composition of Black Sorghum

Fifteen black sorghum cultivars and their pearled products were analyzed for chemical composition to observe the losses in nutrients. Since the main object of pearling is to remove a black surface layer completely, a pearling to the extent of about 25% was essential. Obviously the losses in nutrients in the bhusa due to scoring of some endosperm part are likely. The pearling of sinkers at 650 rpm for 5 min

TABLE 3.31 Distribution of Whole Pearled Grains and Brokens After Pearling of Black Sorghum (Sinkers, 650 rpm 5 min)

S. no.	Cultivar	Whole pearled grain (%)	Brokens (%)
1.	SPV – 96	73.1	26.9
2.	SPV – 462	46.9	53.1
3.	SPV – 1022	65.0	35.0
4.	SPV – 1293	46.7	53.3
5.	SPV – 1330	57.8	42.2
6.	SPV – 1333	57.3	42.7
7.	SPV – 1384	95.1	4.9
8.	SPV – 1385	51.3	48.7
9.	SPV – 1403	89.5	10.5
10.	CSV – 13	56.8	43.2
11.	CSH – 1	67.8	32.9
12.	CSH – 5	77.6	22.4
13.	CSH – 9	92.4	7.6
14.	CSH – 14	23.5	76.5
15.	PVK – 400	49.8	50.2
Range	-	23.5–95.1	4.9–76.5
Mean	-	63.4	36.6
S.D.	-	19.7	19.7

was observed to decrease the protein content by about 5.6%, crude fat by 24.15, ash content by 8.4%, fiber content by about 25.6%, reducing sugars by 30.8%, and nonreducing sugars by 9.5% (Table 3.32). A considerable variation among the cultivars was observed for losses in various nutrients. The losses in chemical constituents can be attributed to the scoring of endosperm surface and loss in germ of grains during pearling. However, the magnitude to nutrient losses is of minor nature as compared to the advantage gained in converting a nearly lost produce into a 60–70% of the edible-grade product.

A pilot-scale trial for pearling of black sorghum was carried out on CIAE-Bhopal make sorghum pearler on 10 kg lots. The produce blackened below about 50% could be directly subjected to pearling. When the produce was found blackened beyond 50% of the grains, the highly infested and damaged light-weight grains were required to be separated

TABLE 3.32 Effect of Pearling of Black Sorghum Sinkers on Nutritional Composition

		Decrease after pearling (%)					
S. no.	**Cultivar**	**Protein**	**Fat**	**Ash**	**Fiber**	**Reducing sugar**	**Non-reducing sugar**
1.	SPV – 96	4.8	25.0	2.3	17.6	27.2	5.9
2.	SPV – 462	6.5	23.1	3.3	18.1	26.6	9.0
3.	SPV – 1022	7.0	31.8	15.3	17.4	37.9	8.9
4.	SPV – 1293	7.5	8.7	7.3	12.1	34.7	11.5
5.	SPV – 1330	2.9	17.6	6.8	14.3	33.3	12.0
6.	SPV – 1333	8.8	44.0	7.8	17.2	31.2	10.4
7.	SPV – 1384	4.7	25.0	4.8	13.7	25.7	12.7
8.	SPV – 1385	5.8	25.0	7.7	13.0	21.2	10.7
9.	SPV – 1403	8.6	17.2	15.2	21.5	45.8	9.8
10.	CSV – 13	4.2	32.1	14.2	10.9	24.1	9.0
11.	CSH – 1	4.9	31.6	8.3	15.2	43.4	8.8
12.	CSH – 5	3.3	24.1	8.5	20.8	27.5	8.3
13.	CSH – 9	5.1	19.0	9.0	11.8	34.2	11.6
14.	CSH – 14	6.0	5.0	9.7	14.2	25.7	5.9
15.	PVK – 400	4.5	13.0	5.9	16.7	23.5	8.0
Range	-	2.9–8.8	8.6–44.0	2.30–15.3	10.9–21.5	21.2–45.8	5.9–12.7
Mean	-	5.6	24.1	8.4	15.6	30.8	9.5
S.D.	-	1.7	8.3	3.9	4.9	7.3	1.9

either by gravity separator or by floating in 40% NaCl solution. The medium- to normal-weight blackened grains could be effectively pearled for 10 to 20 min, depending on the degree of molding, to obtain a clear white produce with about 75 to 80% recovery (Fig. 3.12).

3.8.1.1.7 Chemical Dehulling of Black Sorghum

Different solvents, viz. ether, alcohol, acetone, dilute acid (HCl, H_2SO_4, $HClO_4$), and dilute alkali (NaOH, KOH, $Ca(OH)_2$), were tested at various concentrations (1 to 5%) and grain:solvent ratios (1:1 to 1:5 w/v) at ambient conditions and at 50 or 98°C to solubilize and remove the brown to black pigments of blackened grains. Among the

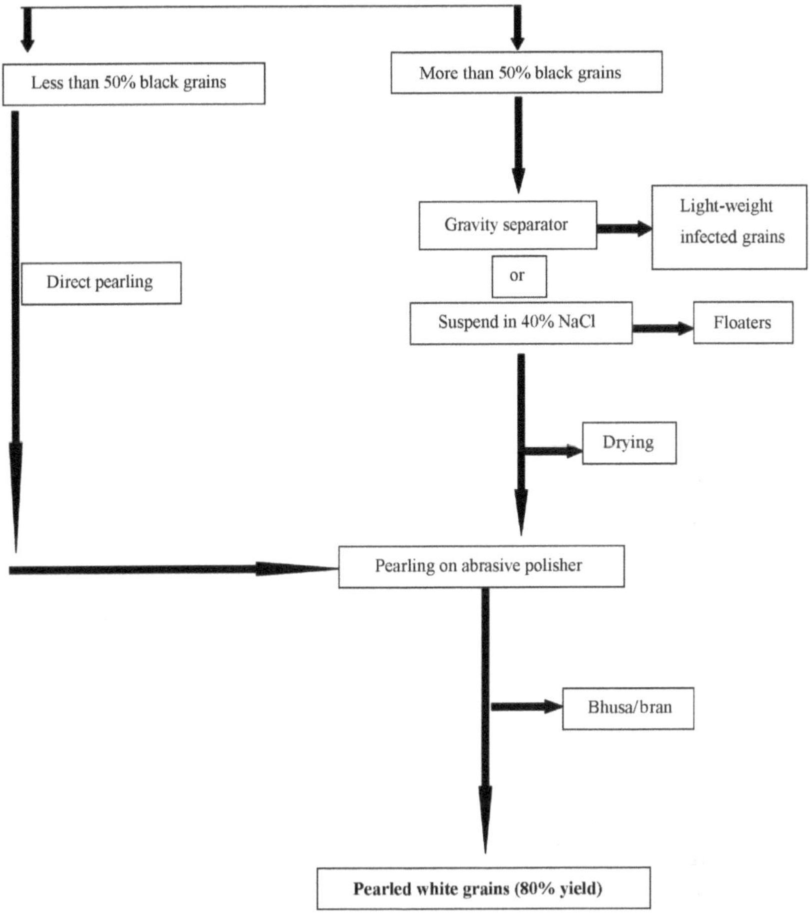

FIGURE 3.12 Pearling process for black sorghum with abrasive rice polisher (lab model).

chemicals tested only dilute HCl and dilute NaOH were found effective in extraction of pigments. However, even with dilute acid or alkali, a complete pigment extraction could not be achieved. Hence, a complete dehulling of grains was attempted. Among the dilute acid or dilute alkali, alkali treatment of grains was observed to be beneficial to loosen the bran layer and its separation. An attempt was therefore made to standardize alkali dehulling treatment to discolored sorghum grains.

Experiment No.1: Standardization of alkali dehulling treatment:

1. Concentration: 2.0, 5.0, 10.0 and 15% (w/v).
2. Temperature: 100°C + 2.0
3. Time of soaking: 2.5, 5, and 10 min.
4. Ratio of grain to solvent: 1:1, (w/v).

After soaking, the grains were washed under running tap water by rubbing in hands to separate the bran, residual alkali was neutralized with 1% acetic acid for 2 min, and grains were dried at 60°C. Among the treatment combinations, a complete removal of the blackened layer was observed with soaking of grains in 5% NaOH solution for 5 min at 100°C.

In a second experiment, various levels of NaOH (2, 3, 4, and 5%) were tested at 100°C for 5 min soaking. It was observed that a complete removal of black layer could be removed when the grains were soaked in 4% NaOH solution.

Experiment No. 2: Varietal evaluation of alkali dehulling for six cultivars, viz., CSH – 5, CSH – 13, CSH – 14, CSH – 15, CSH – 9, and SPV – 462, that had suffered blackening was carried out. The alkali treatment was found to be quite effective in removing the black surface layer completely in all the cultivars tested. The dehulling treatment caused a 15% loss in dry matter without any appreciable breakage of grains. A flow diagram outlines the steps in chemical dehulling of black sorghum (Fig. 3.13).

Experiment No. 3: Nutritional composition and roti quality of alkali-dehulled sorghum.

The alkali dehulling of blackened sorghum resulted in a slight increase in the contents of protein and an appreciable decrease in other constituents like fiber, ash, fat, sugars, and polyphenols (Table 3.33). The roti prepared from dehulled sorghum exhibited superior and acceptable sensory properties, except for the sweetness (Table 3.34).

Problems with alkali dehulling: Although only about 15% loss in dry matter without any breakage occurs with the treatment, the following major constraints exist:

1. Cost of chemical is about Rs. 2/kg of grain processed.
2. A significant quantity of water is required to wash the grains.
3. Heat energy is necessary to soak the grains and subsequent drying.
4. Wetting of grains during soaking and washing causes the grains to be fragile and may exhibit poor storability.
5. A significant loss in sugars will lower the sweetness of the product like *roti*.

Hence, a separation of heavily infested grains by gravity separator or NaCl solution floating technique followed by physical pearling appears to be a more convenient and economical process to pearl the blackened sorghum.

3.8.1.1.8 Identification of Molds on Black Sorghum and Aflatoxin Content

Twelve cultivars including varieties and hybrids grown in *Kharif* and suffering a severe blackening were analyzed by a blotter technique. All

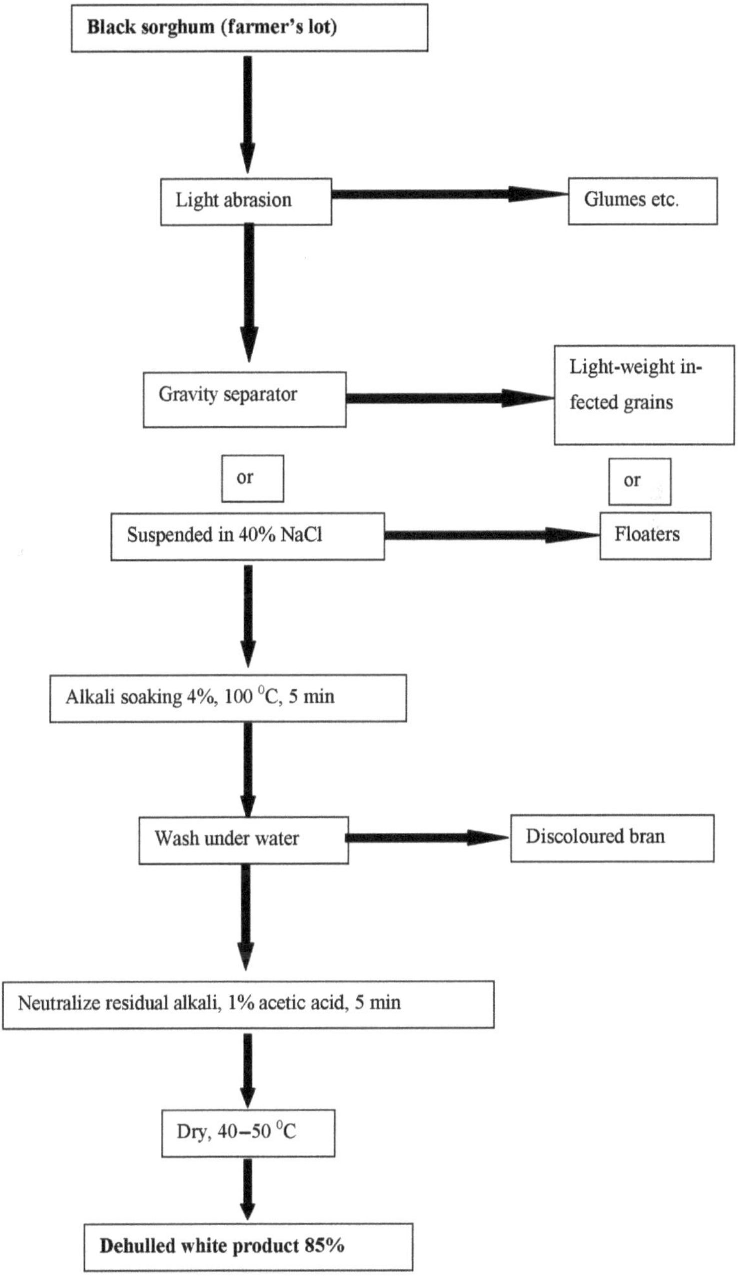

FIGURE 3.13 Process for alkali-dehulling of black sorghum.

TABLE 3.33 Nutritional Composition of Blackened and Alkali-Dehulled Sorghum

Constituent (%)	Blackened grain	Dehulled grain
Crude protein	12.5	13.8
Crude fat	2.1	1.8
Crude fiber	2.7	1.6
Total ash	2.1	1.8
N-free extract	80.3	79.5
Reducing sugars	0.17	0.13
Nonreducing sugars	1.88	1.38
Polyphenols	0.175	0.125

TABLE 3.34 Sensory Properties of *Roti* Prepared From Discolored and Alkali-Dehulled Sorghum

Type of roti	Color	Flavor	Texture	Sweetness/taste	Mean
Discolored sorghum meal	5.6	7.2	7.6	7.8	7.1
Dehulled sorghum meal	8.4	8.0	7.7	6.8	7.7
C.D. at 5%	1.039	NS	NS	NS	-

the cultivars exhibited profuse mold colonies on all 25 seeds tested. The molds identified were species of *Alternaria, Fusarium, Drechslera, Curvularia, Rhyzopus*, and *Aspergillus* (Table 3.35). The aflatoxin content was determined by a thin layer chromatography technique using AFB_1 as a standard check. The mean value for AFB_1 was 100 μg/kg of seed. These results indicated that, in addition to undesirable color, appearance, and taste, the black sorghum is unsuitable for any food or feed purpose due to the presence of aflatoxins.

The black sorghum grains of 12 cultivars were subjected to pearling on Lab-model INDOSAW abrasive pearler for 650 rpm for 5 min and analyzed for the content of aflatoxins. The pearled grains did not contain detectable level of aflatoxin. Very few and poor fungal colonies were observed on pearled grains (Table 3.35). These results indicated that black sorghum grains, after initial cleaning and discarding of severely infested grains, can be effectively processed into white edible product, free of molds and aflatoxins.

3.8.1.1.9 Effect of Storage on Shelf-Life of Black and Pearled Sorghum

Both black and pearled sorghum were stored in plastic and cloth bags at 30 ± 2°C and 80% RH in a controlled humidity chamber for up

TABLE 3.35 Identification and % Fungal Colony Growth on Black and Pearled Sorghum Cultivars

S. no.	Name of variety	No. of seeds showing fungal colonies (25 seeds)		% of seed showing fungal colonies		Prominent types
		Back sorghum	Pearled sorghum	Back sorghum	Pearled sorghum	
1.	SPV – 1403	25	4	100	16	All *Fusarium*
2.	SPV – 462	25	5	100	20	*Curvularia, Fusarium,* bacteria
3.	CSH – 14	25	5	100	20	*Fusarium, Drechslera, Curvularia,*
4.	SPV – 1385	25	6	100	24	*Curvularia, Fusarium*
5.	SPV – 1333	25	3	100	12	*Fusarium, Drechslera,* bacteria
6.	CSH – 5	25	3	100	12	*Curvularia, Fusarium*
7.	SPV – 1384	25	3	100	12	*Fusarium*
8.	CSV – 13	25	3	100	12	*Drechslera*
9.	PVK 400	25	5	100	20	*Rhizophus, Curvularia, Fusarium*
10.	CSH – 6	25	5	100	20	*Fusarium*
11.	SPV – 1022	25	2	100	8	*Fusarium, Aspergillus*
12.	Aflatoxin content (μg/kg)	100	Not detected	-	-	-

to 40 days and evaluated for mold growth and changes in grains odor. The black produce was found to exhibit severe fungal growth and foul odor while the pearled produce showed a very slight off-odor without any fungal growth (Table 3.36). These results indicated that a pearled product exhibited a shelf-life of over 40 days. Obviously, such grains being pearled and containing a proportion of broken grains cannot be stored for several months unlike normal white sorghum.

TABLE 3.36 Effect of Ambient Storage on Sensory Properties of Black and Pearled Sorghum

Storage (days) parameter	Black sorghum		Pearled sorghum	
	Plastic box	Cloth bag	Plastic box	Cloth bag
A. INITIAL				
1. Grain appearance	Black	Black	White	White
2. Molds	Visible	Visible	Absent	Absent
3. Grain odor	Off	Off	Normal	Normal
B. AFTER 40 DAYS				
1. Grain appearance	Black, broken easily	Black, broken easily	White	White
2. Molds	Visible, increased	Visible, moderately increased	Absent	Absent
3. Grain odor	Foul	Extremely foul off	Very slightly	Slightly off

TABLE 3.37 Sensory Properties of the *Roti* Prepared From Pearled Sorghum

Sorghum	Color	Flavor	Texture	Sweetness/taste	Mean
0 DAY STORAGE					
Discolored sorghum	4.6	5.2	4.8	5.2	5.0
Pearled sorghum	8.0	7.2	7.6	7.4	7.6
C.D. at 5%	0.52	0.33	0.44	0.88	–
30 DAYS STORAGE					
Discolored sorghum	4.6	4.2	4.6	4.8	4.6
Pearled sorghum	7.2	6.6	7.4	7.2	7.1
C.D. at 5%	0.69	0.97	1.09	0.79	–

3.8.1.1.10 Utilization of Pearled Sorghum

The *bhakari*/*roti* prepared from pearled sorghum both at 0 day and 30 day storage at $27 \pm 2°C$ and 80% RH, exhibited superior and acceptable sensory properties (Table 3.37). The breads were prepared using 5–30% pearled sorghum flour in wheat maida with the usual baking formula in a pilot bakery unit of the Department of Food Science, MPKV, Rahuri. The breads were evaluated for various sensory

TABLE 3.38 Sensory Properties of Breads Prepared From Wheat: Pearled Sorghum Blends

Flour blend (%) (wheat:pearled sorghum)	Color	Flavor	Texture	Taste
100:00	8	8	8	8
95:05	8	8	8	8
90:10	8	8	8	8
80:20	8	8	8	8
70:30	8	7	6	7
S.D. ±	0.00	0.44	0.89	0.54

properties (Table 3.38). The results indicated that acceptable quality breads can be prepared by using 20% flour of pearled black sorghum with wheat maida.

3.8.1.1.11 Summary

The *Kharif* grown sorghum undergo a mild to severe fungal infestation and blackening of grains. The molds identified on such grains were species of *Alternaria, Fusarium, Drechslera, Curvularia, Rhyzopus,* and *Aspergillus*. The molded and blackened grains were found to contain about 100 μg of AFB_1/kg of grains. A significant genetic variability was observed for the degree of molding and blackening. The blackening was found to be limited only to the surface of grains under mild infestation. A severe infestation, however, damages and spoils the grains nearly completely. The mold infestation results in a marked losses in grain weight, grain hardness, proteins, starch, sugars with an increase in fiber, ash, oil or phenolics to varying levels. The cultivars with higher grain peroxidase activity were found to be resistant to mold infestation and blackening.

A simple procedure based on 40% NaCl solution has been standardized to separate heavily infested and blackened light-weight grains from that of discolored but normal heavy grains. Pearling conditions have been standardized for removing the outer blackened layer to obtain pearly white grains. The sinkers obtained from NaCl solution can be effectively pearled to obtain clear white produce. The yield of pearled product ranged between 56.4 to 90.5% with a mean of 77.2% for 15 cultivars on a laboratory rice polisher. Of the whole pearled material, the clean grains were about 63.4%, while the remainder were broken. Both whole and broken can however be milled together and used to prepare pan breads.

The pearling treatment caused about 25% removal of dry matter with concomitant decrease in proteins (5.6%), crude fat (24.1%), and mineral matter (8.4%). A considerable variation in respect of nutrient losses was also observed between the cultivars. Such losses can be attributed to the scouring of endosperm layers and part of the germ during pearling.

A pilot scale pearling trials were conducted on CIAE-Bhopal model sorghum pearler. The pearling of blackened grains for 10 to 20 min was found to produce a clear white product with about 75 to 80% recovery depending on the initial degree of molding. A produce with less than 50% black grains can be directly pearled, while produce having a higher proportion of molding and blackening is required to be subjected to gravity separation or 40% NaCl floating treatment to separate heavily infested and lightweight grains before pearling. An alkali dehulling of black sorghum may not be practically and economically feasible due to the requirement for a large volume of water, and the cost of chemical and drying.

The pearled produce did not contain surface molds and was free of aflatoxins. The pearled product exhibited a shelf life of about 40 days under laboratory conditions and produced an acceptable quality *bhakari/roti*. The refined flour of pearled sorghum can be blended to the extent of 20% with wheat maida to produce acceptable quality breads.

References

Aisien, A.O., 1989. Utilization of sorghum in brewing lager beer in Nigeria. In: Summary Proceedings of a Symposium on the Current Status and Potential Industrial Uses of Sorghum in Nigeria, Kano, Nigeria, p. 29.

Anonymous, 1993. Descriptors for sorghum [*Sorghum bicolor* (L.) Moench]. International Board for Plant Genetic Resources. Rome, Italy and International Crops Research Institute for Semi Arid Tropics (ICRISAT), Patancheru, India, pp. 12–21.

AOAC, 1995. 16th ed. Official Methods of Analysis of AOAC International, vol. II. AOAC International, Suite, Arlington, USA, Chapter 27.3.6, Chapter 27.4.4.

Azaizeh, H.A., Pettit, R.E., Sarr, B.A., Phillips, T.D., 1990. Effect of peanut tannin extracts on growth of *Aspergillus parasiticus* and aflatoxin production. Mycopathologia. 110, 125–132.

Bandyopadhyay, R., Mughogho, L.K., 1988. Evaluation of field screening techniques for resistance to sorghum grain molds. Plant Dis. 72, 500–503.

Bernfeld, P., 1955. Amylases, α and β. In: Colowick, S.P., Kaplan, N.O. (Eds.), Methods in Enzymology. Academic Press, New York, pp. 149–158.

Betina, V. (Ed.), 1984. Mycotoxins—Production, Isolation, Separation and Purification. Elsevier, Amsterdam.

Bier, M., 1962. Lipases. In: Colowick, S.P., Kaplan, N.O. (Eds.), Methods in Enzymology. Academic Press, New York, pp. 627–642.

Codex Alimentarius Commission, 1989. Report of the Twentieth Session of the Codex Committee on Food Additives and Contaminants. Food and Agriculture Organization, Alinorm 89/12 Rome, 16 pp.

Codex Alimentarius Commission, 2011. Joint FAO/WHO food standards programme. Report for the 5th Session of the Codex Committee on Contaminants in Foods, Hague, Netherlands, 21–25 March 2011. Food and Agriculture Organization of United Nations, Rome, Italy, 10 pp.

Collinge, D.B., Kragh, K.M., Mikkelsen, J.D., Nielsen, K.K., Rasmussen, U., Vad, K., 1993. Plant chitinases. Plant J. 3, 31–40.

Da Silva, J.B., Pozzi, C.R., Mallozzi, M.A.B., Ortega, E.M., Correa, B., 2000. Mycoflora and occurrence of aflatoxin B_1 and fumonisin B_1 during storage of Brazilian sorghum. J. Agric. Food Chem. 48 (9), 4352–4356.

Devi, K.T., Mayo, M.A., Reddy, G., Emmanuel, K.E., Larondelle, Y., Reddy, D.V.R., 2001. Occurrence of ochratoxin A in black pepper, coriander, ginger and turmeric in India. Food Addit. Contam. 18 (9), 830–835.

Egan, H., 1982. Methods 3 (BF method)—quantitative determination of aflatoxins B1, B2, G1, G2 in peanuts and peanut products by thin layer chromatography. In: Stoloff, L., Castegnaro, M., Scott, P., O'Veill, I.K., Bartsch, H. (Eds.), Environmental Carcinogens Selected Methods of Analysis, vol. 5. International Agency for Research on Cancer, Lyon, pp. 147–182.

El-Bazza, Z.E., Zedan, H.H., Toama, M.A., El-Tayeb, O.M., 1983. Factors affecting aflatoxin production by a local strain of *Aspergillus flavus* (Isolate NO.14). Proceedings of the International Symposium on Mycotoxins. National Research Centre, Cairo, Arab Republic of Egypt, pp. 231–242.

Forbes, G.A., Bandyopadhyay, R., Garcia, G., 1992. A review of sorghum grain mold. In: De Milliano, W.A.J., Frederiksen, R.A., Bengstan, G.D. (Eds.), Sorghum and Millets Diseases, a Second World Review. ICRISAT, Patancheru, AP, pp. 265–272.

Harris, H.B., Burns, R.E., 1973. Relationship between tannin content of sorghum grain and preharvest seed molding. J. Agron. 65, 957–959.

Jambunathan, R., Butler, L.G., Bandyopadhyay, R., Mughogho, L.K., 1986. Polyphenol concentrations in grain leaf and callus tissues of mold susceptible and mold resistance sorghum cultivars. J. Agric. Food Chem. 34, 425–430.

Jeuniaux, C., 1962. Chitinase. Methods Enzymol. 8, 644–650.

Komala, V.V., Ratnavathi, C.V., Vijay kumar, B.S., Das, I.K., 2012. Inhibition of aflatoxin B_1 production by an antifungal component, eugenol in stored sorghum grains. Food Control. 26 (1), 139–146.

Kunitz, M., 1947. Crystalline soybean trypsin inhibitor. II. General properties. J. Gen. Physiol. 30, 291–310.

Lansden, J.A., 1982. Aflatoxin inhibition and fungistasis by peanut tannins. Pean. Sci. 9, 17–20.

Lerch, K., 1987. Monophenol monooxygenase from *Neurospora crassa*. In: Colowick, S.P., Kaplan, N.O. (Eds.), Methods in Enzymology., 142. Academic Press, New York, pp. 165–169.

Lindsey, D.L., Turner, R.B., 1975. Inhibition of growth of *Aspergillus flavus* and *Trichoderma viride* by peanut embryos. Mycopathologia. 55, 149–152.

Lowry, O.H., Rosebrough, N.J., Farr, A.L., Randall, R.J., 1951. Protein measurement with the folin phenol reagent. J. Biol. Chem. 193, 265–275.

Martinelli, S.D., Kinghorn, J.R., 1994. *Aspergillus*: 50 years on. Prog. Ind. Microbiol. 29, 763–788.

Norton, R.A., 1997. Effect of carotenoids on aflatoxin B_1 synthesis by *Aspergillus flavus*. Phytopathology. 87, 814–821.

Odoemelam, S.A., Osu, C.I., 2009. Aflatoxin B_1 contamination of some edible grains marketed in Nigeria. E-J. Chem. 6 (2), 308–314.

Padule, D.N., Salunkhe, D.K., 1984. Effects of diseases on yield and quality of grain sorghum. In: Salunkhe, D.K., Chavan, J.K., Jadhav, S.J. (Eds.), Nutritional and Processing Quality of Sorghum. Oxford and IBH Publishing, New Delhi, India, pp. 231–254.

Punja, Z.K., Zhang, Y.Y., 1993. Plant chitinases and their roles in resistance to fungal diseases. J. Nematol. 25, 526–540.

Ratnavathi, C.V., Sashidhar, R.B., 1998. Microassay for the quantitation of protein precipitable polyphenols: use of bovine serum albumin – benzidine conjugate as a protein probe. Food Chem. 61:, 373–380.

Ratnavathi, C.V., Sashidhar, R.B., 2000. Changes in the enzyme activities and the aflatoxin elaboration in sorghum genotypes following *Aspergillus parasiticus* infestation. J. Sci. Food Agric. 80, 1–9.

Ratnavathi, C.V., Sashidhar, R.B., 2003. Substrate suitability of different genotypes of sorghum in relation to *Aspergillus* infection and aflatoxin elaboration. J. Agric. Food Chem. 51, 3482–3492.

Ratnavathi, C.V., Sashidhar, R.B., 2004. Induction of chitinases in response to *Aspergillus* infection in sorghum (*Sorghum bicolor* (L.) Moench). J. Sci. Food Agric. 84, 1521–1527.

Ratnavathi, C.V., Sashidhar, R.B., 2006. Inhibitory effect of polyphenols on the growth of *Aspergillus parasiticus* (NRRL 2999) and aflatoxin production. J. Sci. Food Agric. 87, 1140–1148.

Ratnavathi, C.V., Komala, V.V., Vijay Kumar, B.S., Das, I.K., Patil, J.V., 2012. Natural occurrence of aflatoxin B_1 in sorghum grown at different geographical regions of India. J. Sci. Food Agric. 92 (12), 2416–2420.

Reissig, J.L., Strominger, J.L., Leloir, L.F., 1955. A modified colorimeteric method for the estimation of *N*-acetyl amino sugars. J. Biol. Chem. 217, 959–966.

Sashidhar, R.B., Sudershan, R.V., Ramakrishna, Y., Nahdi, S., Bhat, R.V., 1988. Enhanced fluorescence of ergosterol by iodization and determination of ergosterol by fluorodensitometry. Analyst. 113, 809–812.

Sashidhar, R.B., Ramakrishna, Y., Bhat, R.V., 1992. Moulds and mycotoxins stored in traditional containers in India. J. Stored Prod. Res. 28, 257–260.

Seetharaman, K.A., Waniska, R.D., Rooney, L.W., 1996. Physiological changes in sorghum antifungal proteins. J. Agric. Food Chem. 44, 2435–2441.

Seetharaman, K.A., Whitehead, E., Keller, N.P., Waniska, R.D., Rooney, L.W., 1997. In vitro activity of sorghum seed antifungal proteins against grain mold pathogens. J. Agric. Food Chem. 45, 3666–3671.

Snedecor, G.W., Cochran, W.G., 1968. In Statistical Methods. Oxford and IBH Publishing, New Delhi, India, pp. 120–132, 172–195, 258–296.

Southgate, D.A.T., 1976. On Determination of Food and Carbohydrates. Applied Science Publishers, Oxford, UK, pp. 52–55.

Thirumala-Devi, K., Mayo, M.A., Reddy, G., Reddy, S.V., Reddy, D.V.R., 2000. Recent advances in mycotoxin dignostics. In: Chandrashekar, A., Bandyopadhyay, R., Hall, A.J. (Eds.), Technical and Institutional Options for Sorghum Grain Mold Management: Proceedings of an International Consultation, Pages 116–123, 18–19 May 2000, ICRISAT, Patancheru, India. International Crops Research Institute for the Semi-Arid Tropics, Patancheru, Andhra Pradesh, India.

Toteja, G.S., Mukherjee, A., Diwakar, S., Singh, P., Saxena, B.N., Sinha, K.K., et al., 2006. Aflatoxin B_1 contamination in wheat grain samples collected from different geographical regions of India: a multicenter study. J. Food Prot. 69 (6), 1463–1467.

Trudel, J., Asselin, A., 1989. Detection of chitinase activity after polyacrylamide gel electrophoresis. Anal. Biochem. 178, 362–366.

Turner, R.V., Lindsey, D.L., Davis, D.D., Bishop, R.D., 1975. Isolation and identification of 5, 7-dimethoxy isoflavone, an inhibitor of *Aspergillus flavus* from peanuts. Mycopathologia. 57, 39–40.

Wheeler, E.L., Ferrel, R.E., 1971. A method for phytic acid determination in wheat and wheat fraction. Cereal Chem. 48, 312–316.

Willis, R.B., Montgomery, M.E., Allen, P.R., 1996. Improved method for manual, colorimetric determination of total kjeldhal nitrogen using salicylate. J. Agric. Food Chem. 44, 1804–1807.

Zhang, Y., Haunerland, N.H., Punja, Z.K., 1996. Chitinase profiles in mature carrot (*Daucus carota*) roots and purification and characterization of a novel isoform. Physiol. Plant. 96, 130–138.

CHAPTER

4

Sorghum Uses—Ethanol

C.V. Ratnavathi, V.V. Komala and U. Lavanya
ICAR-Indian Institute of Millets Research, Rajendranagar, Hyderabad, India

OUTLINE

Sorghum Biochemistry: An Industrial Perspective.
DOI: http://dx.doi.org/10.1016/B978-0-12-803157-5.00004-6

Abstract

Sweet sorghum has sugar-rich stalks containing approximately 16–18% fermentable sugar, which can be directly fermented into ethanol by yeast. Technical challenges of using sweet sorghum for biofuels are a short harvest period for highest sugar content and fast sugar degradation during storage. Sweet sorghum juice mainly contains sucrose, glucose, and fructose. Sucrose accumulation initiates at the flowering stage and increases up to maturity. The highest green cane was observed in June planting in all genotypes, followed by February, December, and October. The presence of high reducing sugars and invertase activity make the juice unable to produce good quality jaggery and sugar. Therefore, sweet sorghum juice is only suitable for ethanol and syrup production. Harvesting at a physiological maturity was found to be optimum in terms of juice volume, percent extractability, and percentage of total soluble sugars. A pilot study was conducted and recovery of ethanol was approximately 9% of the juice.

4.1 INTRODUCTION

A group of sorghums called sweet sorghums, which are distinct for their sweet and juicy stalk, are considered as high energy sorghums. They are known to exist sparsely in the sorghum growing regions of Africa and India. They had been of less significant importance in the man-animal food chain and hence had received very little attention from agronomists and plant breeders. The global energy crisis of the early 1970s brought this group of sorghums into focus for their potential as a bioenergy crop.

Among the cultivated annual crops, sorghum has one of the highest dry matter accumulation rates, exceeded only by the Napier grass (Loomis and Williams, 1963). A highest productivity rate of about 50 g dry matter/m^2/day was recorded in sorghum in a number of world locations (Zelitch, 1973). It is also reported to surpass sugarcane, another C_4 plant, in its efficiency in converting cultural energy (Heichel, 1976). The dry matter yield of sweet sorghum is positively correlated with the accumulated solar energy received (Ferraris, 1981). Therefore, the sweet sorghums have all the essential attributes of an efficient, and possibly low-cost bioenergy crop suited to the tropics where sunlight is abundant. Another unique advantage is its ability to provide a good dry matter yield even in the low-fertility farming situations of the semiarid tropics. Sweet sorghums can also be grown during the warmer season in the highest latitude, where sugarcane cannot be grown because of its longer crop duration. Intensive evaluation and research on sweet sorghum are taking place in several countries, notably the USA, Mexico, Brazil, Venezuela, South Africa, Australia, the former Soviet Union, and India. The research in the

USA and Brazil has led to the development of several improved sweet sorghum cultivars and processing and fermentation technologies to produce bioenergy and sugar products.

Sweet sorghum is the only crop that provides grain and stem that can be used for sugar, alcohol, syrup, jaggery, fodder, fuel, bedding, roofing, fencing, paper, and chewing. Sweet sorghum can be used as an alternative feedstock for ethanol production because it has higher tolerance to salt and drought compared to the sugarcane and corn that are currently used for biofuel production. In addition, the high content of carbohydrates and fermentable sugars in sweet sorghum stalk makes it more suitable for fermentation to ethanol. Therefore, it is suggested to plant sweet sorghum for biofuel production in hot and dry countries to solve problems such as increasing the octane of gasoline and to reduce greenhouse gases and gasoline imports (Almodares and Hadi 2009). Sweet sorghum is currently being looked at for the production of bioethanol due to its several advantages over other crops: (1) it contains large amounts of sugars in its stalk that are directly fermentable; (2) it has a short growing period of 4–5 months; (3) it not only tolerates drought but also grows in colder regions of the temperate zones and hence has wider growing areas; and (4) the whole plant (grain, stalk juice, and lignocellulosic biomass) can be used for fuel ethanol production. Previous barriers to commercialization of sweet sorghum to ethanol have primarily been the high capital cost involved in building a central processing plant that may be operated only seasonally. Technical challenges of using sweet sorghum for biofuel are a short harvest period for the highest sugar content and fast sugar degradation during storage.

4.2 HISTORY OF SWEET SORGHUM

Sweet Sorghums appeared to have invited the attention of early plant domesticators and taxonomists thanks to their sweet stalk plants resembling small sugar canes with loose panicles under *Milium indicum sacchariferum*, and their origin is thought to be from Anadabad (presently Ahmadabad) in Gujarat, India. Linnaeus (1753) placed this taxonomic group under *Holcus saccharatus* L. and also described its habitat as India.

The primitive food gatherers might have found sweet sorghums more advantageous over the grain crops for their instant chewability. With the progress in settled agriculture and cooking skills, grain types might have become more important. Therefore, there is a good reason to consider that sweet sorghums might be one of the earliest domesticated plants in their native land. There is circumstantial evidence to believe that sweet sorghums were introduced to Africa, most probably in the south, from India at a very early period through trade channels. According to Doggett (1970), "Chinese Amber" type sorghum reached North America in 1853 through France. It is also recorded that sweet stalked sorghums were the first sorghums introduced to North America and the grain type sorghums gained importance here only in the beginning of 19th century. Sweet sorghums are also called *sorgho* and *imphee*. "Sorgho" is an Italian word for sorghum. The Italian *sorgho* had (Lobel, 1576) as well as loose panicles.

4.3 DISTRIBUTION OF SWEET SORGHUM

In Africa and India, the traditional habitats of sweet sorghums received no attention during the recorded period of crop improvement. While efforts were being constantly made to improve the land races for higher grain yield, many sweet stalked sorghums faced extinction because of their poor grain yielding ability and quality. Today, only a few hundred sweet sorghums are found in Africa, India, and elsewhere. About 983 sweet stalked sorghums were identified by chewing test from over 9000 Sorghum World Collections maintained at the International Crops Research Institute for Semi-Arid Tropics (ICRISAT), India (Seetharama et al., 1987). The majority of these sorghums were originally collected from different parts of Africa and India, the USA, and Lebanon contributing the rest. Among them, fewer than a hundred have a fairly higher level of sugar in the stalk (Seetharama and Prasada Rao, 1988). But for its preference for chewing fodder and purposes, sweet sorghums might have almost become extinct. It is traditionally planted sparsely mixed with grain sorghums to improve the fodder quality and to meet the limited needs of home confectionary in some parts of India (Prasada Rao and Murty, 1982). Interestingly, some of the elite breeding lines of grain sorghum developed in India were found to have a juicy stalk with moderate to high levels of sugar content. However, it is not surprising as the sorghum in India largely has to meet the dual needs of food and fodder. In Africa, sweet sorghum is sparsely encountered in a wide geographical range and is commonly preferred for chewing. In Ethiopia, it is known as "Tinkish" and is used in confectionaries. It is called "Ankolibs" in Sudan, "Misale" or "Njiho" in Malawi, "Kamutu halli" and "Misale" in Zambia, and "Takanda" in Nigeria (Apparao, 1980; Prasada Rao and Murty, 1982). While grain of sweet sorghum is not sweet, another group of sorghums, commonly called "Vani" sorghum, which has sweeter grain, is also found in India.

4.4 PRODUCTIVITY OF SWEET SORGHUM

Yield potential of sweet sorghum varieties in the USA is far better because of the research input for the improvement of fresh stalk yield and sugar content. The more recent Brazilian bioenergy program has also developed some high yielding commercial sweet sorghums. Average fresh weight yield realized in improved varieties ranges between 45 and 60 t/ha (Reeves et al., 1978; Broadhead, 1974; Schaffert and Borgonovi, 1980). Higher experimental stalk yields to the order of 80–126 t/ha have also been recorded (Reeves et al., 1979). In Brazil, the farm yield level of stalk is reported to be above 40 t/ha (Schaffert, 1988). Preliminary evaluation of sweet sorghum in New Zealand recorded fresh weight yields between 53 and 80 t/ha at soft dough stage (Piggot et al., 1980). In dry matter or energy productivity, sweet sorghum in Australia and South Africa was found to be comparable with sugarcane (Ferraris, 1981; Inman-Bamber, 1980).

4.5 ACCUMULATION OF DRY MATTER IN SWEET SORGHUM

Sweet sorghums, in general, differ from the grain sorghums for the rate of dry matter accumulation. Rapid elongation of their stalk makes major contribution to the biomass yield. Thickening of stalk is equally important as optimization in stalk length. Rate of dry matter accumulation and the biomass yield are directly related to phenology (Coleman, 1970; Ferraris, 1981; Schaffert and Gourley, 1982; Freeman, 1979). The majority of sweet sorghum genotypes are photosensitive with a longer maturity period ranging between 85 and 120 days, and agronomically superior hybrids are available with this range of maturity period (Bapat et al., 1987; Ruppa et al., 1980; Ferreira et al., 1982).

Long maturity-linked high productivity is not necessarily associated with high productivity rate. Increase in sorghum productivity rate may be associated with reduced vulnerability of the plant processes to drought and other stress conditions and its increased efficiency in utilizing the solar and cultural energies.

Eastin (1971), based on his 14 experiments in grain sorghum, concluded that elongating culm and developing leaves formed the primary sink for assimilates prior to anthesis and the primary sink was shifted to developing grains after anthesis. As the sweet sorghums were continuously selected for their carbohydrate yield from the stalk rather than for their grain yield, the postanthesis polarity of assimilate flow found in the grain sorghum does not appear to operate here.

This is evident from the findings of Lingle (1987) that the panicle in sweet sorghum represents a less competitive sink than the stalk and that a negative correlation exists between the yields of soluble sugars in the stalk and grain (Ferraris, 1981). This study showed that while 75 percent of the recently formed assimilate in the grain type was exported to the developing grains, only about 25 percent of the same was translocated to the grains in sweet sorghum. Although some of the photosynthate first held in the stalk was later mobilized to the grain, the sweet type retained a high percentage of sugar in the stalk all through the grain filling stage. The interrelationship between the stalk and grain sinks, at least in one direction, was also demonstrated by the deheading experiments. While removal of panicle from a poor yielding sweet sorghum variety made no difference on the concentration of sugars in the stalk (Lingle, 1987), it significantly enhanced the concentration of total solids, particularly sucrose in some other varieties of sweet sorghum. Sweet stalk yield and good grain yield can be coupled in the high energy sorghums provided an inherent ability is bred in to attain an optimal equilibrium for the sharing of the assimilate between the stalk and head.

A high dry matter yield in sweet sorghum is always not accompanied by a high yield of juice and fermentables (Seetharama et al., 1987). Composition of carbohydrates, which form the bulk of stalk dry matter, varied widely among genotypes and at different stages of maturity in a genotype. While juiciness of stalk significantly contributed to the fresh weight, the solid extent in the juice influenced the stalk dry matter yield. All soluble and insoluble solids including sucrose and lignocelluloses steadily increased in the stalk during the post anthesis period (Coleman, 1970; Lingle, 1987; Ferraris, 1981; Schaffert and Gourley, 1982).

Since the leaf is the major site of carbon assimilation and dry matter production, foliage characteristics and its functional duration have a critical effect on the productivity of high energy sorghums. Plants with stay-green leaves are reported to accumulate sucrose at a rapid rate even after physiological maturity (McBee et al., 1983). Removal of panicle at anthesis is reported to delay leaf senescence (Ferraris, 1981). Total dry matter yield in sweet sorghum is positively correlated with total leaf area, leaf area index (LAI), at anthesis (Ferraris, 1981). LAI also appeared to have a direct influence on Brix (Choudhari et al., 1987). In a crop canopy, higher LAI may come from a higher number of culms per unit area, higher plant height with more nodes, and slow senescencing leaves. The ability to produce productive, synchronous tillers, therefore, assumes importance in improving the LAI. Tillers are reported to be as good as the primary culm in juice yield and juice quality. It may be possible to identify genotypes with three to four good tillers (Coleman, 1970). Longer leaf area duration is also useful for higher fodder yield and its nutritive value.

4.6 SUCROSE METABOLISM IN SWEET SORGHUM

Indications were available from earlier studies that large increases in stalk sucrose concentration occurred after anthesis (Ventre et al., 1948; Coleman, 1970; McBee and Miller, 1982). Concentrations of sucrose and reducing sugars along the length of stalk were found to be variable. The four terminal internodes had higher starch, sucrose, and acidity (Ventre and Byall, 1937), while the basal internodes had higher percentage of invert sugars. A better understanding on the metabolism of the stalk sugar is available from a recent study by Lingle (1987).

Total concentration of sugars was lowest until the boot stage and it reached highest at the soft dough stage. While sucrose is the predominant sugar during all developmental stages, it constituted only about 50% of soluble sugar at the boot stage, glucose and fructose making up the remainder. Concentration of glucose was always higher than that of fructose. Glucose and fructose concentration all along the stalk was highest at flag leaf stage, which approximately leveled off in all internodes (to about 8 and 4 g glucose and fructose/Kg FW, respectively) at anthesis and declined during successive stages. Accelerated accumulation of sucrose during the period between the boot stage and soft dough stage was largely responsible for the rapid rise in sugar concentration. On the whole stalk basis, this rate rose from 154 to 278 mg/Kg FW/h in the sweet sorghum variety Rio. In other words, the increase in the tempo of sucrose accumulation took place only after elongation of all internodes was completed. According to Lingle (1987), sweet sorghum, like sugarcane, has soluble invertase activity at two pH optima, 5.0 and 7.0. Invertases showed highest activity in the developing internodes. A strong positive correlation ($r = 0.98$) was found between the internode elongation rate and the acid invertase activity. Acid invertase activity of whole stalk reached a peak at boot stage and declined to barely detectable levels after anthesis. Neutral invertase activity on the other hand was present during all developmental stages with an activity maxima around the boot and anthesis stages. Acceleration of sucrose accumulation occurred only after a decline in acid invertase activity.

It is presumed that while most of the acid invertase activity in sweet sorghum is largely located in the vacuole of parenchyma tissue similar to the storage roots of sugar beet, it may also be located in the apoplast as is in the case of sugarcane. In sugarcane, sucrose concentration is positively correlated with the neutral invertase activity (Hatch and Glasziou, 1963). A similar relationship was not found in sweet sorghum (Lingle, 1987). In sugarcane, accumulation of sucrose involves its cleavage in the apoplast with the help of acid invertase and uptake to parenchyma tissue across the tonoplast (Hatch and Glasziou, 1963;

Glasziou and Gaylor 1972; Bowen and Hunter, 1972). Sucrose-phosphate synthetase (SPS) is believed to play a key role in the uptake and resynthesis of sucrose before or during transport across the tonoplast (Hatch, 1964). According to Lingle (1987), SPS may not be playing an active role in the sucrose storage per se in the sweet sorghum. Therefore, involvement of an active sucrose transporter located at plasma lemma or tonoplast or a passive sucrose uptake by parenchyma cells is suggested.

4.7 AGRONOMY OF SWEET SORGHUM

When a crop system is commercially exploited for bioenergy production, its efficiency in converting solar and cultural energy into biomass assumes critical importance. When solar energy is not limiting, the crop productivity is limited by the ability and efficiency of the crop in using the cultural energy. Crop response to fertilizer and irrigation is hence important. One of the unique advantages of sorghum, whether grown for its grain, biomass, or both, is its ability to tolerate moisture stress and to provide a reasonably higher dry matter yield under marginal fertility and adverse soil conditions. In its traditional semiarid tropical tracts, sorghum is neither irrigated nor even moderately fertilized. In this setting lies the unique strength of sweet sorghum as a bioenergy crop. Nevertheless, its emerging importance as a cheap bioenergy crop makes it imperative to examine whether its dry matter productivity could be further enhanced through better nutrient supply.

Response of sweet sorghum to major nutrients has mostly been studied in the USA and Brazil using local improved cultivars. Sweet sorghum was found to respond to increased doses of N up to 112 Kg/ha in the presence of phosphorus and lime (Wiedenfeld et al., 1984). However, increased biomass yield at higher levels of N tended to decrease the level of carbohydrate in the stalk (Wiedenfeld et al., 1984). Studies in Brazil (Rosolem et al., 1983) also underlined the positive role of lime in grain and stalk yield and found the differential crop response to N and P in clay and sandy soils.

The response of stalk yield and sugar content to irrigation under normal rainfall season was not significant, but water stress decreased the biomass yield and increased the sugar percentage (Seetharama et al., 1987). Biomass productivity of sweet sorghum could be improved under rain fed conditions with the incorporation of better drought tolerance. Thus, among the major bioenergy crops like sugarcane, maize, sugar beet, and cassava, sweet sorghum has the lowest requirement of cultural energy in the form of irrigation and fertilizer to produce a comparable yield of dry matter or fermentables.

Sowing density and population stand are important in optimizing the solar energy conversion efficiency of sweet sorghum. The highest stalk and total sugar yield was obtained by planting an estimated population of 1,24,000 plants/ha in 90 cm wide rows. Increasing planting density by narrow rows causes individual plants to be lean and tall with reduced leaf, leaf area, leaf specific weight, stalk weight, and concentrations of sucrose and fermentable sugars (Broadhead and Freeman, 1980). Increased planting density with narrow rows, however, significantly increased the yield of fermentables and ethanol (Broadhead and Freeman, 1980). Under typical farming conditions, a planting density of 1,50,000 plants/ha may be optimal.

Sweet sorghums tend to show significant variations in sugar content and yield across season and locations and in some genotypes this variation is relatively less. Biomass and sugar yield tended to vary between rainy and post rainy crop season in the tropics (Seetharama et al., 1987). Short day conditions of the post rainy cool season in northern latitudes appear to affect the dry matter accumulation rate. During the warmer season in higher latitudes, the biomass and sugar yields are influenced by the number of frost-free days.

4.8 MAXIMIZATION OF BIOMASS IN SWEET SORGHUM THROUGH GENETIC ENHANCEMENT OF SWEET SORGHUM

Efforts have been made to improve the genetic potential of sweet sorghum. As part of this program, a total of 75 base entries of sweet sorghum were pooled from different participating centers and tested for their performance during *Kharif* 2000 on various breeding parameters and biochemical parameters. Twenty-six superior genotypes from these 75 entries were promoted for advanced trial. A total of 31 entries consisting of 26 entries from previous trial and five hybrids from NARI were tested for their performance on various breeding parameters and biochemical parameters during *Kharif* 2001. Fourteen genotypes showed superior performance and were promoted for an advanced trial. Fourteen genotypes in the advanced trial and 11 genotypes in the initial trial along with the check SSV 84 were tested for their performance in *Kharif* 2002. Based on their performance, 10 varieties and one hybrid were promoted to an advanced trial. Six genotypes in the initial trial along with 11 genotypes in the advanced trial and check SSV 84 were tested for their performance on various breeding and biochemical parameters during *Kharif* 2003. All the genotypes were superior over the check SSV 84 in terms of biomass and percentage extraction. Therefore, all the entries were promoted for an advanced

TABLE 4.1 Overall Performance of the Sweet Sorghum Genotypes

S. no.	Genotype	Plant height (cm)	Biomass (t/ha)	Rank	Grain yield (q/ha)	Extraction (%)	Brix	Total sugars (%)	Reducing sugars (%)	Sucrose (%)
1.	RSSV 57	300.5	42.34	1	16.93	31.36	17.9	14.28	1.41	12.62
2.	RSSV 44	244.6	40.52	2	12.21	35.24	16.3	13.56	1.58	11.96
3.	RSSV 46	229.9	40.06	3	12.05	37.11	16.6	13.82	1.83	11.76
4.	NSSV 219	221.5	39.43	4	15.98	33.3	17.5	14.09	2.13	11.74
5.	RSSV 59	261.8	37.97	5	12.41	34.1	18.3	14.71	1.97	12.31
6.	RSSV 106	259.4	35.05	6	14.06	40.13	17.1	14.16	1.37	12.62
7.	RSSV 91	302.8	34.73	7	14.03	26.39	17.9	14.88	1.44	13.25
8.	RSSV 120	266.7	34.41	8	10.98	33.22	18.3	15.37	1.44	13.47
9.	NSSV 218	222.6	31.53	9	9.46	34.01	17.3	13.51	2.36	10.5
10.	RSSV 58	225.7	31.51	10	8.6	34.36	18.3	14.88	1.56	13.05
11.	RSSV 24	275.4	30.83	11	9.25	32.81	15.9	12.92	1.93	10.76
12.	NSSV 254	240.4	30.57	12	15.95	35.15	17	14.86	1.85	12.48
13.	NSSV 13	276.7	29.95	13	21.64	36.77	17.5	13.77	2.09	11.51
14.	NSSV 216	247.4	29.29	14	15.74	34.69	18.4	15.16	2.07	12.87
15.	RSSV 45	271.1	28.83	15	12.06	34.27	16.9	14.77	1.75	12.77
16.	NSSV 253	273.9	26.24	16	16.06	35.99	18.7	14.03	1.18	12.39
17.	SSV84 (C)	241.5	25.83	17	21.91	32.9	17.6	14.46	2.02	12.23
18.	NARISSH 43	248.3	25.38	18	24.58	39.01	14.7	11.65	2.37	9.52
	C.V (%)	8.99	17.69		36.12	10.91	8.21	10.61	19.4	12.41

trial. The entries in the advanced trial were sown during *Kharif* 2004 and the final data are presented in Table 4.1.

4.9 MAXIMIZATION OF BIOMASS IN SWEET SORGHUM

4.9.1 Time of Sowing Trial

Agronomic trial performance was studied by planting throughout the year at bimonthly intervals, at four locations, viz., Hyderabad, Akola, Parbhani, and Rahuri. The data from all the locations were pooled and statistically analyzed. The highest green cane yield was obtained in Keller (1.24 kg/plant) in the fourth date of planting (December) followed by BJ-248 in the fifth planting (February) at trial location. The genotypes Keller, NSSV-04, Wray, SSV-84, RSSV-09, and BJ-248 have high green cane yield and suitable planting dates are June, October, December, and February. June planting yielded high juice compared to other plantings. Brix recording was found to be highest in the June planting date, followed by October, December, and February plantings. Highest total sugars were observed in the October and December planting dates.

Of the six plantings, planting during August and April were found to be not suitable for sweet sorghum sowing. Therefore, in 2002–2003 four plantings were taken up, that is, June, October, December, and February, at six locations, viz., Hyderabad, Akola, Parbhani, Rahuri, Anakapalle, and Virinjipuram. The highest green cane was observed in the June planting in all genotypes followed by February, December, and October. The percentage juice extractability was observed to be highest in the June planting, followed by February, December, and October planting dates. The highest juice Brix percent was observed in the February planting, followed by June planting. In the February planting, the total sugars percentage was highest followed by the June planting (Fig. 4.1).

June and February plantings are ideal to obtain maximum biomass (Figs. 4.2 and 4.3) and high sugar content in sweet sorghum. Of the ten genotypes NSSH 104, RSSV 9, RSSV 47, and NSSV 208 were superior for green cane yield, juice extractability, and juice yield.

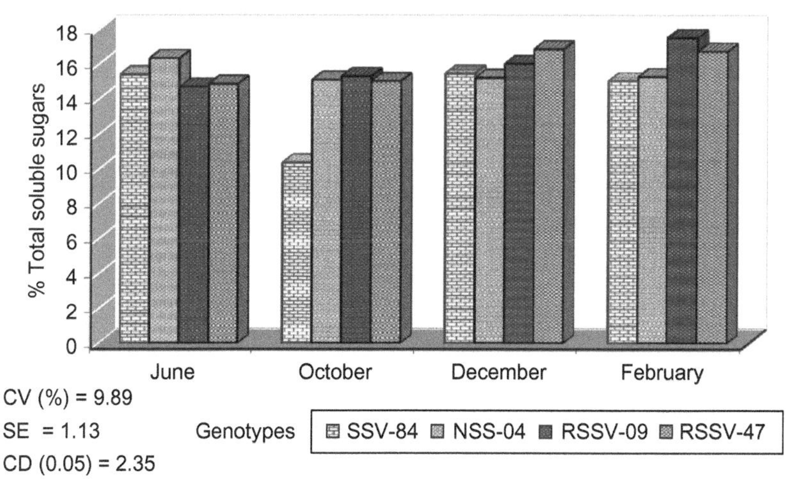

FIGURE 4.1 Influence of planting dates on total soluble sugars (2003–04).

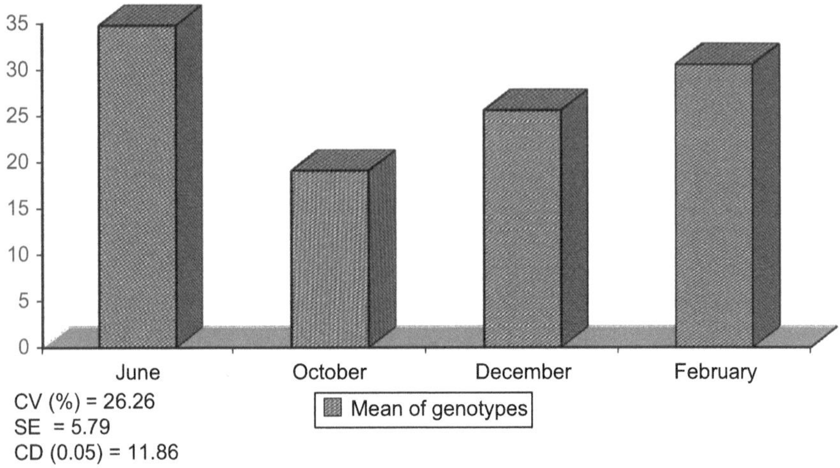

FIGURE 4.2 Influence of planting date on green cane yield.

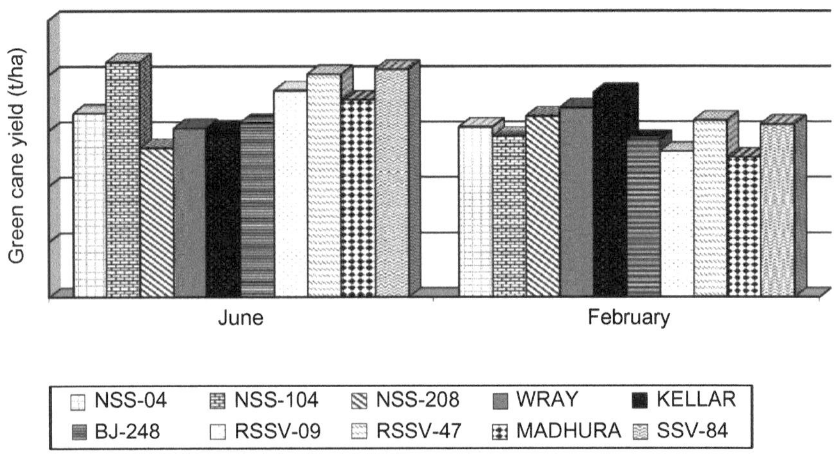

FIGURE 4.3 Interaction of planting date and genotypes.

4.10 RESISTANCE TO DISEASES AND PESTS

Sweet sorghum is vulnerable to different diseases and pests. Higher concentrations of sugar in the culm and leaves as well as the relatively low concentration of tannins in some genotypes contribute to the relatively higher vulnerability of sweet sorghums to diseases and pests. Important among the foliar diseases are the green leaf spot (*Cercospora sorghi* and *Colletotrichum graminicola*) and rust (*Puccinia purpurea*). These diseases seriously affect the biomass productivity of plants. Red rot and leaf anthracnose (*C. graminicola*) and downy mildew (*Peronosclerospora sorghi*) affect the stalk to reduce its yield and sugar content. Charcoal rot (*Macrophomina phaseolina*) and *Fusarium* stalk rot are serious during the post rainy cool crop season in the tropics.

Major virus diseases affecting sweet sorghum are the maize dwarf mosaic (MDMV) and sugarcane mosaic (SMV) when cultivated in rotation or along with sugarcane.

The principal insects causing serious reductions in juice and sugar yield in sweet sorghum are the borer (*Chilo partellus*), sorghum shoot fly (*Atherigona soccata*), and the sugarcane borer (*Diatraea* sp.). Field resistance to borers is an essential agronomic attribute for extracting good quality juice.

About nine genera of stylet-bearing nematodes, more commonly *Helicotylenchus dihystera*, *Aphelenchoides* sp., *Meloidogyne javanica*, have been described as infesting sweet sorghums.

4.11 EVALUATION OF SWEET SORGHUM GENOTYPES FOR RESISTANCE TO KEY PESTS (SHOOT FLY, CORN PLANT HOPPER, AND STEM BORER) IN RELATION TO DIFFERENT DATES OF PLANTING

Ten sweet sorghum genotypes along with susceptible (DJ 6514) and resistant checks (IS 18551 and IS 2312) were evaluated in the year 2001–02 for their reaction to key pests at four planting dates, viz., June (1), September (2), November (3), and February (4), at one location, that is, NRCS, Hyderabad. RSSV 9 was found relatively less susceptible to shoot fly when compared to the rest of the genotypes in terms of both oviposition and expression of dead hearts (Fig. 4.4). Stem borer infestation was low in June planting in comparison to November planting suggesting that delayed plantings are prone to higher stem borer infestation under natural conditions. June planting indicated that the population of both the adults and nymphs of corn plant hopper have gradually increased, reaching a peak in the September planting and subsequent decline in the November planting. Among the test genotypes, NSS 208 was relatively less susceptible for both colonization and establishment of adults and nymphs

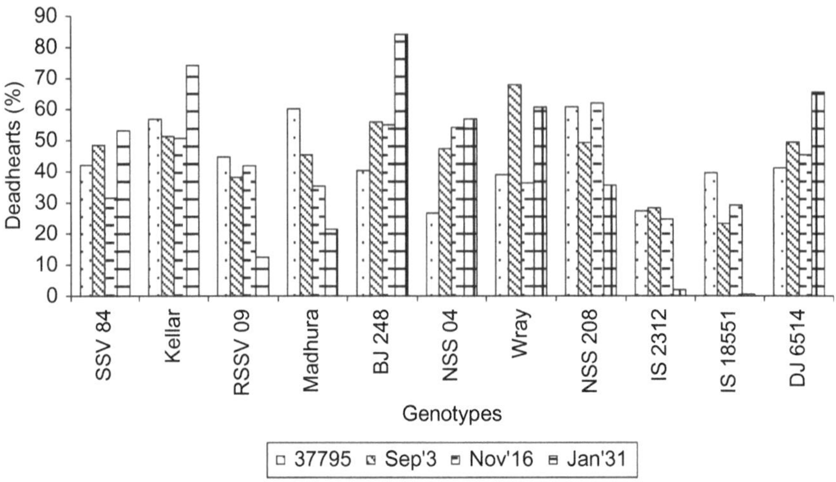

FIGURE 4.4 Incidence of shoot fly damage in relation to different planting dates in sweet sorghum (2001–02).

as well as plant damage at all the planting dates. The results of this trial summarized that genotypes planted early escaped from infestation of shoot fly, stem borer, and head bug, and the delayed plantings are prone to higher susceptibility to these key pests. However, the corn plant hopper showed a different response showing a slow increase in the infestation in June planting, with a gradual increase and rapid decline in the September planting and the November planting, respectively.

Eight sweet sorghum genotypes along with resistant (IS 2312 and IS 18551) and susceptible (DJ 6514) checks for shoot fly were tested for resistance to shoot fly (*A. soccata* Rond.), corn planthopper (shoot bug) (*Peregrinus maidis* Ashm.), and the spotted stem borer (*C. partellus* Swin), under four plantings, viz., June, September, November, and January, in the year 2002–2003. It is inferred that June and November plantings are relatively safer for shoot fly incidence and RSSV 9 was relatively less susceptible across the planting dates. Corn plant hopper infestation was found to be lower during the November planting and the February planting. Stem borer infestation was low during the first planting in comparison to other plantings, while the planting in November recorded greater stalk tunneling by the stem borer, and there was intermediate incidence in the January planting (Fig. 4.5). Among the test genotypes, BJ 248 was significantly less susceptible for stalk tunneling and significantly not different from resistant genotypes IS 2312 and IS 18551 across three planting periods.

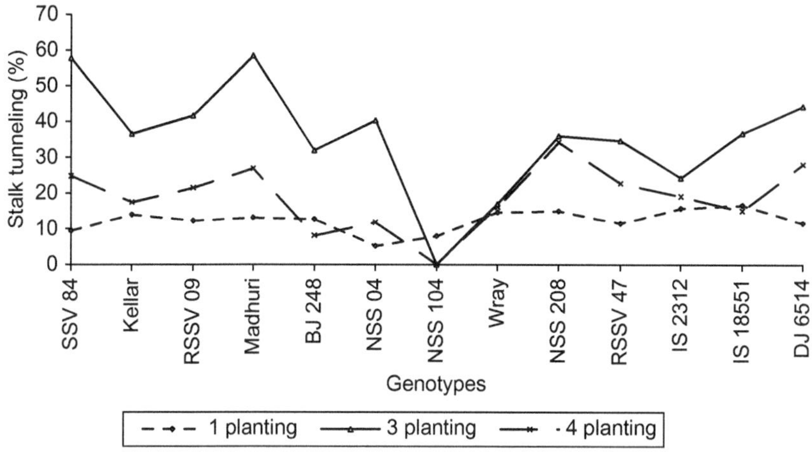

FIGURE 4.5 Stalk tunneling caused by the spotted stem borer in relation to different dates of plating in sweet sorghum.

4.12 GENETIC IMPROVEMENT OF SWEET SORGHUM

Earlier breeding efforts to improve sweet sorghum in the United States of America were directed to syrup quality and yield. While sweet sorghum is being reexamined for its potential for commercial extraction of sugar (Schaffert and Gourley, 1982), the emphasis on this crop as a bioenergy source is much more recent. In Brazil, the world's largest bioenergy producer, an ambitious National Alcohol Programme, was initiated in 1975 to produce alcohol from sugarcane, sweet sorghum, and cassava through a network of microdistilleries at a daily turnover of 25 million liters. United States Department of Energy projects an annual production of 11.4 billion liters ethanol by 1990 and 31.4 billion liters by 2000 AD from sweet sorghum.

4.13 GENETIC VARIABILITY

Genetic improvement of sweet sorghum for an increased yield of biomass and fermentables and increased hectarage under this crop are important aspects of the emerging bioenergy scenario. There are syrup type cultivars of sweet sorghum in USA with the potential to yield as high as 125 t fresh weight/ha. Only very limited breeding efforts have been made elsewhere to develop it as a high energy crop. The breeding base initially available in the New World was narrow with much of the variability belonging to syrup types (Schaffert and Gourley, 1982). A significant broadening of this base was made in the 1940s with the introduction of about 1200 varieties collected from Africa by Carl O. Grassl. Since 1980 sweet sorghum has invoked much interest among plant breeders world over. This helped to broaden the base of information on the genetic and nongenetic variability available for most of the important plant characteristics which contribute to the sugar and biomass yield, their development rhythms, and correlated response with other agronomic attributes, futuristic crop models, and breeding methodologies (Miller and Creelman, 1980a,b; Schaffert and Gourley, 1982; Choudhari et al., 1987; Seetharama et al., 1987).

Sweet sorghums present a fair genetic variability in several characters of breeding interest. These include plant height, stalk girth, tillering ability, LAI and duration, stalk juiciness, juice extractability, juice purity, concentration of starch, sucrose, and reducing sugars in juice and juice acidity, stalk yield, stalk fiber content, maturity period, photoperiodic response (Seetharama et al., 1987; Bapat et al., 1987; Nimbkar et al., 1987; Schaffert and Gourley; 1982; Broadhead, 1981; Freeman, 1979; Ferreira et al., 1982), grain yield, and quality (Bapat et al., 1987; Choudhari et al., 1987), resistance to important diseases and insect pests (Broadhead, 1981; Freeman, 1979; Wiedenfeld et al., 1984), and adaptability across agronoclimatic conditions.

4.14 CULM CHARACTERISTICS

Variability for plant height usually ranges from less than 1 to 5 m. Most sweet sorghums tend to grow very tall. Height together with higher girth and fleshiness of stalk determine the stalk yield. Stalk diameter at mid height of selected sweet sorghum types varied from 8 mm to 23 mm. Maximization of fresh stalk yield is possible through optimization of stalk height and thickness. Resistance to lodging and suitability for mechanical harvesting are desirable agronomic attributes. A significance tendency for lodging was noted when the height exceeded 3 m. A moderate height, nonbrittle nodes, flexible stalk, and a strong root system prevent lodging tendency. Resistance to lodging is inherited as a simple dominant character and can be easily incorporated. Culm thickness is a consistent character (Smith et al., 1987) and appears to have high heritability. Because stalk is the primary sink in sweet sorghum, its structure is important for efficient accumulation and retention of sugars. Sweetness of stalk is a recessive character controlled by two genes (Coleman, 1970; Clark, 1981). Juicy stalk may have a firm flesh which when cut across may appear greyish or yellowish brown. Basal internodes are usually fleshier and the terminal ones are pithy and less juicy. Fleshier is the stalk, higher the juice content and yield. Juiciness of stalk is not always related to concentration of sugars (Seetharama et al., 1987). Juiciness is a simply inherited recessive character (Coleman, 1970). High sugar yield per unit weight of fresh stalk is possible only when it has higher juice yield and superior juice quality.

4.15 JUICE YIELD AND QUALITY

Sweet sorghum genotypes show wide variability in juice quality and juice extractability (Schaffert and Gourley, 1982; Seetharama et al., 1987; Choudhari et al., 1987; Bala Ravi et al., 1997). Recovery of juice during crushing determines the relative juiciness of sweet sorghum genotypes. It is discerned with the help of an experimental mill having two or three rollers with an adjustable roller gap. Evaluation of 160 sweet sorghum genotypes for juice extractability at physiological maturity stage using a two discharge roller mill indicated variability from 14% to 68%. Juice yield and clarity not only depend on the genotype but also on the growing conditions and maturity stages. Their peak yields were reported at soft dough stage (Ventre and Byall, 1937; Coleman and Stokes, 1964) as well as at post physiological maturity stage (Coleman, 1970).

Juice quality is determined by the relative concentration of soluble solids like sugars and starch and other components like aconitic acid. Brix refractometry is widely used to screen variability in the percentage of soluble solids in juice. Brix values up to 26.6 were recorded in sweet sorghum identified from sorghum World Collections (Seetharama et al., 1987). Sweet sorghums show large differences in the amount of sucrose and reducing sugars in the stalk (Ventre et al., 1948; Ferreira et al., 1982; Schaffert and Gourley, 1982; Bapat et al., 1987). Brix and sucrose content determined by colorimetric methods do not show high positive correlation (Seetharama et al., 1987). Genotypic differences also appear to exist for invertase activity which hydrolyzes sucrose in the stalk and extracts juice to glucose and fructose.

Total concentration of sugars was lowest until the boot stage and it reached its highest at the soft dough stage. Sweet sorghum juice mainly contains sucrose, glucose, and fructose. While sucrose is the predominant sugar during the whole developmental stages, it constituted only about 50% of soluble sugar at the boot stage, glucose and fructose making the remainder. Concentration of glucose was always higher than that of fructose. A study on changes of the chemical quality of juice was done in terms of its inversion at different periods of time after harvesting of the cane. The parameters such as Juice Brix, pH, total sugars, reducing sugars, and sucrose were evaluated. Canes from CSH 22SS were harvested, samples were drawn, and juice quality was determined every 6 h up to 48 h. After 48 h, there is a 17% reduction in sucrose content. Inversion of sucrose results in the higher production of aldehydes during fermentation. This is higher compared to sugar cane. The presence of high reducing sugars and invertase activity make the juice unable to produce good quality jaggery and sugar. Therefore, sweet sorghum juice is only suitable for ethanol and syrup production.

Percentage of sucrose in total solids decides the juice purity. Proportion of reducing sugars to sucrose is termed as "sugar ratio." Sweet sorghums show a large variability in juice purity and sugar ratio (Nimbkar et al., 1987). Two broad groups of sweet sorghum are commonly identified on the basis of sugar ratio and juice purity. Varieties with relatively higher levels of reducing sugars are called syrup type sorghums. Crystallization sugar called "sugaring" is not easily achieved in their juice. Sweet sorghums having high juice purity and low sugar ratio are called sugar types. Concentration of their juice easily crystallizes the sucrose. Juice purity is adversely affected by the presence of starch. The amount of starch extracted with juice was found to vary from less than one to 7.6 percent in an evaluation made on 160 sweet sorghum genotypes. While the genetic principles governing the differences in the amount of sucrose and reducing sugars is not delineated, the nongenetic factors influencing them are better discerned.

Aconitic acid, apart from influencing juice acidity, is important because of its adverse influence on the crystallization of sucrose (Accorsi et al., 1981). Juice acidity usually varies from pH 4.5 to 5.5. Lack of a rapid technique for the quantitative determination of aconitic acid is posing problems in determining variability in sorghum for this critical juice constituent.

The color of juice differs depending on the pigments extracted from stalk and leaves. Important among them are chlorophyll and anthocyanins. Sorghum stalks are commonly classified as tan (nonreddening) and purple, on the basis of the types of anthocyanin pigments present. These pigments usually get discolored on exposure to light, air, or heat and on glucosidase or polyphenol oxidase mediated degradation.

Extracted juice undergoes a change in color during storage. Starch dispersed in the juice may also influence the color.

Most of the sweet sorghum identified from the Sorghum World Collection were found to belong to the Snowdenean races Caudatum and Durra with sparse distribution across Bicolor and Guinea and some of their mixed races (Seetharama et al., 1987).

In general, the amount of sucrose in sweet sorghum stalk is lower than in sugarcane. There are, however, a few genotypes which have sugar yield comparable to sugarcane (Ferraris, 1981; Inman-Bamber, 1980). Considering the long years of intense selection and breeding efforts that have been done to sugarcane to improve its Brix and juice quality, the marginal inferiority of sweet sorghum in these respects may not remain a disadvantage with intensified research. Although it is realized that improvement of juice quality in sweet sorghum would be more difficult, the different studies on variability in juice quality provide a little hope in this respect.

4.16 STUDIES ON JUICE QUALITY (NRCS, HYDERABAD)

Sweet sorghum genotypes show wide variability in juice quality and juice extractability (Bala Ravi et al., 1997). Evaluation of 160 sweet sorghum genotypes for juice extractability at physiological maturity using a two-discharge roller mill indicated variability from 14% to 68%. Recovery of juice during crushing determines the relative juiciness of sweet sorghum genotypes. Sweet sorghum juice mainly contains sucrose, glucose, and fructose. While sucrose is the predominant sugar during the whole developmental stages, it constituted only about 50% of soluble sugar at the boot stage, glucose and fructose making the remainder. Concentration of glucose was always higher than that of fructose. A study on changes of the chemical quality of juice was done in terms of its inversion at different periods of time after harvesting the cane. The parameters such as Juice Brix, pH, total sugars, reducing sugars, and sucrose were evaluated. Cane from NSS-104 was harvested, and samples was drawn and juice quality was determined every 6 h up to 48 h. After 48 h, there is a 17% reduction in sucrose content. Inversion of sucrose results in a higher production of aldehydes during fermentation. This is higher compared to sugar cane. The presence of high reducing sugars and invertase activity make the juice unable to produce good quality jaggery and sugar. Therefore, sweet sorghum juice is only suitable for ethanol and syrup production.

TABLE 4.2 Comparison of the Juice of Sweet Sorghum and Sugarcane

S. no.	Character	Sweet sorghum	Sugar cane
1.	Appearance	Juice is thick and turbid.	Juice is clear and transparent.
2.	Contents	High starch & chlorophyll.	Has no starch.
3.	Density	High.	Low almost equal to water.
4.	Sugars	Sucrose and high amount of reducing sugar (2–4%).	Contains more sucrose and less reducing sugar (0.5–1.0%).
5.	Invertase activity	High invertase.	No invertase.
6.	Time of sugar accumulation	Starts after flowering.	Sucrose accumulates at early stage.
7.	Purity of juice	70–90%.	90–98%.

4.17 IMPROVEMENTS ACHIEVED IN JUICE QUALITY OF SWEET SORGHUM: DETERMINATION OF SUCROSE IN SWEET SORGHUM

The key analytical factors that contribute to sweet sorghum juice quality are mainly juice extractability, sucrose, reducing sugars content, and purity of juice. In this context, juice from sweet sorghum and sugarcane are compared (Table 4.2). Sucrose was being quantified with an indirect chemical assay using phenol and sulfuric acid. This method is not accurate and a more precise method of quantifying sucrose in sweet sorghum was developed at NRCS. This method is being widely adapted in sugarcane; however, it is not applicable directly to sweet sorghum juice. Therefore, modification of the method in terms of the clarifying agent lead acetate and the time of clarification were studied. The basic principle of the method is using a saccharimeter. The study on spiking with external sucrose added to the juice showed 98% recovery. Percentage of added sucrose in sugarcane and sweet sorghum are similar confirming the fact that the other factors in the juice are not interfering in the quantification.

The detailed data on the spiking studies of external sucrose are presented in Table 4.3. This study was conducted with different stages of juice of sweet sorghum with varied Brix and sucrose content. The study showed that sucrose can be quantified accurately by clarifying with lead acetate in the juice of sweet sorghum. This method is a fast and accurate method compared to the chemical method. No interference is observed in sweet sorghum for sucrose estimation.

TABLE 4.3 Spiking With External Sucrose and Interference Study in Sweet Sorghum & Sugarcane

		Brix				Sucrose				Sucrose			Sucrose recovered		
		Pure juice	After addition of Sucrose (g)			Pure juice	After addition of sucrose (g)			Recovered (g)			After addition of sucrose (g)		
S. no.	Genotype		2	4	6		2	4	6	2	4	6	2	4	6
1.	SSV-84	13	15	17	18	7.2	9.1	11.15	13.05	1.9	3.95	5.9	95	98.75	98.33
2.	SSV-84	14	16	18	20	8.6	10.53	12.57	14.43	1.93	3.97	5.63	96.5	99.25	93.8
3.	SSV-84	13	—	—	—	6.7	8.8	—	—	—	—	—	100	—	—
4.	SSV-84	6	8	10	12	1.5	3.68	5.45	7.44	2.18	3.95	5.94	100	98.75	98.99
5.	SSV-84	7	8.5	10	12	1.5	3.8	5.5	7.7	2.3	4	6.2	100	100	100
6.	SSV-84	6	8	10	12	1.5	3.63	5.43	7.35	2.14	3.96	5.92	100	99	98.7
7.	Sweet sorghum bulk	18	—	21	23	12.7	—	16.4	18.3	—	3.7	5.6	—	92.5	93.33
8.	Sugar cane bulk	20	22	23	25	14.9	16.71	18.69	20.56	1.81	3.79	5.66	90.5	94.75	94.52

Each value is mean of three independent analyses.

4.18 JUICE QUALITY IN SWEET SORGHUM AT DIFFERENT CROP GROWTH STAGES

Sweet sorghum requires a period of 4 months for maturity. However, to minimize crop growth period, study on juice quality evaluation at different crop growth periods was done. The study shows that at physiological maturity of the grain, in the majority of the genotypes, the crop has the maximum cane yield and juice yield.

In a study to determine the suitable stage of harvest to optimize cane yield, juice extractability and yield, crop was harvested at three different grain maturity stages (ie, milky stage, physiological maturity, and harvesting stage). Harvesting at a physiological maturity was found to be optimum in terms of juice volume, percent extractability, and percent total soluble sugars. The cane yield showed a significant difference with stage of harvest. Most of the genotypes recorded higher green cane yields at harvesting stage (Fig. 4.6). Exceptionally, NSSV 219, NSSV 253, RSSV 24, and RSSV 58 recorded higher green cane yield during the milky stage, while RSSV 106, RSSV 120, and SSV 84 recorded higher green cane yield at physiological maturity.

A significant difference was observed in juice volume with stage of harvest. The highest juice volume was recorded (during physiological maturity) in most of the genotypes, while genotypes such as NSSV 13, RSSV 46, and RSSV 57 recorded highest juice volume during harvesting stage and RSSV 24 during milky stage (Fig. 4.7). Extraction percentage showed a significant difference with stage of harvest. Maximum extraction percentage was obtained during physiological maturity in all genotypes, followed by the harvesting stage. Hence, crop can be harvested at physiological maturity, as juice yield is a key factor for high ethanol recovery.

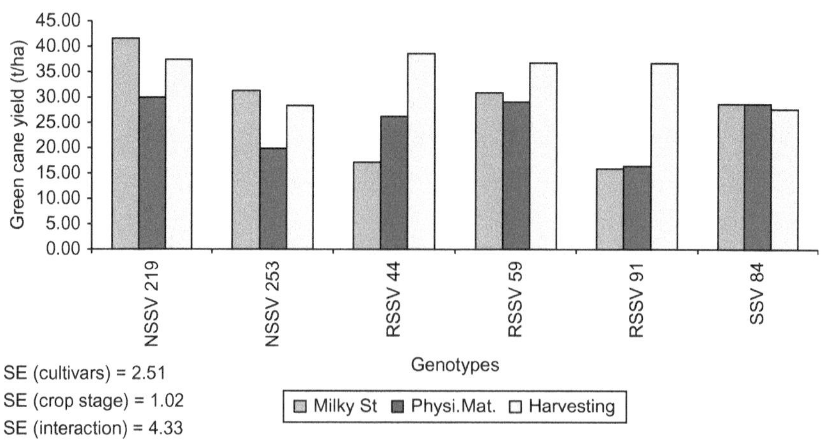

FIGURE 4.6 Effect of crop growth stage on green cane yield.

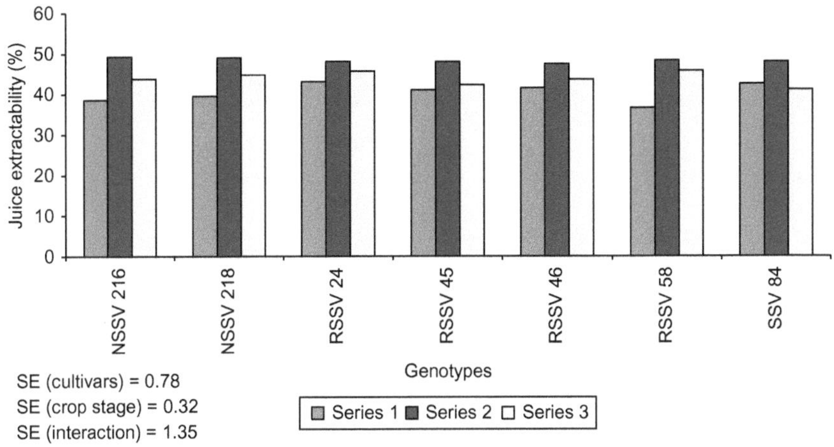

FIGURE 4.7 Effect of crop growth stage on juice extractability.

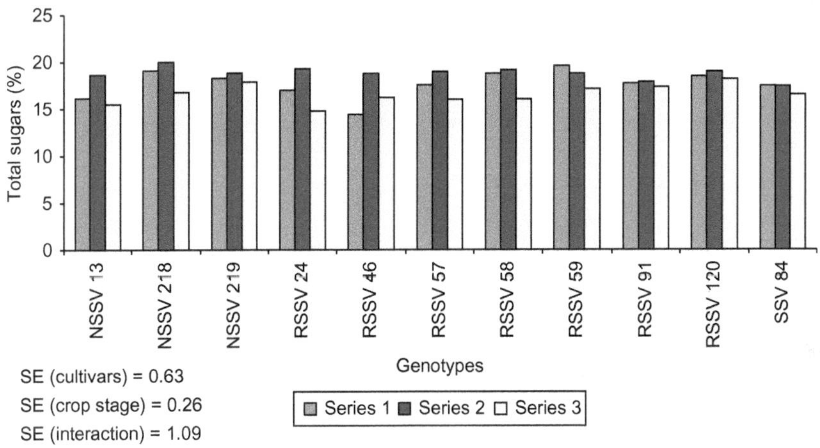

FIGURE 4.8 Effect of crop growth stage on per cent total sugars of juice.

A significant difference was observed in percentage of total soluble sugars with stage of harvest. Most of the genotypes recorded maximum percentage of total soluble sugars at physiological maturity. However, maximum percentage of total soluble sugars during the milky stage was obtained in genotypes RSSV 44, RSSV 45, RSSV 59, and RSSV 106, while in genotypes NSSV 216 and NSSV 253 it occurred during the harvesting stage (Fig. 4.8).

Correlations were made between Brix versus total sugars and Brix versus sucrose at different crop growth stages of sweet sorghum. The correlation coefficient (*r*) is significant at all the stages, milky stage, physiological maturity, and maturity stages. The correlation coefficient (*r*)

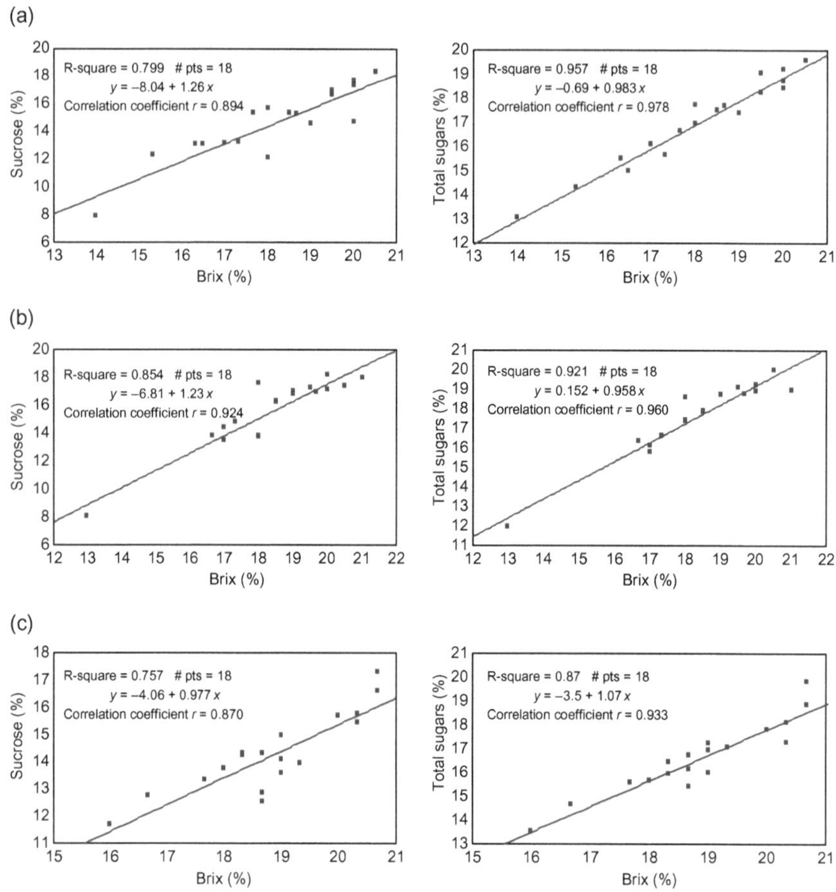

FIGURE 4.9 Scatter graphs of Brix, sucrose and Brix, total sugars at different crop growth stages in 18 sweet sorghum genotypes. (a) Grain milky stage. (b) Grain physiological maturity stage. (c) Grain maturity stage.

was always higher for total sugars compared to sucrose. However, the correlation coefficient r of Brix versus total sugars was maximum ($r = 0.978$) at the milky stage of the grain (Fig. 4.9).

4.19 JUICE QUALITY IN SWEET SORGHUM AS A VEGETATIVE CROP

Sweet Sorghum is grown normally for a period of 120 days. Sugar accumulation starts at flowering stage and peaks at physiological maturity and maturity stages. Though the Brix readings are high, sucrose levels are at their maximum at physiological maturity and maturity stages of grain. To reduce the crop growth period and to match the lean periods of sugarcane harvest, an attempt was made to evaluate the juice quality characteristics by removing the panicle before emergence. The experiment was performed with a control, that is, a normal crop with panicle/earhead and another crop without panicle in six sweet sorghum genotypes, which were under testing in the AICSIP advanced varietal trial.

From the milky to the physiological maturity stage, there was no significant difference in any of the parameters studied between the control and deheaded plants. The mean percentage increase in Brix is 3% and 1% at milky and physiological stages, respectively. However, this increase was not reflected in sucrose accumulation. The mean percentage increase in sucrose was 1.3% at the milky stage and 0.9% at the physiological maturity stage (Fig. 4.10). The negative factor for a deheaded plant is growth of a large number of side tillers. Due to the tillering, the sugars being accumulated would be distributed/diluted

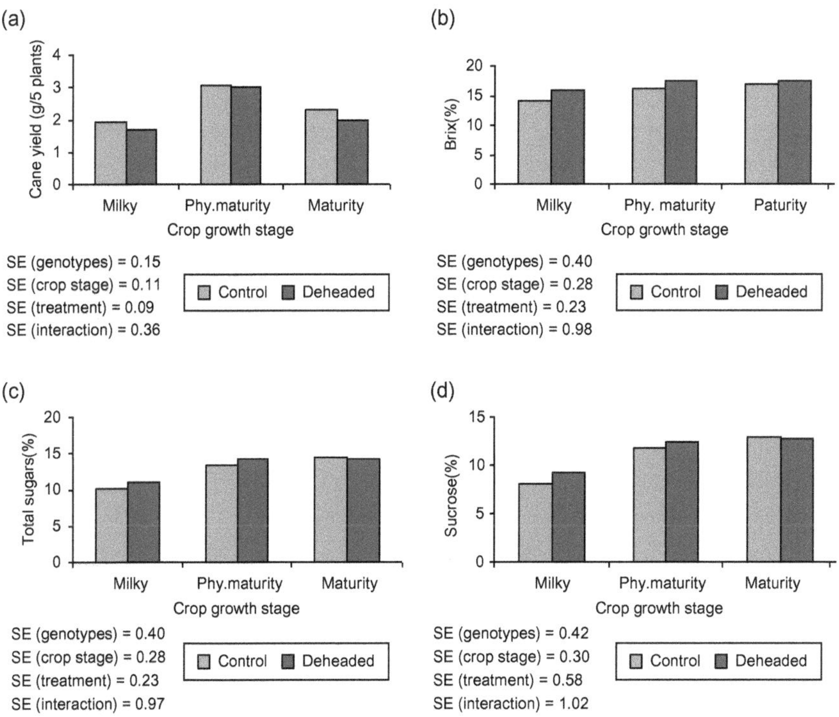

FIGURE 4.10 Effect of deheading on biomass and juice quality in sweet sorghum genotypes (*Kharif*, 2004). (a) Green cane yield, (b) Brix of stalk juice, (c) Total sugars in stalk juice, (d) Sucrose of stalk Juice. SE, Standard Error.

and hence an increase in sugar accumulation was not reflected in deheaded plants. In addition to that, the normal crop gives a good amount of grain and the total net return was higher in control plants compared to deheaded plants.

4.20 PHENOLOGY AND BIOMASS PRODUCTIVITY

Biomass yield in sweet sorghum is predominantly influenced by the phenological behavior. Nevertheless, genotypes with similar phenology widely differ in their biomass productivity. These differences in the productivity appear to have a linear relationship with their ability to assimilate solar and cultural energies. An evaluation of 90 agronomically superior sweet sorghum genotypes, which showed maturity range between 85 and 125 days at ICRISAT, showed the variation in biomass productivity rate from 7.4 to $55.6/gm^2/d$ with a mean value of $22.8/gm^2/d$ (Seetharama and Prasada Rao, 1988). Photosensitivity seriously limits progress in varietal improvement in sweet sorghum and its latitudes. Two high yielding varieties, Wray and Honey, are good sources of photo insensitivity (Schaffert and Gourley, 1982). Although maturity in sorghum is known to be governed by four genes, breeding cultivars of diverse maturity period is not easy because the character is complexly interrelated to several physiological–genetic components involving photo period–temperature–genotype interactions.

As photosynthetically active leaf area constitutes the major source, LAI and LAD are important in sweet sorghum productivity. LAI at maturity is positively correlated ($r = 0.57^{**}$) with Brix (Choudhari et al., 1987) and biomass yield. Some of the grain sorghum cultivars with stay-green characteristics were found to have moderate sugar content in stalk and would be useful for sweet sorghum improvement. Tillering ability is another character which could be used for higher biomass yield. Sweet sorghum shows variability for this character (Broadhead, 1981). Genotypes capable of producing two to four synchronous tillers which would develop into larger diameter stalks usually produce higher yields (Coleman, 1970). Similarly, the variability available among sweet sorghum for the ratio of stalk and foliage (Nimbkar et al., 1987) could be taken advantage of to increase the stalk yield.

4.21 GRAIN YIELD

Sweet sorghums are believed to have a poor grain yielding ability. Recent studies on sweet sorghum collected from Africa and Asia, however, revealed the wide variability among them for grain yield. Variation ranging from 16.6 to 515.4/gm^2 with a mean value of 156.6/gm^2 among 90 agronomically superior sweet sorghum stocks evaluated at ICRISAT, India, was reported by Seetharama and Prasada Rao (1988). Differences existing between grain and sweet sorghums for their sink sites (Seetharama and Prasada Rao, 1988; Lingle, 1987) do not appear to cause limitations for attaining higher grain yield in sweet sorghum. It seems the higher biomass productivity of sweet sorghum comes as an advantage for higher grain also. While high yield in grain sorghum is achieved at 35–45% harvest index, equivalent grain sorghum in high energy sorghums may be attained at 5–10% harvest index. Several sweet sorghum genotypes having fresh biomass productivity potential above 100 t/ha in 100 to 150 days in the tropics and higher latitudes have been identified (Schaffert and Gourley, 1982; Smith and Reeves, 1979; Seetharama and Prasada Rao, 1988). Discovery of high concentration of sugar in the stalk of elite high yielding grain sorghums (Choudhari, 1986) provides strong support for the possibility of combining high grain yield and stalk sugar yield. Major grain yield components like the number of grains in the panicle and 1000 grain weight (TGW) showed wide variability among several sweet sorghum stocks evaluated at the National Research Centre for Sorghum.

4.22 GRAIN QUALITY

Substitution of grain sorghums with high energy sorghum in their present habitats in Africa and Asia would be possible only if they possess the required grain quality to suit the different food, feed, and beverage preparations. Most of the current commercial cultivars of sweet sorghum have very limited grain value as they have small and poor quality grains. During their selection history, emphasis was not on grain yield and quality. However, land races of sweet sorghum from Africa and Asia offer wide variability on different aspects of grain quality. Examination of 176 sweet sorghum types selected from sorghum World Collections showed that their TGW varied from 14.1 g to 51.6 g. About 30% of this varieties have TGW higher than 30.5 g, predominantly in white and yellow endosperm varieties. Chalkiness of the grain was, however, more frequent among these grain types. Nevertheless, about 28% of genotypes possessing white and yellow endosperm were pearly. Availability of acceptable grain characters among the grain sorghums which have high juice and sugar yield further brightens the possibility of breeding sweet sorghum with suitable grain quality.

4.23 RESISTANCE TO DISEASES AND PESTS

Stalk and foliage damaging pests and pathogens may cause serious loss on the total biomass and stalk yield in sweet sorghum. They may also lead to quantitative and qualitative changes in sugar concentration. Hence, resistance or field tolerance in sweet sorghum against major pests and diseases is important for realizing its biomass productivity potential. A fair variability for resistance against most of these diseases and pests is available in sweet sorghum. In addition, the sources of resistance available in grain sorghums could also be used in sweet sorghum breeding. These sources of resistance and the screening techniques in breeding nurseries are described elsewhere (Agrawal and House, 1982; Jotwani, 1976; Rao and Goud, 1977).

Possibilities of incorporating resistance for green leaf spot, anthracnose, stalk rot and downy mildew (Freeman, 1979), and MDMV into high yield background have been demonstrated. Cultivation of sweet sorghum on a wider scale around sugarcane tracts in the tropics may require bred-in-resistance against SMV and sugarcane borer. There is a need to widely screen sorghums to identify different sources of resistance against SMV and cane borer. The possibility of creating new resistance sources through somaclonal variation also needs investigation. In situations where sweet sorghum is also used for grain, resistance against clinch bug, head bug, and head molds is important.

4.24 RESISTANCE TO STRESS CONDITIONS

A unique advantage of sorghum unlike most other cereals and sugar yielding crops is its ability to withstand different stress situations and still provide a modest yield. Its ability to grow in problem soils, low fertility, and moisture-stress conditions as well as under too much water is well known. It appears that in more instances drought tolerance and stay-green foliage in sorghum is either a consequence or cause of a higher percentage of stalk sugar. For instance, the stocks IS-12611, IS-21005, IS 2050, and DA-70 identified for drought tolerance have moderate to high stalk sugar accumulation ability. Therefore, sweet sorghum offers a rare opportunity in crop design to conjunctively manipulate its reduced vulnerability for drought and other stress conditions to maintain a higher biomass productivity rate.

4.25 EASY TO STRIP STALK

Another agronomic character of importance in mechanized agriculture is suitability for mechanical stripping, harvesting, and billeting. Sheath usually encircles the stalk, extending about $1^{1/2}$ times around the base of the internodes (Artschwager, 1948) with only little variation in length and looseness. While a loose sheath may make the stripping process easier, it is not known whether it may lead to increased vulnerability to pests and pathogens.

4.26 FUTURISTIC CROP MODEL

Modifications in crop phenology and growth pattern are important considerations for achieving a significant increase in the efficiency of biomass productivity rate in sweet sorghum. These changes have to be in conjunction with other favorable physiological attributes, such as tolerance to stress and adaptability under varying soil conditions and management milieu. Therefore, the major aspects of a futuristic crop model of high energy sorghum are:

1. An inherent ability for high biomass productivity with about 70–75% of fresh biomass as stalk.
2. Tall, stout, and lodging resistant stalk with structural make up and metabolic efficiency to accumulate and retain high levels of sucrose and other fermentable sugars with low levels of starch.
3. Higher juice content, purity, and extractability.
4. An ability to produce a moderate grain yield with acceptable grain quality, wherever required.
5. A crop phenology that suits the length of growing season and ensures high biomass productivity rate. The ideal maturity period in the tropics may be 90 to 130 days and 120 to 160 days in the higher latitudes, depending on the duration of frost-free warm days.
6. Stay-green foliar characteristics for protracted photosynthetic assimilation and an ability to prevent postmaturity diversion of sugars on brace roots and auxiliary branches.
7. Drought tolerance and ability to give good yield under low fertility and problem soil conditions.
8. Resistance to major diseases and pests.
9. Better stability in biomass and sugar yield across broader environmental lines.
10. Suitability for mechanized harvesting and processing.

The earlier sweet sorghum varietal models suggested by Coleman (1970) and Schaffert and Gourley (1982) mentioned some of the above characters.

4.27 STATUS OF SWEET SORGHUM BREEDING

Recent interest in sweet sorghum as a bioenergy crop has led to its larger scale introduction from the USA and evaluation in several countries. Their performance in tropical countries is less impressive mainly due to reasons of phenology and susceptibility to pests and diseases. Successes were, however, reported in identifying high yielding introductions and their hybrids from Brazil (Schaffert and Gourley, 1982) Australia (Ferraris, 1981), and New Zealand (Piggot et al., 1980). From India there were reports that high yielding US varieties like "Rio" and "Wray" failed to give good yield. They, however, found some of the US introductions had a good combining ability in crosses with cultivars of Indian grain sorghum and were useful for developing sweet sorghum with high biomass yield and modest grain yield. Some of the sweet sorghums introduced from parts of tropical Africa like Sudan, Ethiopia, and Nigeria were found to perform better under Indian conditions (Bapat et al., 1987). This diverse variability in sweet sorghum along with several elite grain sorghum selections having high levels of extractable juice and sugar content offers a fair opportunity to combine most of the characters listed under the varietal model described.

There are indications that the genetic pattern of major components of biomass yield and the stalk sugar content are amenable for transfer with fair success in selected and specific cross combinations. Therefore, breeding approaches already available for developing grain sorghum varieties and hybrids could be followed in sweet sorghum. As in the case of grain sorghum, hybrids are reported to be more productive in sweet sorghum (Schaffert and Gourley, 1982) and they have higher juice yield (Choudhari, et al., 1987). Superiority of hybrids in juice yield and sugar yield/ton of fresh stalk was also evident from the evaluation of a few hybrids and several selections of sweet sorghum. Discovery of high juice and sugar yield in some of the B-lines used in the Indian grain sorghum breeding programme (Choudhari et al., 1987) provides an immediate opportunity to develop high energy sorghum hybrids. Hybrids produced using A-lines derived from combined kaffir tend to show poor vigor, stability, and juice quality (Schaffert and Gourley, 1982). They stressed the need for developing new series of female parents to produce sweet sorghum hybrids. Enhanced juice quality in hybrids was found to be possible when they were sterile (Coleman, 1970). Wherever the grain is not required, sterile hybrids with high biomass yield and juice quality is a possibility.

4.28 UTILIZATION OF SWEET SORGHUM

The primary potential of sweet sorghum is as a source of fermentable carbohydrates for the production of bioenergy. It also has the potential for yielding sugar. Its use for the production of syrup is an old recent history of the crop. The sorgos of the United States are largely bred for their syrup yield and silage value. In countries like India, it can serve as an alternative source for the production of noncentrifugal sugars, such as jaggery conventionally prepared from sugarcane. In countries where grain sorghum is cultivated for the food or feed needs, the sweet sorghum could be developed as a multiutility crop to meet the food, feed, and fuel needs. Sweet sorghums possessing a high rate of dry matter accumulation that may lead to high stalk and grain yield are termed as high energy sorghums (Miller and Creelman, 1980a,b). In addition to these important uses, the by-products derived from the industrial processing of sweet sorghum for energy, sugar, or jaggery production can be utilized as fodder, solid fuel, raw material for biogas, biofertilizer, paper, and organic solvents.

4.29 STALK PROCESSING

Utilization of sweet sorghum either for production of bioenergy, syrup, jaggery, or sugar depends on the quantity and quality of juice extracted from the stalk. Juice extraction depends on the efficiency of the milling method used. Milling is either by a mechanical process involving crushing of the stalk between roller mills or by a physical process of diffusion of sugars into hot water. In the latter process, extractability of sugars is improved when stalk is chopped into small pieces. Efficiency of this process can be greater than 90% (Schaffert and Gourley, 1982). A combination of mechanical and diffusion methods would substantially enhance extraction efficiency (Nathan, 1978).

Mechanical variables influencing juice expression are the number of rollers and crushing pressure applied through setting clearance between rollers. Extraction optimization studies (Monroe et al., 1984) with stalks of different diameters and discharge roller clearances showed that highest recovery of total sugars was possible when the rollers were set at 0.6 mm clearance irrespective of the differences in stalk diameter. According to Schaffert and Borgonovi (1980) quoted by Schaffert and Gourley (1982), juice extractability between 45 and 76%

was possible from several varieties by applying 250 kg/cm pressure for one minute, using a hydraulic press. It appears that a medium diameter of stalk and a reasonable uniformity of its size may favor maximization of juice and total sugar yield during mechanical crushing.

A crushing rate between 0.3 and 0.6 t stripped stalk/h can be obtained when rollers, whether vertical or horizontal, were set at the lowest practical gap (Monroe et al., 1984). Stripping the stalk free of leaves before milling is a labor intensive process. Whole plant crushing, on the other hand, may save labor and improve yield of fermentables because the plant top, and leaves constitute about 16–32% of total mass and contain about 5% sugar (Monroe and Bryan, 1983). Crushing along with leaves, however, reduces the milling throughput (Monroe and Bryan, 1983) and imparts a darker green color and distinct odor to the juice. Billeting helps to automate the feeding process and maximize the throughput and higher extraction at the lowest rate of energy consumption.

Mechanization of sweet sorghum cutting was started with the use of corn binders. Now simple mechanized systems to strip the standing stalks, harvest and load them in a self-loading wagon, billet the stalks at metered sizes, and feed them to a juice extraction mill are available (Monroe et al., 1983). A major disadvantage of mechanical harvest and billeting, however, is the poor stability of stalk and rapid loss of fermentable sugars. Storage aspects of stalk are discussed in the next section.

4.30 CROP RETENTION AND STALK STORAGE

Processing of sweet sorghum either on a large or small scale may often entail delayed harvesting or storage of cut stalks for a few days before crushing. Therefore, the ability of the crop to stay a few days beyond its normal harvest time as well as the storability of harvested stalk for a few days without deterioration in juice yield and quality become important agronomic attributes. The optimal time for the harvest of a specific variety may have to be determined for different climatic zones. In higher latitudes, the physiological maturity of stalk may lag behind that of grains.

Only very limited information is available on the ability of the crop to stay beyond maturity without a reduction in sugar yield. Inman-Bamber (1980) reported that at Natal Midlands of South Africa sugar yield could be kept fairly constant during delayed harvest and the delay helped to increase juice purity and sugar recovery. Varietal differences in stalk retention ability were observed when two sweet sorghums were continuously monitored for concentration of sugars and starch at weekly intervals for seven weeks from anthesis. Varieties with a tendency for rapid dehydration of stalk showed significant reduction in juice and sugar yield when harvest was delayed beyond grain maturity stage. Rapid dehydration was observed from terminal part. The sponginess of stalk substantially reduced juice extractability. Stay-green habit, inability of the plant to produce brace roots and axillary branches, and resistance to red rot and stem borer would all help in preventing sugar loss during delayed harvest.

Two major events taking place in the stalk during storage are dehydration and inversion of sucrose. Inversion of sucrose initiates almost immediately after harvest. The extent of inversion depends on the genotype and the physical injury to the stalk. Inversion is relatively more rapid in syrup type sorghums than in sugar-type varieties, and in certain syrup types it takes only 3–4 days storage to complete the inversion (Coleman and Stokes, 1964). Inversion is faster in billeted stalks than in intact stalks. Dehydration of stalk leads to gradual drying and decrease in extractable juice. This is usually accompanied with an increase in Brix and such juice reduction could be minimized by storing stalk in moist conditions (Coleman, 1970). Microbial infection and destruction of stalk sugars is one possible danger under such conditions. According to Eiland et al. (1983) microbial damage of sugars was to the extent of 49% during storage for a week of mechanically cut and billeted stalks. Loss of sugars in manually cut and billeted stalks was relatively lower. Prolonged storage of whole or billeted stalk at ambient temperature for 2 to 4 months without any significant loss of fermentable sugar yield is possible through treatment with sulfur dioxide at 3000 and 4000 ppm (Eiland et al., 1983).

Juice extracted is more vulnerable to microbial spoilage and loss of fermentable sugars. Daeschel et al. (1981) found that freshly extracted juice of sweet sorghum contained about 100 m microorganisms/mL. The dominant species were *Leuconostoc mesentoroides*, gram negative rods, *Lactobacillus*, yeast, and nonfuel coliform bacteria. Among the juice spoiling microflora, the most important ones were *L mesentoroides* at 20°C and *Lactobacillus plantarum* at 32°C (Daeschel et al., 1981). Treatment of juice, stored at very low to 32°C ambient, with 0.5% sulfur dioxide could significantly prevent microbial deterioration.

4.31 FERMENTATION

Conversion of hexose sugars to ethyl alcohol is mediated by the enzyme zymase of yeast. Under aerobic conditions, hexoses are converted to pyruvate through the glycolysis pathway and the pyruvate under anaerobic conditions, or high sugar concentration (Crabtree effect) is converted to ethyl alcohol and carbon dioxide. The reaction is exothermic and releases 31.2 K cal energy for every molecule of sucrose converted. In addition to hexoses, sucrose, maltose, and maltotriose are also taken up by the yeast. More complex carbohydrates should be hydrolyzed to simple sugars before fermentation. In addition to ethanol and carbon dioxide, several other compounds, many of them in traces are produced during fermentation. They are of immense importance for flavor and taste of alcoholic beverages.

Saccharomyces cerevisiae is the most commonly used species of yeast in alcohol fermentation. Several of its strains exhibit varying fermentation ability and tolerance to different fermentation conditions.

The rate of fermentation increases almost linearly between 20°C and 35–40°C and rapidly decreases thereafter. The most rapid rate of ethanol production may take place in the temperature range of 30–34°C. Temperature inhibition mainly operates on the growth of the yeast population, which is necessary to maintain the fermentation rate. Advantages of using yeast strains capable of efficient fermentation at 40°C in view of the cost of the cooling facility required to maintain the fermentation temperature are discussed by Hacking et al. (1984).

Concentration of sugars and minerals in the broth also determines the maximization of the ethanol yield. A concentration around 12% fermentables may represent a best balance between economic distillation and fermentation inhibition.

The specific growth rate of yeast is severely depressed when ethanol concentration in the fermentation broth rises above 7–10%. Yeast strains differ in their tolerance to alcohol concentration. In the commercial parlance, an ethanol concentration of 8–9% is considered as an acceptable level (Hacking et al., 1984). However, yeast strains capable of continuing fermentation under higher concentrations of alcohol are important to scale down the cost of alcohol production. While the energy cost of distillation accounts for about 70–85% of total cost of energy required for alcohol production (Ladisch and Dyck, 1979), a 15% reduction in this cost is possible by raising the level of alcohol in fermentation broth from 8–9% to 12% (Patura, 1968). Yeast strains capable of raising alcohol concentration over 12% (v/v) have been identified (Ghose and Tyagi, 1979). Some of these strains are capable of growth and fermentation at 40°C.

There are other characteristics which make different yeast strains suitable to different systems of fermentation. Fermentation systems available are the most common batch type, continuous type, vacuum type, tower type, gradient tube type, and systems employing immobilized yeast cells. The ability of yeast cells to flock has a specific advantage in some of these fermentation systems. Fermentation systems also differ in their optimum concentration of sugar, required growth rate, and concentration of yeast, residence time, for realizing high efficiency. Matching the fermentation system with these fermentation conditions and specific characteristics of the yeast strain is essential for enhancing the efficiency of fermentation and alcohol yield. Although the batch system is most widely favored for the production of beverages, continuous fermentation systems offer many-fold increases in efficiency. The introduction of immobilized cells in continuous fermentation system may further enhance the fermentation efficiency (FE) and yield, in addition to other operational advantages.

FE of yeast could be improved through mutations and a biotechnological approach. A Gram-negative bacterium, *Zymomonas mobilis*, is found to be more efficient in converting sugar to alcohol. Compared with yeast, this bacterium isolated from fermenting sugarcane juice showed higher substrate uptake, rapid volumetric rate of ethanol production, and tolerance to higher concentration of ethanol. Application of this bacterium, however, is limited because it can take up only hexoses and sucrose and not pentoses, galactose, maltose, or maltotriose. Its deployment in ethanol production is reported from Brazil and Australia.

In situ fermentation of sugars in stalk is reported to give higher yields of ethanol than fermentation of extracted juice. Alcohol yield was higher from stalks chopped to 1 cm size billets than from shredded stalks (Coble et al., 1984). Sweet sorghum juice contains starch varying from 0.6% to 7.6% (Schaffert and Gourley, 1982). Prefermentation cooking of juice at 60–85°C is suggested to hydrolyze the starch to fermentable carbohydrates.

Fermentation of feedstocks to bioethanol is carried out by microorganisms, both prokaryotes (bacteria) and eukaryotic yeasts and fungi. The most commonly used organism for ethanol production is the yeast *S. cerevisiae* which is known to generate high ethanol yields apart from being more resistant to fermentation inhibitors. So far, the research is only focused on the fermentation of hexose sugars which is because many organisms utilizing them naturally exist in nature. However, the pentoses (which form the major fraction of the hemicellulosic hydrolysate of pretreated lignocellulosic biomass)-degrading microbes (except for the bacterium *Z. mobilis* which can use xylose and ferment it to alcohol; Zhang et al., 1995) were not explored, and the research on pentose-fermenting organisms is required for more economical utilization of lignocellulosic feedstocks and to increase overall ethanol yield.

There are several reports of various aspects of ethanol production from sweet sorghum which include: yeasts were commonly used for ethanolic fermentation and various aspects related to yeast were studied including the selection of high ethanol producing yeast strains, use of thermotolerant yeasts, and immobilized *S. cerevisiae* for fermentation (Table 4.4).

TABLE 4.4 Microorganisms Used in the Fermentation for Ethanol Production From Sweet Sorghum

S. no.	Microorganisms used in the fermentation of juice	Ethanol yield	References
1.	*Saccharomyces cerevisiae CFTR 01 and SG*	0.39–0.48 (g/g consumed sugar)	Ratnavathi et al. (2010)
2.	Fermax yeast (*S. cerevisiae*)	77.07–79.58 (gL^{-1})	Kundiyana et al. (2001)
3.	SuperStart yeast (*S. cerevisiae*)	73.18–76.95 (gL^{-1})	Kundiyana et al. (2001)
4.	*S. cerevisiae* TISTR 5048	0.42–0.48 g/g	Laopaiboon et al. (2007)
5.	*Saccharomyces* strains	29–87% (sugar conversion efficiency)	Bulawayo et al. (1996)
6.	*S. cerevisiae* (Nanyang)	91.61%	Liu et al. (2008)
7.	*S. cerevisiae*	0.42–0.45 (g/g sugar) (80–90%)	Sipos et al. (2009)
	Bagasse		
8.	*S. cerevisiae* 424 A (LNH-ST	EtOH Conc.-42.3 g/L, EtOH yield: 82.4–96.9%)	Li et al. (2010)
9.	*Pachysolen tannophilus, S. cerevisiae*	0.147 g/g dry material	Ban et al. (2008)

Shigeru Chohnan et al. (2011) produced ethanol from three varieties of sweet sorghum using repeated-batch fermentation. *Saccharomyces cerevisiae* cells could be recycled in 16 cycles of the fermentation process with good ethanol yields. This technique gives scope for using a broader range of sweet sorghum varieties for ethanol production.

To reduce the high investment cost involved in the production of ethanol from sweet sorghum juice, Kundiyana et al. (2001) investigated in-field production of ethanol with an objective to determine whether the fermentation process can occur without process control. They tested the effect of yeast strains, pH, temperature, and nutrients (urea) on the efficiency of the fermentation process and concluded that the addition of urea and lowering the pH did not have any significant effect on ethanol conversion efficiency of two yeast strains (Superstart and Fermax yeasts) tested. Maximum ethanol produced was 7.9% (w/v) in 120 h under ambient temperature conditions.

Areesirisuk et al. (2010) in their study to improve the GC analysis for organic acids and solvents in ABE fermentation from sweet sorghum juice by *Clostridium beijerinckii* JCM1390 found that the highest butanol concentration produced was 7.56 g/L with a butanol yield of 0.33 g/g sugar utilized.

Genotypic variation for ethanol production from sweet sorghum juice was studied by Ratnavathi et al. (2010). Five sweet sorghum genotypes (Keller, SSV 84, Wray, NSSH 104, and BJ 248) were studied. A maximum ethanol production of 9.0% (w/v) was obtained with the Keller genotype (with 20% sugar concentration) when *S. cerevisiae* (with a FE of 94.7%) was used. Under similar experimental conditions, when unsterile sweet sorghum juice (15% sugar concentration) was used, 6.47% (w/v) ethanol was produced. High biomass of yeast was obtained with BJ 248 variety. Two genotypes, SSV 84 (a variety) and NSSH 104 (a hybrid), were identified as superior genotypes for ethanol production with NSSH 104 having higher sugar conversion efficiency than SSV 84.

Hunsigi et al. (2010) tested several sweet sorghum genotypes for biomass yield, quality parameters, and ethanol production from stalk. Twelve and 14 genotypes, respectively, were tested during *Kharif* season of 2004 and 2005. Highest biomass yields were obtained from NSSH 104 (45.0 t/ha) followed by RSSV 56 (41.2 t/ha) and PAC 52093 (35.3 t/ha) during the year 2004, while PAC 52093 (48.00 t/ha) and ICSV 700 (45.8 t/ha) gave higher yields during 2005. With regard to juice quality, total sugars were highest in NSSV 255 (14.6%) followed by NSSV 6 (14.2%) during 2004, while RSSV 56 (14.2%) and NSSV 254 (13.2%) showed higher total sugars in 2005. Highest reducing sugars were observed in SSV 84 (2.73%), NSSV 6 (2.5%) (AICSIP, 2004) and NSSV 254 (2.15%) (AICSIP, 2005). They also tested for the ideal sowing time for sweet sorghum and suggested sowing in the month of June is ideal and delayed sowing invites shoot fly and thereby results in reduced biomass yield. They also suggested that summer sowing, January–February were better in yield and quality. A separate experiment was conducted to study the performance of three ruling varieties (SSV-74, RSSV-9 and NSSH-104). The stalk yields varied from 22 to 27 with an average of 25 t/ha. The fiber content is 13%. The ethanol yield is 1500 L (60 L/ton) with power generation of 3.25 MW/ha (0.5 MW/ton). A big mill test was made in sugar mill by crushing 356 tons of Sweet Sorghum stalks. The extraction was 50% and total reducing sugars was 5.64% with ethanol yield of 38.89 L/ton stalk.

Khongsay et al. (2010) investigated the effects of initial cell concentrations and nitrogen supplementation on growth and ethanol production from sweet sorghum juice using *S. cerevisiae* under normal gravity (fermentation of mashes containing 20–24% of dissolved solids) and very high gravity conditions (mashes containing $\geq 27\%$ of dissolved solids). Under conditions of 30°C, 100 rpm in a 2-L fermenter, the initial cell concentration of 1×10^8 cells/mL, and absence of nitrogen supplementation gave maximum ethanol production efficiencies under both normal and very high gravity conditions. The productivity (2.92 ± 0.04 g/L/h) and yield (98.33 ± 1.25% of theoretical yield) of ethanol at 36 h fermentation time were higher at normal gravity conditions than at very high gravity conditions. However, the ethanol concentration (105.12 ± 1.34 g/L) was similar under both conditions.

Laopaiboon et al. (2007) reported ethanol production by *S. cerevisiae* TISTR 5048 in sweet sorghum juice in which the ethanol concentration produced, yield, and productivity were 100 g/L, 0.42 g/g, and 1.67 g/L/h, respectively, in batch fermentation, whereas in fed-batch fermentation they were 120 g/L, 0.48 g/g, and 1.11 g/L/h, respectively. These findings suggest that fed-batch fermentation improves the efficiency of ethanol production in terms of ethanol concentration and product yield. Liu and Shen (2008) reported the effect of main factors (fermentation temperature 37°C, agitation rate 200 rpm, particles stuffing rate 25%, and pH 5 on ethanol yield and CO_2 weight loss rate) on bioethanol fermentation from stalk juice of sweet sorghum by Immobilized *S. cerevisiae* (CICC 1308). He concluded that the determined optimal condition was suitable and reasonable for the ethanol fermentation.

Bulawayo et al. (1996) explained the sugar conversion efficiency with the 10 strains of *Saccharomyces* sp. that produced ethanol from the sorghum juice or from yeast extract/phosphate/sucrose (YEPS) media. Bioethanol from sweet sorghum juice by immobilized *S. cerevisiae* with optimum inorganic salts K_2HPO_4 0%, $(NH_4)_2SO_4$ 0.2%, $MgSO_4$ 0.05% increased the ethanol yield to 91.61% (Liu et al., 2008).

In ethanolic fermentation of sweet sorghum juice using *S. cerevisiae*, when sucrose was used as an additional component, maximum efficiency of ethanol production with concentration, productivity, and yield of 120.68 ± 0.54 g/L, 2.01 ± 0.01 g/L/h, and 0.51 ± 0.00 g/g, respectively, was obtained. However, when sugarcane molasses was used, only 109.34 ± 0.78 g/L, 1.52 ± 0.01 g/L/h, and 0.45 ± 0.01 g/g, were obtained, respectively (Laopaiboon et al., 2007) (Fig. 4.11).

Jianliang et al. reported the ethanol production from sweet sorghum juice by immobilized yeast in optimized media conditions (Yu et al., 2009). They observed the maximum ethanol production rate (119.12 g/L h) for a medium consisting of 0.77 g/L phosphorus, 2.15 g/L nitrogen, and pH = 6.39. Under this condition, the ethanol fermentation rate was 122.85 g/L h.

There are two common categories in fermentation: (1) Separate hydrolysis and fermentation (SHF) where, enzymatic hydrolysis and fermentation are carried out separately. (2) Simultaneous saccharification

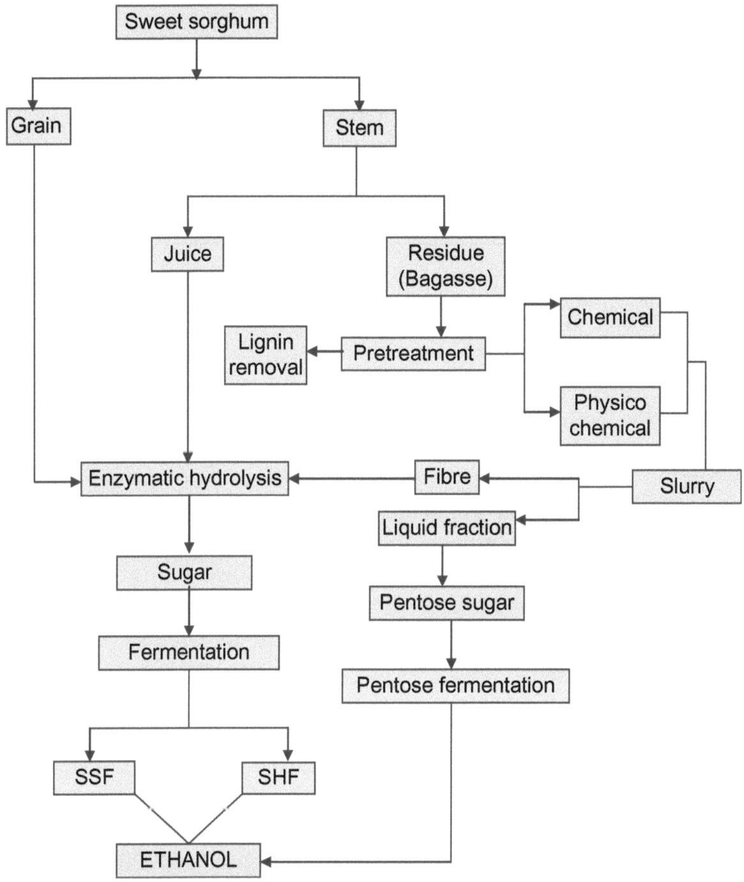

FIGURE 4.11 Bioethanol production from sweet sorghum.

and fermentation (SSF) where hydrolysis is done in the presence of fermentative microbes. SSF has advantages over SHF in that product inhibition is avoided as the sugars released in the hydrolysis are immediately utilized by the fermentative microbes present.

SSF reports in sweet sorghum include the studies conducted by Yu et al. (2010) for ethanol production from H_2SO_3-steam-pretreated fresh sweet sorghum stem. The conditions used for pretreatment were 10% (w/v) solid loading, acid dosage of 0.25 g/g DM H_2SO_3, and at 100°C for 120 min. At the fermentation reaction conditions of 37°C, pH 7.0, 180 rpm, 50 h, cellulose loading of 60 FPU/g DM and 5 g dry yeast cells/L, a high ethanol yield of 110% was achieved.

In a study by Sipos et al. (2009), they used the whole sweet sorghum plant as a feedstock for bioethanol production and concluded that both juice extracted from the stem and residual bagasse can be used for fermentation to ethanol. At 190°C, 10 min and 200°C, 5 min pretreatment parameters, 89% and 92% cellulose conversion of the separated washed fiber fraction could be achieved, respectively.

4.32 ETHANOL YIELD

Bioconversion of hexoses should theoretically yield 51.1% ethanol and 48.9% carbon dioxide. On this basis, every kilogram of glucose or fructose, sucrose, and starch would yield 510, 539, and 568 g or 646, 680, and 720 mL alcohol, respectively. Sucrose, glucose, and fructose are the fermentable sugars in sweet sorghum: it has been suggested to use an average of values of these sugars for computing the theoretical yield of alcohol. On this basis, every kilogram of sweet sorghum fermentable sugars would yield 663 mL ethanol. It is estimated that 5% of sugar is used by the yeast to produce new cells and minor products like acetic acid, lactic acid, glycerol, fuel oil, and a mixture of higher alcohols (Bothast and Destroy, 1981). Together with other losses associated with fermentation and distillation, 80% of the theoretical yield (530 mL) is considered practical in large-scale processing.

When this ethanol yield is scaled up to the unit area of sweet sorghum crop, other factors like extractable juice yield and amount of fermentable sugars in juice also influence the alcohol yield. Because of these variable factors, widely divergent yields of ethanol/ha have been reported from laboratory and microlevel fermentation studies. More frequently, reported ethanol yield ranged from 1200 to 5600 L/ha (Schaffert and Gourley, 1982) but yields as high as 7400 and 8700 L/ha have also been reported on the basis of stalk yield and stalk sugar analyses.

4.33 ALCOHOL FROM SWEET SORGHUM (NRCS, HYDERABAD)

Initially five sweet sorghum genotypes (Keller, SSV 84, BJ 248, NSSH 104, and Wray) were evaluated for total sugar, alcohol production, and FE using batch fermentation with *S. cerevisiae*. Higher amounts of total sugar per 100 mL of extractable juice were obtained from Keller. High juice extractability was observed with SSV-84 (34368 L/ha). Total sugar yield was found to be high in SSV-84 (7.35 Mg/ha) and Keller (4.66 Mg/ha). The highest alcohol yields of 4502.2 L/ha were observed in SSV 84, whereas the highest FE of 91.0% was observed with Keller.

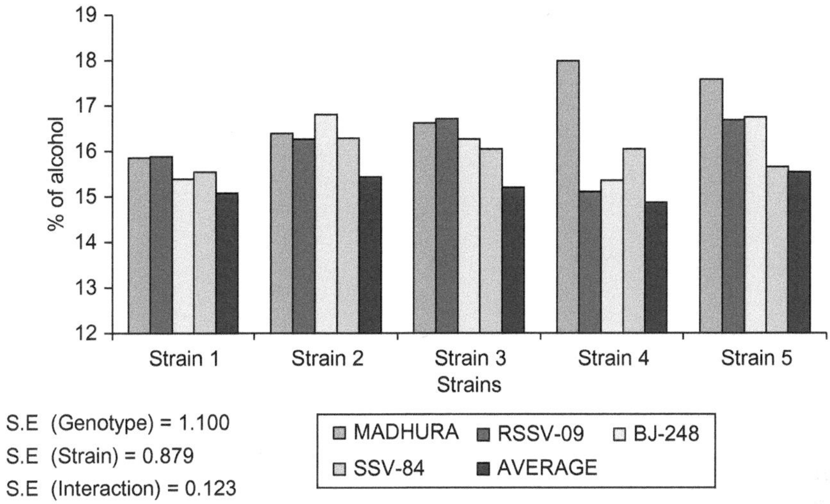

FIGURE 4.12 Percent alcohol production using five different yeast strains.

Ten strains of yeast (*Saccharomyces* sp.) were isolated from sorghum and evaluated for their alcohol production and tolerance to alcohol. Four strains were found to be superior for alcohol production, out of which one strain (Strain 2) had the highest alcohol tolerance (20%). These strains were evaluated for their performance in the production of alcohol and FE in comparison to the strain (*Schizosacchoaromyces pombe*) obtained from distilleries. Strains 2 and 3 recorded the highest percentage of alcohol production. Strains 1 and 2 have the superior mean FE (81.42% and 79.82% respectively) as compared to the FE of strain 5 (75.7%) (Distiller's strain) (Fig. 4.12).

4.34 RECOVERY OF ETHANOL AT DIFFERENT CROP GROWTH STAGES WITH DIFFERENT YEAST STRAINS (NRCS, HYDERABAD)

Ethanol was produced in laboratory with four different yeast strains, viz., (1) *S. cerevisiae*, (2) *Candida tropicalis*, (3) *Saccharomyces pombe*, and (4). *Cryptococcus albidus*, by fermenting the juice obtained at different crops' growth stages, that is, milky stage, physiological maturity, and harvesting stage or normal maturity stage. The study was conducted with 18 promising sweet sorghum genotypes. The total soluble sugars left out in the medium were also estimated after 72 h of fermentation.

At harvesting, high recovery of ethanol was obtained with all four yeast strains followed by the physiological maturity and milky stages,

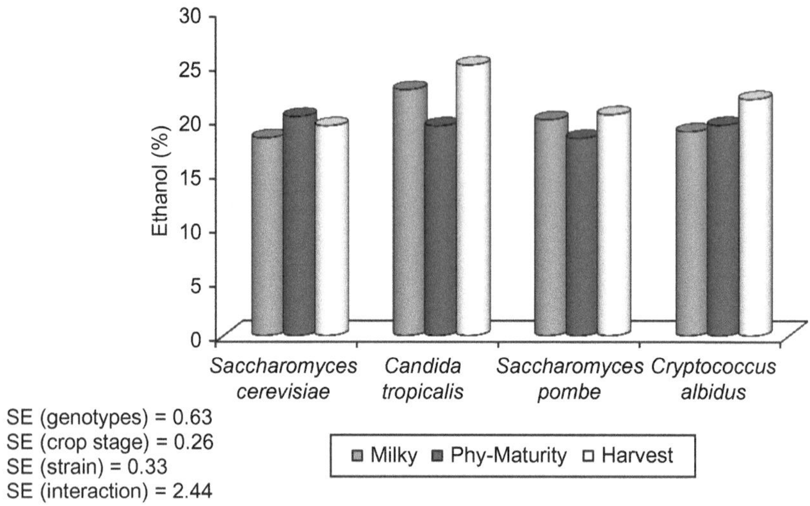

FIGURE 4.13 Recovery of ethanol in different crop stages.

respectively (Fig. 4.13). The yeast strain *C. tropicalis* and *C. albidus* have produced higher ethanol (25 and 21.9%) compared to the distillery strain *Saccharomyces pombe* (20%) and *S. cerevisiae* (20%). Among the genotypes, NSSV-21, NSSV-254, NSSV-216, and NSSV-219 from NRCS, and RSSV-59, RSSV-46, and RSSV-91 from MPKV Rahuri are promising for ethanol recovery.

Aldehyde is a by-product produced during fermentation. The production of aldehydes and fused oil should be minimal. In the fermentor experiment using two genotypes (NSSH 104 and BM 09), the percentage of aldehyde content was in the range of 0.0005 to 0.0008%. The lowest amount of aldehyde content was reported in strain Svt after 48 h of fermentation. The content of aldehyde in alcohol from sweet sorghum is minimal.

4.35 TOTAL ALCOHOL RECOVERY (STALK AND GRAIN) IN SWEET SORGHUM (PDKV, AKOLA)

Advance sweet sorghum genotypes from breeding performance trials were evaluated for the total alcohol recovery from sweet sorghum stalk juice and grain. The data presented in Table 4.5 show that the genotypes RSSV-91 (1709.28 L/ha) and NSSV-218 (1572.73 L/ha) were found to be highest in recovery of alcohol from stalk juice; however, maximum grain alcohol was obtained from NARISSH-43 (2014.65 L/ha) followed by AKSSV-21 (2006.54 L/ha) and NSSV-219 (1969.51 L/ha). As regards of total alcohol yield from stalk juice and

TABLE 4.5 Average Alcohol Yield (L/ha) From Advanced Sweet Sorghum Genotypes

Cultivars	Alcohol yield from stalk juice (L/ha)	Alcohol yield from grain (L/ha)	Total alcohol yield (L/ha)
RSSV-24	684.61	1021.63	1706.24
RSSV-44	670.77	1729.69	2400.46
RSSV-45	490.72	783.6	1274.32
RSSV-46	890.69	1720.87	2611.56
RSSV-57	1018.91	1009.72	2028.63
RSSV-58	987.37	1635.23	2622.6
RSSV-59	1439.82	877.59	2317.41
NSSV-216	568.54	1547.13	2115.67
NSSV-218	1572.73	1532.43	3105.16
NSSV-219	857.51	1969.51	2827.02
RSSV-91	1709.28	1452.07	3161.35
RSSV-106	1275.38	1286.08	2561.46
RSSV-120	877.6	1359.89	2237.49
NSSV-13	994.12	767.33	1761.45
NSSV-253	1091.47	950.81	2042.28
NSSV-254	1502.55	967.99	2470.54
NARISSH-43	1122.17	2014.65	3136.82
AKSSV-22	1492.66	1862.01	3354.67
AKSSV-21	868.61	2006.54	2875.15
SSV-84	1020.07	854.16	1874.86

grain AKSSV-22 recorded highest recovery (3354.67 L/ha) followed by RSSV-91 (3161.35 L/ha) and NARISSH-43 (3136.82 L/ha) (Table 4.5).

4.36 PILOT SCALE EVALUATION OF ETHANOL PRODUCTION FROM SWEET SORGHUM STALK JUICE (NRCS, HYDERABAD)

A pilot study was conducted during the year 2002 with the Sri Renuka sugar factory (Fig. 4.14) located at Munoli, Belgaum District, Karnataka, to determine the feasibility of ethanol production from sweet sorghum

FIGURE 4.14 Sri Renuka sugar factory located at Munoli, Belgaum District, Karnataka.

juice. Two varieties of sweet sorghum (SSV 74 and SSV 84) and one hybrid (Madhura) were grown during *Kharif* 2002 in more than 600 acres of dryland in and around the factory. In 2002, being a drought year, green cane yield and grain yield were 25 tons/ha and 2 tons/ha respectively. The juice Brix was 18% and brought down to 12% for fermentation. About 112 tons of cane was used for one fermentor of capacity 60 kL. Fermentation was carried out for 48 h. The recovery of ethanol was approximately 9% of the juice. The bagasse (46.44% of stalks with 2.58% sugars) obtained was used for the cogeneration of electricity. The comparative economy of ethanol production is presented in Table 4.6.

In another pilot study, carried out with Sagar Sugars and Allied Products at Nelavoy in Chittoor District of Andhra Pradesh, SSV 84 was grown in *Kharif* 2004 on about 378 ha. In Nallatur village in Tamil Nadu, SSV 84 was sown during the last week of May and harvested during the first week of October. The green cane yield and grain yield were 40 tons/ha and 2 tons/ha, respectively. However, the diffuser machinery was listed for sweet sorghum in *Rabi* season and it was tested suitable.

TABLE 4.6 Comparative Economy of Ethanol Production

S. no.	Item	Sweet sorghum	Molasses
1.	Manpower	0.5	0.25
2.	Steam	1	1
3.	Electricity	1	1
4.	Yeast	0.1	0.1
5.	Management/Admin	0.1	0.25
6.	Pollution Control	Nil	0.25
7.	Raw material	10.41	12.13
	Total	13.11	14.98

4.37 SWOT ANALYSIS ON PRODUCTION OF ETHANOL FROM SWEET SORGHUM (NRCS, HYDERABAD)

A SWOT analysis was performed for the technology on the strengths, weaknesses, and opportunities for using sweet sorghum for ethanol production.

Following are the strengths of sweet sorghum: minimum irrigation (2 to 3 times) requirement for sweet sorghum cultivation, minimum requirement of purchased inputs, short duration (4 months) compared to sugarcane of 12 months, possibility of mechanizing the harvest of sweet sorghum (to mitigate labor demand at peak harvest period), clean grain yield, superior fodder and by-product yield compared to sugarcane, additional employment to farm women, superior quality of ethanol that is low in aldehydes and free from sulfates, high biological value of bagasse to use as animal feed and cogeneration of electricity from bagasse.

The factors rendering weaknesses are inadequate seed supply, inadequate extension efforts for creating awareness about the advantage of sweet sorghum among farmers, unclear picture of industry's demand, lack of suitable package of practices for cultivation of sweet sorghum in different agroclimatic zones, less labor efficiency in harvesting compared to sugarcane, high transportation costs due to light weight stalks, difficulty in de-trashing of leaves before crushing, time-consuming stalk crushing process, lack of suitable industry-farmer linkages, and weaker marketing side aspects.

Opportunities for sweet sorghum are as follows: sweet sorghum can be cultivated with less water to utilize installed capacity of sugar mills effectively. Advantage of widening the raw material base from single source (sugarcane molasses) to multiple feedstocks to impart stability in ethanol production expected sharp rise for ethanol demand for blending with petrol/diesel, saving of forex (Rs. 830 crore/year) as it reduces the import of crude oil and less effluent problem.

The threats considered are (1) import of cheaper ethanol from foreign countries, (2) promoting sweet sorghum as supplementary rather than a substitute to sugarcane molasses, and (3) exploitation of marginal and small farmers by the industry, etc., could undermine the prospects of sweet sorghum in its industrial utilization. Sugar beet with higher yield potential, high sugar yield, and medium crop duration can also be a threat to sweet sorghum.

In the wake of many industries coming forward to use Sweet sorghum for the production of biofuel, NRCS and our cooperating center MPKV, Rahuri have taken up large-scale seed production of the state released variety SSV 84 in an area of 4 ha and 0.5 ha, respectively, during *Rabi* 2004. A total of 13.5 tons of seed from NRCS and 1 ton from MPKV, Rahuri were supplied to several industries located at Andhra Pradesh, Tamil Nadu, and Maharashtra.

A pilot experiment on biofuel production from sweet-stalked sorghum was extended to two major sugar factories located in Andhra Pradesh and Maharashtra (Sagar Sugars and allied products, Nelavoy, Chittoor District, (Fig. 4.15) and Godavari sugars Limited, Shakarwadi) to work out the feasibility and farmers' acceptance during *Kharif* 2004.

FIGURE 4.15 Harvesting of sweet sorghum cane at Chittoor District.

4.38 ETHANOL PRODUCTION FROM DAMAGED GRAIN

Laboratory evaluation studies on grain alcohol from eight genotypes at Seagram Distilleries Ltd, Nasik, showed that CSH-16 is a promising hybrid for the grain alcohol production.

In India, 480 sugar factories produce 1.3 billion liters from the installed capacity of 2.5 billion liters using molasses as the chief raw material (substrate for 97% of the potable alcohol production). The gap in capacity utilization is mainly because the seasonal availability of molasses (raw material) is restricted to 6 months (Yadnik and Gadkari, 2003). The molasses production came down from 9.0 million tons in 2002–2003 to 6.5 million tons in 2003–2004 exerting pressure on the ethanol industry. Simultaneously, the molasses price skyrocketed to Rs. 5000/t representing an increase of 400% during 2004 (AMB, 2004). The trend in state-wise production of sugarcane and molasses is shown in Tables 4.7 and 4.8, which indicate a highly fluctuating trend. Table 4.9 shows the fluctuation in availability of molasses for alcohol production, which may adversely affect the alcohol-based industries, such as polymers, rubber, pesticides, plastics, dyes, and synthetic fiber industries. The increasing molasses price has a potential negative impact on the industrial sector in terms of the availability of the ethanol. It is imperative to keep the uninterrupted supply of the industrial alcohol for its smooth functioning. Diversion of the molasses alcohol for potable alcohol production makes it unavailable for the important industrial purposes and adversely affects alcohol industries. Moreover, in the wake of the new policy of the government of India to make it

TABLE 4.7 State-Wise Sugarcane Production in India During 1995–2000

S. no.	State	95–96	96–97	97–98	98–99	1999–2000	2000–01	2001–02	2002–03
1.	Andhra Pradesh	15.12	15	14	16.5	18.7	17.69	18.08	15.39
2.	Gujarat	10.5	11.4	11.8	13.6	14.1	1269	12.46	14.7
3.	Karnataka	24.9	23.4	28.3	34.8	36.5	43.25	33.02	32.48
4.	Maharashtra	469.7	41.8	38.2	47.2	53.1	50.22	45.14	37.02
5.	Tamil Nadu & Pondicherry	32.9	25.9	30.2	33.8	35.1	34.06	32.62	30.28
6.	Uttar Pradesh	11.98	125.3	129.3	116.5	115.4	106.9	117.98	116.38
	All-India	281	277.3	279.7	288.8	299	299.21	297.21	281.57

From Agricultural Statistics at a Glance, 2003 and 2004. Various Issues of 'Cooperative Sugar', vol. 34, February 2003.

TABLE 4.8 State-Wise Molasses Production in India During 1995–2000

S. no.	State	95–96	96–97	97–98	98–99	1999–2000
1.	Andhra Pradesh	0.418	0.323	0.369	0.506	0.513
2.	Gujarat	0.505	0.418	0.379	0.473	0.515
3.	Karnataka	0.581	0.345	0.369	0.556	0.542
4.	Maharashtra	2.168	1.228	1.354	1.921	2.31
5.	Tamil Nadu & Pondicherry	0.937	0.544	0.698	0.912	0.868
6.	Uttar Pradesh	2.699	2.195	1.879	1.971	2.408
	All-India	8.285	5.936	5.607	6.977	7.923

From Various Issues of 'Cooperative Sugar', vol. 34, February 2003.

TABLE 4.9 Production of Molasses and Ethanol in India During 1996–2003

S. no.	Year	Molasses production (million tons)	Molasses available for distillation (million tons)	Alcohol production (million liters)
1.	1996–97	5.93	5.34	1201.9
2.	1997–98	5.54	4.99	1123.2
3.	1998–99	6.97	6.28	1413
4.	1999–2000	8.16	7.35	1654
5.	2000–01	8.32	7.49	1686.9
6.	2001–02	7.87	7.08	1594.7
7.	2002–03	8.32	7.49	1685.8

From National Conference on 'Prospects of Ethanol as Sustainable Fuel' Held at Hyderabad, September 2003 by APITCO Ltd.

mandatory to blend ethanol with petrol, demand for ethanol production would soar and molasses is expected to completely cater to the needs of ethanol production for biofuel, neglecting potable alcohol production. Severely damaged sorghum grain is neither preferred for food nor for animal feed. Thus the sorghum grain especially the blackened one can find a niche here. In 2001–2002, about 0.092 million tons of sorghum grain was used for alcohol production and it is projected to increase to 0.22 million tons in 2010, based on assumption of the annual potable alcohol production growth of 10% (Dayakar Rao et al., 2003). It is reported that 33% of the total alcohol is produced from grain sources alone; typical grain sources include corn, wheat, and sorghum

(Praj Industries, 2004). A few past studies have proved that sorghum could form a supplementary or alternate source of raw material for these industries.

In light of the above discussion, we believe there is enormous scope for this grain to be used as a valuable feedstock for potable alcohol production, besides its traditional use as a food, feed, and fodder. The present work examines the importance, utility, techno-economic feasibility, and scope of sorghum grain for ethanol production.

4.39 SORGHUM GRAIN-BASED POTABLE ALCOHOL – GLOBAL EXPERIENCES

In the USA, the world's major sorghum grain producer, about 12% of the total sorghum grain is absorbed by the ethanol industry (AMB, 2004). It is being attempted to utilize even the chemically treated sorghum seed grain as an ethanol feedstock as the disposal of the bulk of chemically treated seeds poses environmental hazards with the intact chemicals and is also a costly affair. The chemically treated sorghum grains could be safely used as an ethanol feedstock, and ethanol did not contain traces of the seed-treating chemicals. However, the stillage of this feedstock contains toxic chemicals higher than permissible limits and cannot be used as animal feed. The traditional opaque beer and clear beer made from sorghum grain are popular in African countries (Nigeria, Burkino Faso, Ghana, Mali, Uganda, Zimbabwe, Togo, and South Africa), whereas sorghum wine and various types of distilled spirits (alcohol content of 20–65%) are consumed in China. The well-known distilled spirits (eg, Maotai, Fenjiu, and Wuliangye) of China are produced with varying amount of sorghum grain (Smith and Frederiksen, 2000). The nonalcoholic fermented beverage, Kunun-zaki, is produced in Nigeria using Red Guirita sorghum which is used both as malt and adjunct (Taylor and Belton, 2002). The major breweries in Mexico, Africa (especially Nigeria and South Africa), and Asia, use sorghum grits as an inexpensive source of fermentable carbohydrates in brewing barley beer. The red coloured sorghum is preferred, but white sorghums with tan plant color and straw-colored glumes are preferred due to their improved milling properties. In Nigeria, sorghum and/or maize grits, some with high oil content, are used with industrial enzymes to produce a clear, lager beer.

Worldwide consumption of alcoholic beverages averages is about five liters of pure alcohol per person per year, and has been relatively constant due to higher production rates, although consumption levels are rising in many developing countries. Importantly, the data on developing countries most likely underestimate actual alcohol consumption, since they generally ignore artisanal production and smuggled beverages (WHO, 2004).

4.40 DYNAMICS OF ETHANOL UTILIZATION IN VARIOUS DEMAND QUARTERS

Ethanol has three distinct end uses as potable alcohol, biofuel, and other industrial utilities. However, the most important single factor that determines the intersectoral allocation of molasses is the price of the molasses and the economic value of the corresponding end-product. Obviously, the potable alcohol production pays off a high premium; therefore, this industry is absorbing sizeable quantum of molasses as raw material. Temporally, it is observed that the price of the molasses against the price of potable alcohol/industrial alcohol/ biofuel mix dictates the allocation terms of the molasses between different uses. Therefore, during the peak supply period, ethanol from molasses is diverted to fuel alcohol production that pays more. It is observed (Table 4.10) that molasses production deviates considerably from the average production over the years, whereas the deviation in production of sorghum is comparatively less (ie, almost half of that value) using a measure of standard deviation. It is obvious that the potential raw material should have a flat production trend rather than a fluctuating one.

TABLE 4.10 Statistical Deviation in Raw Materials Production During 1995–2003

State	Molasses[a]	*Kharif* sorghum production[b]
Andhra Pradesh	83.5	59.47
Gujarat	58.14	55.32
Karnataka	112.2	85.26
Maharashtra	483.78	205.87
Tamil Nadu	166.98	83.01
Uttar Pradesh	332.84	60.68
All-India	1179.13	608.36

From Agricultural Statistics at a Glance, 2004. Various Issues of 'Cooperative Sugar', vol. 34, February 2003.

[a]*Molasses price data for 1995–2000 as the latest available data (except for All-India for which 1995–2003 data is available).*

[b]Kharif *sorghum production data during 1995σ2003 is the only latest available data.*

4.41 FEEDSTOCKS FOR MANUFACTURING OF ETHANOL—AN OVERVIEW

Although sugarcane is the known ethanol feedstock, other sources such as corn, cellulose, grains, and tubers can also be used as feedstock for ethanol manufacturing. There are three basic raw materials groups under which almost entire range of ethanol feedstocks could be classified.

4.41.1 Saccharine

Simple carbohydrates in the form of 6- or 12-carbon sugar molecules. Sugarcane, sugar beets, fruits, molasses, and sweet sorghum are classified under this category.

4.41.2 Starchy Materials

Have complex carbohydrates such as starch but which could be broken into simple sugars by acid hydrolysis or by enzymatic action. Feedstocks such as corn, sorghum, barley, potatoes, and sweet potatoes, etc., fall under this category.

4.41.3 Cellulose Materials

Using either acid or enzymatic hydrolysis, the complex structure of these materials is broken down to simple sugars. Wood, wood-waste, paper, straw, corn cobs, corn stalks, cotton, etc., are examples.

Table 4.11 shows the alcohol yield as a percentage of total fermentable sugars content. The sorghum grain feedstock has the higher FE (91%) compared to the 88% of molasses feedstock. It highlights the superiority of sorghum grain feedstock on a FE basis.

TABLE 4.11 Comparative Study of Fermentation Efficiency of Ethanol Feedstocks

S. no.	Feedstock	Fermentable sugar (kg/t)	Maximum alcohol yield (Kg/t)	Fermentation efficiency (%)
1.	Sorghum grain	610.0–660.0	300–325 (370.0–407.01/t)	91.0
2.	Molasses	450–500	229.95–255.5 (29323.1/t)	88.0

From Somani, R.B., 1996. Project appraisal titled 'An over view of the potential of the sorghum and millets for industrial uses in India'. Agro Products Development Research Center, PDKV, Akola, Maharashtra, India, pp. 269.

4.41.4 Grain-Based Ethanol Production Technology

Alcohol can be produced from any starchy cereal grain by hydrolyzing the grain starch by a combination of enzymes. The hydrolysate containing the free sugar is fermented to produce alcohol.

Ethanol production from cereal grain typically involves processing steps such as cleaning and milling of grains, hydrolysis of starch to release fermentable sugars, followed by inoculation with yeast. The grain starch is broken down to simple glucose by a combination of two enzymes, viz., amylase and amyloglucosidase, before adding yeast for fermentation. Thus, the alcohol produced is distilled from fermented broth. The remaining broth is processed to produce distillers dried grain and solubles (DDGS).

4.42 TECHNICAL SUITABILITY OF MOLDED SORGHUM GRAIN IN ETHANOL PRODUCTION

For alcohol production, grains with high starch content (65–72%) and low protein content (8–12%) are desirable. The preferred grain quality requirements for producing the distilled spirit are 65–70% starch content, of which above 80% should be amylopectin, 7–9% protein content, and 0.3–0.5% tannin content. Rain-damaged sorghum grain (molded grain) is usually blackened, fetching low market price. However, the percentage degradation of starch in the molded grain is not significant. Further, the degradation is mostly confined to the simple forms of sugars and therefore does not affect the recovery of alcohol. A study by Ratnavathi et al. (2003) revealed that among the six selected cultivars, SPV-462 and CSH-14 were found to be more desirable for ethanol production as its starch content and total sugar content remains high with less standard deviation values even after severe mold infestation (Table 4.12). Other desirable cultivars include CSV-13 and SPV-462. However, the percentage reduction in starch is reflected by the total soluble sugars (%). Therefore, the percentage recovery of alcohol does not get affected due to grain mold. On the other hand, recently released cultivars CSH-18 and CSH-16 recorded the highest alcohol recovery percentage. The study also reported that grains of sorghum hybrids had a relatively higher alcohol yield compared to varieties.

The high quantity of by-product generated is a positive aspect of sorghum usage in alcohol industry. The by-product contains mostly protein, fiber, and added yeast, which can be dried to make a protein supplement in cattle feed. During the fermentation process, the grain starch gets utilized and the other components such as cellulose and protein remain unutilized in the stillage. The dried stillage is not only rich in protein and fat but also has vitamins produced by the yeast fermentation.

TABLE 4.12 Proximate Composition of Molded Grain and Normal Grain

	Starch (%)			Total soluble sugars (%)		Fat (%)	
Cultivar	**Normal grain**	**Molded grain**	**Mold score**	**Normal grain**	**Molded grain**	**Normal grain**	**Molded grain**
CSH 14	70.8 (0.106)	68.4 (0.212)	4	1.48 (0.06)	2.25 (0.116)	1.78 (0.19)	1.67 (0.05)
CSH 16	69.9 (0.53)	66.3 (0.509)	3.5	1.98 (0.049)	2.29 (0.063)	3 (0.33)	2.2 (0.04)
CSV 11	69.9 (2.25)	67.5 (1.87)	4	1.27 (0.116)	2.34 (0.088)	2.6 (0.167)	1.8 (0.171)
CSV 13	71.7 (2.36)	68.3 (2.83)	3.3	1.09 (0.115)	1.82 (0.084)	3 (0.211)	2.81 (0.206)
CSV 15	68.7 (0.212)	65.4 (1.484)	3	1.89 (0.017)	2.78 (0.038)	1.84 (0.23)	1.5 (0.24)
SPV 462	69.9 (0.954)	68 (0.424)	03:03	2.04 (0.049)	2.78 (0)	2.5 (0.23)	1.84 (0.23)
Mean	70.1	67.3	3.5	1.62	2.37	2.45	1.97

Note: Figures in the parentheses indicate standard deviation

TABLE 4.13 Typical Ethanol Yield From Various Feedstocks

Raw material	Ethanol yield (L/t)
Sorghum	422–448
Black jowar (molded sorghum)	390–435
Broken Rice	420–445
Corn	370–410
Wheat	375–425

From Praj Industries Ltd, 2004. Cereals in alcohol industry-industrial perspectives. In: Presented by Praj Industries Ltd, Pune at International Conference on Biofuels, Winrock International India, New Delhi, September, 2004.

The recovery of the stillage (DDGS) is about 240–260 kg/ton of grain flour used for fermentation. It is reported by Miller and Creelman (1980a,b) that the fermentation of the sorghum grain results in a high quality distills grain, which when dried has a protein content of approximately 30%. However, the stillage is equally contaminated with mycotoxins and level of toxins should be checked before its use as feed.

It is shown in Table 4.13 that sorghum grain offers an excellent ethanol feedstock with a high alcohol recovery of 422–448 L/t compared to

TABLE 4.14 Bio-Chemical Characteristics of Different Grain-Based Ethanol Feed Stocks

S. no.	Contents (% w/w)	Sorghum	Blackened jowar (molded sorghum)	Maize
1.	Moisture	9.0–13.0	9.0–13.0	8.0–12.0
2.	Total solids	87.0–91.0	87.0–91.0	88.0–92.0
3.	Starch	66.0–70.0	60.0–64.0	62.0–66.0
4.	Proteins	9.0–11.0	9.0–11.0	7.0–11.0
5.	Fats/oils	2.2–3.2	2.2–3.2	3.5–5.0
6.	Crude fibers	1.3–2.1	1.3–2.1	1.5–2.0
7.	Inorganic ash	1.2–2.0	1.2–2.0	1.0–1.5
8.	Other organics	8.0–12.0	8.0–12.0	7.0–9.0

From Praj Industries Ltd, 2004. Cereals in alcohol industry-industrial perspectives. In: Presented by Praj Industries Ltd, Pune at International Conference on Biofuels, Winrock International India, New Delhi, September, 2004.

other feedstocks listed. The higher ethanol yield can be attributed to its relatively higher starch (60–70%) and total solids (9–11%) contents of sorghum grain compared to other feedstocks (Table 4.14).

4.43 ECONOMIC PROSPECTS OF ALCOHOL PRODUCTION FROM SORGHUM GRAIN

Based on our industrial survey, the cost of production of ethanol from molasses as raw material and from the alternate source of grain sorghum has been compared (Table 4.15). The cost incurred on steam/coal and electricity components and high water requirement escalates the cost of ethanol-based on sorghum grain. The sum of the variable cost excluding raw material cost, going in to the ethanol production, is almost at a par for both sources (ie, Rs. 7.50/L for molasses and Rs. 7.20–8.60/L for sorghum grain). However, the comparatively lesser cost of the raw material of sorghum grain makes the grain-based ethanol cheaper by 19–31% to that of molasses-based ethanol. Considering the difference between costs of production of these two raw materials, sorghum grain seems to have a price advantage over molasses. However, it is not intended to replace the molasses entirely; rather molasses is to be supplemented with sorghum grain as an alternate source of raw material for alcohol production. Moreover, the alcohol

TABLE 4.15 Cost of Production of Ethanol From Sorghum Grain and Molasses

		(Rs. per liter of ENA)	
		Ethanol based on	
S. no.	**Particulars**	**Sorghum grain**	**Molasses**
1.	Manpower	1.00	1.00
2.	Steam/coal	1.50–1.75	1.25
3.	Electricity	2.00–2.50	2.5
4.	Yeast/enzymes	1.20–1.35	0.25
5.	Water utilization/maintenance	1.50–2.00	0.5
6.	Pollution control	-	2
7.	Total cost (A)	7.20–8.60	7.5
8.	Raw material (B)	8.97–10.25*	15.91*
9.	Gross cost of production (A + B)	16.17–18.85	23.41

*The cost of raw material is calculated as under:

1. Price of sorghum grain @Rs. 3500–4000/ton and recovery of 390L/ton = Rs. 8.97 to 10.25/L of ENA.
2. Price of molasses @Rs.3500/ton (including central excise duty and 13% sales tax) and recovery of 220L/ton = Rs. 11.36/L of ENA.

produced from sorghum grain is used for making value-added potable liquors by virtue of its superior quality and low sulfates and aldehydes content.

A study conducted by Praj Industries Ltd showed that with the average ethanol yield of 410 L/t, the blackened sorghum grain feedstock costs only Rs. 7.3/L of ethanol production. However, the normal sorghum grain feedstock cost little higher (Rs. 11.5/L of ethanol production) with a slightly higher ethanol yield of 435 L/t. Thus, the economics of alcohol production from sorghum grain depends on factors such as costs of production, crop yield, and grain and alcohol prices, which is again related to factors such as supply and demand of the grain.

4.44 EFFLUENT/WASTE WATER TREATMENT FOR GRAIN ALCOHOL DISTILLERY

The wastewater from grain-based distillery has lesser polluting matter compared to molasses-based distillery's effluent. The prerequisite step of hydrolysis of grain starch with costly enzymes and heat makes the molasses alcohol production cheaper and simpler. However, the treatment of the effluent of the molasses alcohol industry is difficult and costly, leading to environmental pollution. Considering that molasses contains 45% sugars, only about 70–75% of the sugars get fermented to alcohol while the remaining sugars go into spent wash. Therefore, the effluent has BOD as high as 40–60 g/L. Most of the spent wash is being disposed without proper treatment. Though the modern wastewater treatment plant efficiently reduced both the BOD and COD, the effluent remains black in color and has a foul odor. On the other hand, the grain feedstock has 60% starch of which 85% of sugars are converted to alcohol. Moreover, the remaining sugars do not go into effluent as the stillage gets dried (Sheorain et al., 2000). The grain distillery wastewater, called "thin slops," can be treated in two ways. A decanter is used to separate the suspended solids in the effluent initially which is then sent for either biomethanation or evaporation in case of DDGS production.

4.45 BIOMETHANATION, AERATION, AND FERTI-IRRIGATION

Biomethanation involves anaerobic conversion of the organic compounds in the effluent to methane gas and carbon dioxide by a consortium of bacteria. The process achieves approximately 85% of BOD and 65% of COD reduction in the effluent. The biogas production for 40,000 L/day grain-based distillery is in the range of 4000–4500 Nm'/day. Aeration further reduces the BOD levels to less than 100 ppm enabling the safe disposal by ferti-irrigation to the land.

4.46 VALUE-ADDED PRODUCT: DDGS

The concentrated effluent from evaporation section is fed to the dryer and the cake removed in the decantation section is mixed with the concentrated effluent. The effluent is then dried and made in the form of solid powder. This can be converted into pellets and marketed as animal feed.

4.47 MARKETING CHAIN OF SORGHUM GRAIN

The movement of sorghum grain starting from farmer's fields to the distillery has a high degree of fragmentation. The first link in the chain is a consolidator at the village level who collects the produce from the farmers. The consolidator in turn sells the produce to the commission agent. The agent passes on the produce to a trader who operates at the district level and who handles the large quantities of produce acquired from a number of agents. The brokers based in the cities, such as Akola, Nanded, and Jalgaon, act as additional intermediaries between large traders and the distillery units. Thus, too many intermediaries are involved in the system of transaction, which hinders the benefits to the producers and distillers. However, the inefficiencies can be transformed into a huge opportunity by having backward linkage with the farmers.

4.48 ADVANTAGES ACCRUING FROM BACKWARD INTEGRATION

The following are the benefits of contract farming with backward integration in the case of sorghum grain production for alcohol industry, accruing to both farmers and distillers.

To distillers

1. Assured supply of raw materials with assured quality
2. The search and information costs for the right quality can be taken care by contract farming
3. Demand of domestic and export market can be met by having control over production

To farmers

1. Assured market and support price
2. Assured quality of seed
3. Better realization of prices without middlemen

4.49 SWOT ANALYSIS OF SORGHUM GRAIN AS RAW MATERIAL FOR POTABLE ALCOHOL

A SWOT analysis on the prospects of sorghum grain as ethanol feedstock is made and presented below.

4.49.1 Strengths

The utilization of sorghum grain in ethanol industry has strengths as discussed below:

- The rain-damaged, blackened, less market-valued, *Kharif* sorghum grain, unfit for human consumption can be used as feedstock serving twin objectives of low cost raw material and creating market value for grains which otherwise have only a few takers.

- Sorghum grain compares favorably with molasses as a feedstock for ethanol production. Sorghum grain has an economically desirable price advantage over the molasses raw material when alcohol is sold.
- The sorghum grain raw material is available throughout the year and can be easily stored against the seasonal and fluctuating availability of molasses raw material.
- The quality alcohol from sorghum grain is superior as it has less aldehydes and is free from sulfates.
- The effluent from sorghum grain-based ethanol unit causes less environment pollution, while on the other hand the molasses-based ethanol unit is environmentally risky as it produces large quantity of effluent with high biological oxygen demand (BOD).
- The silica content is less in sorghum grain than that of broken rice raw material.
- A strong nonregulated market is catering for the growing demand for sorghum grain more than the regulated market channel.

4.49.2 Weaknesses

Sorghum grain's usage as an ethanol feedstock is fraught with the following weaknesses, and they are interlinked with threats.

- The occurrence of the grain mold in a *Kharif* production season determines the availability of the low-priced, mold-affected sorghum grain for the ethanol industry. In other words, grain mold occurrence is strongly correlated to the economic supply of feedstock. The grain mold occurrence is a regular phenomenon.
- The prerequisite of the processing of the grain (hydration, gelatinization, and enzymatic/acid hydrolysis of grain starch) before fermentation requires additional equipment and machineries, which put a burden on the huge initial capital investments.
- Solid content is greater in sorghum grain as ethanol feedstock compared to that of rice.
- Sorghum grain may not have comparative advantage in nonsorghum growing areas.
- There is an absence of effective supply chain management for sorghum grain.
- In India, trials on proven technologies for the use of sorghum in different industries have not been carried out on a large-scale in order to convince the industry and farmers.
- Though it is argued that the sorghum grain-based ENA offers superior quality potable alcohol, the price realization from its sale does not match expectations of the industry. In the absence of the policy insisting to strictly use cereals-based ethanol for

potable alcohol production, molasses is still continuing as the major raw material for the potable liquor sector though it is inferior for the consumer's health view point.

4.49.3 Opportunities

There are several real and perceived opportunities, for usage of sorghum grain in ethanol industry, and are listed below.

- The continuously declining consumption demand for *Kharif* sorghum is expected to continue in the foreseeable future, this offers opportunity for its future continuance as potential feedstock. Moreover, with the increase in per capita income and the rising standard of living, the demand for quality potable liquor, which could be served with sorghum grain-based ethanol rather than the current molasses-based ethanol, will increase. Moreover, the production of potable alcohol from molasses is paying more than that of either fuel alcohol or industrial alcohol. Potable alcohol fetches a price ranging from Rs. 18 to 36/k as against the fuel alcohol price of Rs. 22/ha. Hence, during the peak supply period, ethanol from molasses is diverted to cheaper industrial alcohol production, whereas at peak demand period the molasses is used for potable alcohol production that fetches more.
- The dwindling supply of molasses feedstock and its increasing price necessitate the seeking of supplementary raw material such as molded sorghum grain.
- The expected large demand for ethanol for blending with petrol from transport sector, initialized with the GOI notification on its execution in a phased manner in India, the demand will be met if excess molasses is allotted for fuel-grade ethanol production at the expense of potable alcohol production. Thus, the gap in potable alcohol production can be effectively plugged by use of sorghum grain (even low cost, rain-damaged blackened *Kharif* sorghum grain). The general rise in the income of the population and standard of living increases the demand for superior quality alcohol where sorghum grain can find a niche.
- Encouraging the sorghum grain-based ethanol production helps Indian made whiskey to be classified as whiskey as per European standards.
- Usage of sorghum malt as an adjunct in brewing industry can be exploited.
- Usage of blackened *Kharif* sorghum grain in ethanol units attached to the existing sugar factories will help to mitigate the problem of utilization of the installed capacity for only a part of a year because of the seasonal availability of the molasses raw material and helps to run factories for most part of the year.

4.49.4 Threats Possible Solutions

The role of sorghum as ethanol feedstock is being threatened by many factors, and if these threats are not taken care, they could pose serious challenges in future. They include:

- Poorly informed industry and farmers on the potentiality of the sorghum grain in its industrial uses. Extension efforts to popularize its suitability in industrial application may also hamper its prosperity in the industrial arena.
- Gradual decline in importance of sorghum grain, especially the *Kharif* grain, as government policies are in favor of fine cereals. This could ultimately affect the *Kharif* sorghum production.
- Import of cheaper ethanol from other countries could prove disastrous to Indian ethanol industry. Therefore, imported ethanol should be subject to suitable duties so that domestically produced ethanol will not be costlier than the imported one.

4.50 NEW INITIATIVES

As a step closer to industry-institute collaborative efforts to address industrial needs, Seagram Distilleries, (Nasik, Maharashtra) came forward to collaborate for the pilot experiments on grain-based alcohol from *Kharif* sorghum hybrids. There is also a proposal from the industries' side to procure grain from the farmers' fields directly at a pre-fixed price with the active support of State Department of Agriculture, State Agricultural Universities (SAU's) located in Maharashtra and other neighboring states like Madhya Pradesh.

4.51 CONCLUSIONS

With the demand for ethanol looming large both in the industrial as well as in the consumption sector, the current raw material sugarcane molasses has to be supplemented with suitable raw material, and this is where the molded sorghum grain can find a niche. Moreover, sorghum grain-based ethanol industry will play an important role in increasing the local competition for sorghum grain thus potentially improving the prices paid to farmers. The economically comparable cost of production of ethanol from molded sorghum grain (Rs. 16.17–18.85/L) with that from molasses raw material (Rs. 23.41/L) adds strength to our argument. However, problems such as inadequate supply of molded grain (grain flowing to industry at an acceptable price only when severe molding occurs), problems in large scale pooling, etc., could be tackled with efforts from concerned agencies. Moreover, sorghum is largely cultivated in rainfed conditions where grain mold is a recurring phenomenon. Encouraging large-scale usage of molded sorghum grain in the ethanol industry will benefit both farmers as well as industry: an assured market and remunerative price for farmers for their molded sorghum grains and alternative cheap feedstock supply on a continuing basis for the industry. Hence, creating industrial demand for mold-affected sorghum grain could be a blessing to the poverty stricken poor farming community in these regions.

4.52 RECOMMENDATIONS ON THE INDUSTRIAL USES OF MOLDED SORGHUM GRAIN

The following actions are suggested for the better utilization of sorghum grain in ethanol industry.

1. As the molded sorghum grain is yet to be fully exploited for its usage in the ethanol industry, efforts need to be strengthened to popularize the suitability and economic viability of this coarse grain in ethanol production.
2. Generally, the rural poor cultivate sorghum. Thus, the government policies should encourage the cultivation of this potential industrial raw material by way of giving incentives and subsidies as in the case of other crops.
3. It may be advocated to strictly declare that grain-based ethanol alone should be used for potable alcohol production on the grounds of consumer's health.
4. Government may reduce the excise duty proportionately to compensate the users of grain-based alcohol for potable purpose since they need to pay a little higher cost (Rs. 5/L extra) than the molasses-based alcohol.

Although this may reduce income from excise duty to state Governments, this may be suitably compensated by GOI. Larger quantity of molasses-based ethanol will be relegated to fuel blending, if the production of grain-based ethanol is increased as the burden of consumption demand is transferred to grain-based alcohol. This will save money for GOI due to the proportionate reduction in import expenditure on auto fuel and meet the social target of reducing environmental pollution. Postharvest operations including grading and standardization of molded grains to match the industrial requirements, effective supply network ensuring timely and adequate supply of molded sorghum grains, price regularization, institutional linkages for need-based research and development, industrial demand as well as consumer demand creation for superior quality ethanol, effective monitoring of all these aspects, would take sorghum a long way as an acceptable industrial raw material.

4.53 ENERGY BALANCE IN BIOENERGY PRODUCTION

The economic viability and attractiveness of sweet sorghum as a bioenergy crop crucially depend on its overall output–input energy ratio, called the net energy value. Ethanol in the fermented beer has to be distilled to at least 90% purity to make it acceptable for gasoline blending or use as industrial raw material or solvent. Alcohol distillation is an energy intensive and costly process. It accounts for about 70 to 85% of the total cultural, solar, and chemical energy used for bioenergy production (Ladisch and Dyck, 1979). When fuels like coal, gasoline, or furnace oil were used for distillation, the net energy values of ethanol production from major starch and sugar crops were negative. In the case of sugarcane, sugarbeet, and cassava, when the bagasse or crop residue was used as boiler fuel, the ethanol production became energy positive (Ladisch and Dyck, 1979). Sweet sorghum bagasse in combination with tillage has excellent fodder value. Hence, its use as a boiler fuel should receive only secondary consideration. Use of alternative renewable sources of energy to distil ethanol to 85 to 95% purity was suggested. Under such distillation, a net energy value of 5.99 was calculated for sweet sorghum. A more favorable energy balance appears feasible with the availability of improved distillation techniques and efficient strains of microbial system which can raise the level of ethanol concentration in the fermented beer. Nevertheless, under the identical fermentation and distillation conditions, ethanol production from sweet sorghum may have higher net energy value than from sugarcane because the fertilizer and irrigation requirements of the latter to provide the equivalent per day sugar yield are almost three times higher.

4.54 SUGAR PRODUCTION

Research on extracting sucrose (sugar) from sweet sorghum was initiated in the USA almost a century back. These efforts, however, failed to make any headway mainly due to the nonavailability of economically viable processing techniques to remove the starch and aconitic acid present in the juice and higher concentration of invert sugars. Subsequent identification of genotypes with a higher amount of crystallizable sucrose and low levels of invert sugars could overcome one of these problems. Diffusion of sucrose from stalks and clarification of juice by carbonization were developed to minimize the interference from starch and to improve the extraction and purity of sucrose. These processes, however, required modification of traditional equipment used by the sugarcane industry. Renewed efforts to refine sugar from sweet sorghum were initiated only very recently (Schaffert and Gourley, 1982). Although major processing problems still beg a solution, the possibility of breeding genotypes with reduced starch and aconitic acid levels in the juice and improved juice purity offers hope of extraction of sugar from sweet sorghum. A minimum juice purity of 75% essential for sugar production is within the reach of available variability for this character. Starch content in juice also shows similar variation among genotypes and at different stages of maturity in a genotype. Analytical limitation for the rapid estimation of aconitic acid is continuing to cause problems in the identification of genotypes with low levels of this constituent.

Different studies have estimated sugar yields varying from 3.2 to 7.3 t/ha from sweet sorghum (Ferraris, 1981; Piggot et al., 1980). These estimates, however, were arrived from the stalk yield and stalk sugar analysis. Development of new genotypes that could be adapted into the sugarcane processing industry has the added advantage of cutting down the overhead charges and improving the economy of sugar production (Coleman, 1970; Smith et al., 1973).

4.55 SWEET SORGHUM FOR GRAIN AND FODDER

Sorghum in many tropical countries is grown under marginal climatic and soil fertility conditions where it serves to meet the food and fodder needs. Development of sorghum for specialized commercial uses like production of ethanol or sugar products may have little immediate appeal in these traditional sorghum tracts, unless it may also provide grain and fodder. Therefore, high energy sorghums with multi-utilization potential have promising prospects to replace grain sorghum. Emphasis on a moderate grain yield along with high biomass productivity is hence important, wherever sweet sorghum is targeted to replace the grain sorghum. Grain yields ranging from 1.5 to 5.5 t/ha along with high yields of fresh stalk were reported in sweet sorghum (Schaffert and Borgonovi, 1980). Feasibility of harvesting 2.2 t grain and 37.7 sweet stalk/ha under farmer manager conditions has also been demonstrated widely in Brazil (Schaffert and Gourley, 1982). A very high grain dry matter yield of 5.7 t/ha from a sweet sorghum cultivar FS 26 was reported by Ferraris, (1981) from Queensland, Australia.

Evaluation of 19 sweet sorghum selections which showed a fresh biomass yield potential between 89 and 132 t/ha indicated that the fraction of stalk yield varied from 68% to 84% of the total with the rest being leaves and panicle. A Brazilian study also estimated that plant top and leaves constituted 10–18% of total biomass yield (Schaffert and Gourley, 1982). It is evident from these studies that about 16–25% of total biomass is available as leaves. According to Monroe and Bryan (1984) about 4.9% of total sugar in the plant is present in these fractions. The sweet sorghum's leaves make excellent silage (Caswell et al., 1983) with acceptable nutritive value (Joshi and Lalwani, 1977). In addition, farm level milling of stalk may provide 30 to 40% of stalk yield as bagasse which also could be developed as an excellent fodder with microbial enrichment or supplementation with urea.

References

Accorsi, C.A., Bartolucci, Zama, F., 1981. Sweet sorghum trials: cultivation and processing. Industria Saccarifera Italiana. 74 (5), 117–123, 126–127.

Agrawal, B.L., House, L.R., 1982. Breeding for pest resistance in sorghum. In: Proceedings of the International Symposium on Sorghum, 2–7 November 1981, pp. 435–446.

Almodares, A., Hadi, M.R., 2009. Production of bioethanol from sweet sorghum: a review. Afr. J. Agric. Res. 4 (9), 772–780.

AMB, 2004. Can sorghum sweeten rising sugarcane molasses prices. Ambrosia. 12 (7), August 2004, SAP Media Worldwide Ltd, 13/D, Laxmi Industrial Estate, Mumbai – 400 053.

Apparao, S., 1980. Germplasm Collecting Mission to Zambia, ICRISAT Genetic Resources Unit. Progress Report-25. Anatomy and Morphology of the Vegetative Organs of Sorghum Vulgare. Artschwager, Ernst (1948). TB957.

Areesirisuk, A., Laopaiboon, L., Khongsay, N., Laopaiboon, P., 2010. Improvement of gas chromatographic analysis for organic acids and solvents in acetone-butanolethanol fermentation from sweet sorghum juice. Afr. J. Biotechnol. 9 (38), 6422–6429.

Artschwager, E., 1948. Anatomy and Morphology of the Vegetative Organs of Sorghum Vulgare. US Department of Agriculture, Washington, DC, Technical Bulletin No. 957.

Bala Ravi, S., Biswas, P.R., Ratnavathi, C.V., 1997. Genetic variability and stability for the major traits of sweet sorghum. In: Proceedings of the First International Sweet Sorghum Conference, pp. 298–303.

Ban, J., Yu, J., Zhang, X., Tan, T., 2008. Ethanol production from sweet sorghum residual. Front. Chem. Eng. China. Vol 2, 452–455.

Bapat, P.R., Jadhav, H.D., Gaur, S.L., Salunkhe, C.B., Ingle, U.M., Kulkarni, D.N., et al., 1987. Sweet sorghum cultivars for production of quality syrup and jaggery in Maharashtra. In: Technology and Applications for Alternative Uses of Sorghum. Proceedings of the National Seminar, Parbhani Maharashtra, India, pp. 203–206.

Bothast, R.J., Destroy, R.W., 1981. What is alcohol? How is it made? In: Alcohol and Vegetable Oil as Alternative Fuels. Proceedings of Regional Workshops, Raleigh, NC, 7–9 April; Sacramento, CA, 21–23 April; Peoria, IL, 28–30 April. Purdue University Press, West Lafayette, IN, 31–37.

Bowen, J.E., Hunter, J.E., 1972. Sugar transport in immature inter-nodal tissue of sugarcane II. Mechanism of sucrose transport. Plant Physiol. 49, 789–793.

Broadhead, D.M., 1974. Effect of planting date and maturity on juice quality of Rio sweet sorghum. J. Agron. 64, 389–390.

Broadhead, D.M., 1981. A comparison between primary and tiller sweet sorghum plants for juice brix and sucrose. In: 73rd Annual Meeting of American Society of Agronomy, Madison, Wisconsin, USA, American Society of Agronomy, pp. 130–131 (En.) USDASEA-ARS Meridian Mississippi, USA.

Broadhead, D.M., Freeman, K.C., 1980. Stalk and sugar yield of sweet sorghum as affected by spacing. J. Agron. 72, 523–524.

Bulawayo, B., Bvochora, J.M., Muzondo, M.I., Zvauya, R., 1996. Ethanol production by fermentation of sweet-stem sorghum juice using various yeast strains. World J. Microbiol. Biotechnol. 12, 357–360.

Caswell, L.F., Kalmbacher, R.S., Martin, F.G., 1983. Yield and silage fermentation characteristics of corn, sweet sorghum and grain sorghums. Proc. Soil Crop Science Soc. Fla. 42, 139.

Chohnan, S., Nakane, M., Rahman, M.H., Nitta, Y., Yoshiura, T., Ohta, H., et al., 2011. Fuel ethanol production from sweet sorghum using repeated-batch fermentation. J. Biosci. Bioeng. 111 (4), 433–436.

Choudhari, S.K., Kulkarni, D.N., Chundrawar, A.D., Rao, N.G.P., 1987. Evaluation and selection of diverse sorghum genotypes for high energy, grain and biomass production. In: Technology and Applications for Alternative Uses of Sorghum. Proceedings of the National Seminar, Parbhani, Maharashtra, India, Marathwad Agricultural University, pp. 160–168.

Choudhary, S.D., Pawar, K.R., 1986. Internode patterns in sorghum and their role in grain yield. Indian Journal of Agricultural Sciences. 56, 453–456.

Clark, J.W., 1981. The inheritance of fermentable carbohydrates in stems of *S. bicolor* (L.) Moench. Dissertation abstract. 42, 8518.

Coble, C.G., Egg, R.P., Shimulevich, I., 1984. Processing techniques for ethanol production from sweet sorghum. Biomass. 6 (1–2), 111–117.

Coleman, O.H., 1970. Syrup and sugar from sweet sorghum. In: Walland, J.S., Ross, W.M. (Eds.), Sorghum Production and Utilization. Avi Publishing Co., Inc., Westport, CT, pp. 416–440.

Coleman, O.H., Stokes, I.E., 1964. Storage Studies of Sorgo. USDA Technical Bulliten No. 1307.

Daeschel, M.A., Mundt, J.O., et al., 1981. Microbial changes in sweet sorghum (*Sorghum bicolor*) juices. Appl. Environ. Microbiol. 42 (2), 381–382.

Dayakar Rao, B., Binu, M., Bharath Kumar, K.A., Karthikeyan, K., Hyma Jyothi, S.S., et al., 2003. Industrial Utilisation of Sorghum in India-Status and Prospects. NATP (NRCS) Series No. 2, Rajendranagar, Hyderabad, AP, India, pp. 20.

Doggett, H., 1970. Sorghum. Longman; Published by Wiley, London, New York.

Eastin, J.D., 1971. Photosynthesis and translocation in relation to plant development. In: Rao, N.G.P., House, L.R. (Eds.), Sorghum in Seventies. Oxford and IBH Publishing Co., New Delhi, India, pp. 214–246.

Eiland, B.R., Clayton, J.E., Bryan, W.L., 1983. Reducing fermentable sugar losses in stored sweet sorghum. In: Paper, American Society of Agricultural Engineers No. 83-3065, 13 pp. [En, 7 ref., 1 fig. 6 tab.] USDA, ARS, Sugarcane Harvesting Res. Unit, Belle Glade, FL, USA.

Ferraris, R., 1981. Early assessment of sweet sorghum as an agro industrial crop 1. Varietal evaluation. Aust. J. Exp. Agric. Anim. Husb. 21 (108), 75–82.

Ferreira, J.M., Viana, A.R., Sousa Filho, B.F., 1982. Preliminary evaluation of cultivars and hybrids of sweet sorghum in North Fluminense. Pesquisa em Andamento Pesagro-Rio. 9, 3.

Freeman, K.C., 1979. Germplasm release of sweet sorghum lines with resistance to downy mildew, leaf anthracnose and rust with adequate combining ability to produce progeny with agronomic characters acceptable for commercial syrup and/or sugar production. Miss. AFES Res. Rep. 4 (2), .

Ghose, T.K., Tyagi, R.D., 1979. Rapid ethanol fermentation of cellulose hydrolysate. I. Batch versus continuous system. Biotechnol. Bioeng. 21, 1387–1400, 1387–1420.

Glasziou, K.T., Gaylor, K.R., 1972. Storage of sugars in stalks of sugar cane. Bot. Rev. 38, 471–490.

Hacking, A.J., Taylor, L.W.F., Hamas, C.M., 1984. Selection of yeast able to produce ethanol from glucose at 400°C. Appl. Microbiol. Biotechnol. 11, 31–33.

Hatch, M.D., 1964. Sugar accumulation by sugar-cane storage tissue: the role of sucrose phosphate. J. Biochem. 93, 521–526.

Hatch, M.D., Glasziou, K.T., 1963. Sugar accumulation cycle in sugarcane. II. Relationship of invertase activity to sugar content and growth rate in storage tissue of plants growth in controlled environments. Plant Physiol. 38, 344–348.

Heichel, G.H., 1976. Agricultural production and energy resources. Am. Sci. 64, 64–72.

Hunsigi, G., Yekkeli, N.R., Kongawad, B.Y., 2010. Sweet stalk sorghum: an alternative sugar crop for ethanol production. Sugar Tech. 12 (1), 79–80.

Inman-Bamber, N.G., 1980. An evaluation of sweet sorghum as a sugar crop in the midlands mistabelt. In: Proceedings of the Fifty Fourth Annual Congress, South African Sugar Technologist. Association, pp. 105–108, South African Sugar Association Experimentation Station, Mount Edgecombe, 4300. Natal, South Africa.

Joshi, D.C., Lalwani, D.D., 1977. Sweet sorghum (*Sorghum bicolor*) as feed for cattle and sheep. Indian Vet. J. 54, 1018.

Jotwani, M.G., 1976. Host-plant resistance with special reference to sorghum. Proc. Natl. Acad. Sci. (Biol. Sci.). 46, 42–48.

Khongsay, N., Laopaiboon, L., Laopaiboon, P., 2010. Growth and batch ethanol fermentation of *Saccharomyces cerevisiae* on sweet sorghum stem juice under normal and very high gravity conditions. Biotechnology. 9 (1), 9–16.

Kundiyana, D.K., Bellmer, D.D., Huhnke, R.L., Wilkins, M.R., Claypool, P.L., 2001. Influence of temperature, pH and yeast on in-field production of ethanol from unsterilized sweet sorghum juice. Biomass Bioenerg. 3 (4), 1481–1486.

Ladisch, M.R., Dyck, K., 1979. Dehydration of ethanol: new approach gives positive energy balance. Science. 898–900.

Laopaiboon, L., Thanonkeo, P., Jaisil, P., Laopaiboon, P., 2007. Ethanol production from sweet sorghum juice in batch and fed-batch fermentations by *Saccharomyces cerevisiae.* World J. Microbiol. Biotechnol. 23 (2007), 1497–1501.

Li, B.Z., Balan, V., Yuan, Y.-J., Dale, B.E., 2010. Process optimization to convert forage and sweet sorghum bagasse to ethanol based on ammonia fiber expansion (AFEX) pretreatment. Bioresour. Technol. 101, 1285–1292.

Lingle, S.E., 1987. Sucrose metabolism in the primary stem of sweet sorghum during development. Crop Science. 27 (6), 1214–1219.

Linnaeus, C., 1753. Species plantarum, exhibentes plantas rite cognitas, ad genera relatas, cum differentiis specificis, nominibus trivialibus, synonymis selectis, locis natalibus, secundum systema sexuale digestas. Holmiae, Impensis Laurentii Salvii. [L. salvius, Stockholm.]. [1 May 1753] [Starting point for Spermatophyta, Pteridophyta, Sphagnaceae, Hepaticae, Fungi (incl. slime moulds and lichen-forming fungi) and Algae (pro parte).]

Liu, R., Shen, F., 2008. Impacts of main factors on bioethanol fermentation from stalk juice of sweet sorghum by immobilized *Saccharomyces cerevisiae* (CICC 1308). Bioresour. Technol. 99, 847–854.

Liu, R., Li, J., Shen, F., 2008. Refining bioethanol from stalk juice of sweet sorghum by immobilized yeast fermentation. Renew. Energy. 33 (5), 1130–1135.

Lobel, 1576. Observations, pp. 32.

Loomis, R.S., Williams, W.A., 1963. Maximum crop productivity. Crop Sci. 3, 67–72.

McBee, G.G., Miller, F.R., 1982. Corbohydrates in sorghum culms as influenced by cultivars, spacing and maturity over a diumal period. Crop Sci. 22, 381–385.

McBee, G.G., Waskom, R.M., Miller, F.R., Creelman, R.A., 1983. Effect of senescence and nonse-nescence on carbohydrates in sorghum during late ker-nel maturity states. Crop Sci. 23, 372–376.

Miller F.R., Creelman, R.A., 1980a. Sorghum – a new fuel. In: Texas Agricultural Experiment Station, College Station, Texas, Presented in 35th Annual Corn and Sorghum Research Conference.

Miller, F.R., Creelman, R.A., 1980b. Sorghum – a new fuel. In: Proceedings of 35th Annual Corn and Sorghum Conference. American Seed Trade Association.

Monroe, G.E., Bryan, W.L., 1983. Design factors for milling sweet sorghum. In: Paper, American Society of Agricultural Engineers 83-3070, 14 pp.

Monroe, G.E., Bryan, W.L., 1984. Design factors for milling sweet sorghum. T. Am. Soc. Agric. Eng. 27 (4), 1211–1216, 1221.

Monroe, G.E., Sumner, H.R., Hellwig, R.E., 1983. A leaf-removal principle for sweet sorghum. T. Am. Soc. Agric. Eng. 26 (4), 1006–1010, 1015.

Monroe, G.E., Nichols, R.L., Bryan, W.L., Sumner, H.R., 1984. Sweet sorghum juice extraction with 3-roll mills. T. Am. Soc. Agric. Eng. 27 (3), 651–654.

Nathan, R.A., 1978. Fuels from Sugar Crops: Systems Study for Sugarcane, Sweet Sorghum, and Sugar Beets. United States Department of Energy, Technical Information Centre, Oak Ridge, Tennessee, USA.

Nimbkar, N., Ghanekar, A.R., Joseph, R.D., 1987. Development of improved cultivars and management practices in sweet sorghum as a source of alcohol. In: Ingle, U.M., Kulkarani, D.N., Thorat, S.S. (Eds.), Proceedings of the National Seminar on Technology and Application for Alternative Uses of Sorghum, 1987. Marathwada Agricultural University, Parbani, India, pp. 180–188.

Patura, J.M., 1968. By Products of the Cane Sugar Industry. Elsevier, Newyork.

Piggot, G.J., Farrell, C.A., Stebleton, G.L., Shannon, P.W., 1980. Summer brassica forages in Northland. Proc. Agron. Soci. NZ. 10, 13–15.

Praj Industries Ltd, 2004. Cereals in alcohol industry-industrial perspectives. In: Presented by Praj Industries Ltd, Pune at International Conference on Biofuels, Winrock International India, New Delhi, September, 2004.

Prasada Rao, K.E., Murty, D.S., 1982. Sorghum for special uses. Proceedings of the International Symposium on Sorghum Grain Quality. ICRISAT, India, pp. 129–134.

Rao, M.J.V., Goud, J.V., 1977. Inheritance of grain-yield and its components in sorghum. Indian J. Genet. Pl. Br. 37, 31–39.

Ratnavathi, C.V., Dayakar Rao, B., Seetharama, N., 2003. Sweet Sorghum Stalks: A Suitable Raw Material for Fuel Alcohol Production, NATP (NRCS) Series No.1, Rajendranagar, Hyderabad, AP, India, pp. 8.

Ratnavathi, C.V., Suresh, K., Vijay Kumar, B.S., Pallavi, M., Komala, V.V., Seetharama, N., 2010. Study on genotypic variation for ethanol production from sweet sorghum juice. Biomass Bioenerg. 34, 947–952.

Reeves, S.A., Hipp, B.W., Smith, B.A., 1978. Sweet Sorghum Biomass – A Renewable Energy Source. Technical Report 78-1. Texas Agricultural Experiment Station. Texas A&M University, College Station, Texas, USA.

Reeves Jr., S.A., Hipp, B.W., Smith, B.A., 1979. Sweet sorghum biomass. I. Agronomic data. Sugar y Azucar. 74 (2), 55–58 [Es, 5 refs.] Agricultural Experimental Station Texas A and M Univ., Weslaco, Texas, USA. See FSTA 11 L464 for original En version.

Rosolem, C.A., Malavolta, E., Machado, J.R., 1983. Comparative effects of nitrogen, phosphorus and potassium fertilization on two sweet sorghum cultivars. I. Effect on grain and stalk production. Revista Brasileira de Ciencia do Solo. 5, 124–128.

Ruppa, A.A., Cordeiro, D.S., Petrini, J.A., Porto, M.P., Brancao, N., Santos Filho, B.G. dos, 1980. [Sweet sorghum cultures in the southern region of Rio Grande do Sul.] A culture do sorgo sacarino na regiao sudeste doRio Grande do Sul. Circular Technica, UEPAE de Pelotas No.12, 15 pp. [pt] Unidade de Execucao de peaquisa de Ambito Estadual, 6. 100 pelotas, Rio Grande do Sul, Brazil.

Schaffert, R.E., 1988. Sweet sorghum substrate for industrial alcohol. In: Paper Presented in the International Workshop on Policy and Potential Relating to Uses of Sorghum and Millets, June 8–12, 1988. SADCC/ICRISAT, Bulawayo, Zimbabwe.

Schaffert, R.E., Borgonovi, R.A., 1980. Sorghuma Opcao Para a Producao de alimentos energia. A Granja. 36, 60–64.

Schaffert, R.E., Gourley, L.M., 1982. Sorghum as an energy source. In: Sorghum in the Eighties: Proceedings of the International Symposium on Sorghum, 2–7 November 1981, ICRISAT Center, India. International Crops Research Institute for the Semi-Arid Tropics, Patancheru, AP, India, pp. 605–623.

Seetharama, N., Prasada Rao, K.E., 1988. Sweet Sorghum Germplasm at ICRISAT. Report of Cereal Programme ICRISAT, Patancheru, AP, India, 16 p.

Seetharama, N., Prasad Rao, K.E., Vaidyanathan, S., Murthy, D.S., 1987. Screening for sweet stalk sorghum and environmental effect on stalk sugar concentrations. In: Procedings of National Seminar Technology and Application for Alternate Uses of Sorghum, 2 – 3 February 1987, MAU, Parbhani, Maharahtra, India, pp. 169–179.

Sheorain, V., Rakesh, B., Chavan, M., 2000. Ethanol production from sorghum. In: Chandrashekar, A., Bandyopadhyay, R., Hall, A.J. (Eds.), Technical and Institutional Options for the Sorghum Grain for Sorghum Grain Mold Management: Proceedings of an International Consultation, 18–19 May 2000, ICRISAT, Patancheru, India. International Crops Research Institute for the Semi-Arid Tropics, Patancheru, AP, India

Sipos, B., Réczey, J., Somorai, Z., Kádár, Z., Dienes, D., Réczey, K., 2009. Sweet sorghum as feedstock for ethanol production: enzymatic hydrolysis of steam-pretreated bagasse. Appl. Biochem. Biotechnol. 153, 151–162.

Smith, B.A., Reeves Jr., S.A., 1979. Sweet sorghum biomass. IL plant composition. Sugar y Azucar. 74, 3V35.

Smith, C.W., Frederiksen, A.R., 2000. Sorghum-Origin, History, Technology and Production, eds. Texas A & M University, Texas, pp. 713.

Smith, B.A., Romo, R.V., Smith, R.C., Cruz, R.A., de la and Lime, B.J., 1973. Production of raw sugar from sorghum juices. Sugar J. 35 (12), 22–27 [En 22 refs.] Food Crops Util. Res. Lab., USDA, Weslaco, Texas, USA.

Smith, M.O., Bagby, R.T., Lewellan, D.L., Doney, P.H., Moore, E.J., Hills, L.G., et al., 1987. Evaluation of sweet sorghum for fermentable sugar production Potenti 2.V.A. Crop Sci. 27, 788–793.

Somani, R.B., 1996. Project appraisal titled 'An over view of the potential of the sorghum and millets for industrial uses in India'. Agro Products Development Research Center, PDKV, Akola, Maharashtra, India, pp. 269.

Sorghum. In: Taylor, R. N. John, Belton, S. Peter (Eds.), Pseudocereals and Less Common Cereals – Grain Properties and Utilization Potential. Springer-Verleg, Berlin Heielberg, Germany.

Ventre, E.K., Byall, S., 1937. Distribution and variation with maturity of dissolved solids, sucrose and titratable acidity in the sorgo stalk. J. Agric. Res. 55, 553–562.

Ventre, E.K., Byall, S., Catlett, L.L., 1948. Sucrose, dextrose and levulose contents of some domestic varieties of sorghum different stages of muturity. J. Agric. Res. 76, 141–151.

WHO (World Health Organization), 2004. http://www.who.int/substance_abuse/publications/alcohol.

Wiedenfeld, P.R., et al., 1984. Nutrient requirements and use efficiency by sweet sorghum. Energ. Agric. 3, 49–59.

Yadnik, D.S., Gadkari, S.S., 2003. Etho technological challenge. In: Alfa Laval Ltd, Pune, in National Conference on Prospects of Ethanol — A Sustainable Biofuel, APITCO Ltd, Hyderabad.

Yu, J., Zhang, X., Tan, T., 2009. Optimization of media conditions for the production of ethanol from sweet sorghum juice by immobilized *Saccharomyces cerevisiae*. Biomass Bioenerg. 33, 521–526.

Yu, J., Zhong, J., Zhang, X., Tan, T., 2010. Ethanol production from H2SO3-steam-pretreated fresh sweet sorghum stem by simultaneous saccharification and fermentation. Appl. Biochem. Biotechnol. 160 (2), 401–409.

Zelitch, I., 1973. Alternate pathways of glycolate synthesis in tobacco and maize leaves in relation to rates of photorespiration. Plant Physiol. 51 (2), 299–305.

Zhang, M., Eddy, C., Deanda, K., Finkelstein, M., Picataggio, S., 1995. Metabolic engineering of a pentose metabolism pathway in ethanologenic *Zymomonas mobilis*. Science. 267, 240–243.

Further Reading

Agricultural Statistics at a Glance, 2003. Directorate of Economics and Statistics, Department of Agriculture and Cooperation, Ministry of Agriculture, GOI, New Delhi, p. 113.

Agricultural Statistics at a Glance, 2004. Directorate of Economics and Statistics, Department of Agriculture and Cooperation, Ministry of Agriculture, GOI, New Delhi, p. 87.

Anonymous, 2003. Cooperative Sugars, vol. 34, February 2003.

Avgerinos, G.C., Fang, H.Y., Biocci, I., Wang, D.I.C., 1980. A novel single step microorganism conversion of cellulose biomass to a liquid fuels ethanol. In: Proceedings of the 6th International Fermentation Symposium, Toronto, pp. 80.

Broadhead, D.M., Freeman, K.C., Zummo, N., 1981. 'M81E' – a new variety of sweet sorghum. Miss. Agr. Expt. Sta. Inform. Sheet 1309.

Broadhead, D.M., Freeman, K.C., Zummo, N., 1982. The use of trichlorfon to identify insecticide- resistant sweet sorghum cultivars. J. Am. Soc. Sugarcane Technol. 1, 5–6 [En, 9 ref.] USDA Sugar Crops Field Sta., Meridian, Miss., USA.

Brown, L.R., 1980. Food or Fuel: New Competition for the World's Cropland. World Watch Paper 35. World Watch Institutė, Washington, DC, USA.

Chandrashekar, A., Bandyopadhyay, R., Hall, M., 2000 Technical and institutional options for sorghum grain mold management: proceedings of an international consultation. 18–19 May 2000, ICRISAT, Patancheru, India. (In En Summaries in En, Fr). International Crop Research Institute for Semi Arid Tropics, Patancheru, Andhra Pradesh, India, 299 p. ISBN 92-9066-428-2.

Choudhary, S.D., Pawar, K.R., 1986. Internode patterns in sorghum and their role in grain yield. Indian Journal of Agricultural Sciences. 56, 453–456.

Directorate of Millets Development, 2004, Personal Communication, Jaipur, India.

Du Preez, J.C., Prior, B.A., 1985. A quantitative screening of some xylose-fermenting yeast isolates. Biotechnol. Lett. 7, 241–246.

Du Preez, J.C., De Jong, F., Botes, P.J., Lategan, P.M., 1985. Fermentation alcohol from grain sorghum starch. Biomass. 8 (2), 101–117.

Du Preez, J.C., Bosch, M., Prior, B.A., 1986. Fermentation of hexose and pentose sugars by *Candida shehatae* and *Pichia stipitis*. Appl. Microbiol. Biotechnol. 23, 228–233.

Gaessler, W.G., Hixon, R.M., 1941. Levulose from chicory, dahlias and artichokes. Iowa State Coll. J. Sci. 16, 291–297.

Hills, F.J., Johnson, S.S., Geng Abshahi, A., Peterson, G.R., 1983. Comparison of four crops for alcohol yield. Calif. Agric. 37 (3/4), 17–19.

<http://vvww.praj.net>.

<http://www.ethanol.net/>; 2004, Situation and Outlook, India.

<http://www.petroleum.nic.in/psbody/html>; 2003, Ministry of Petroleum, India.

Kresovich, S., 1980. Sweet sorghum (*Sorghum bicolor* (saccharam group)). In: McClure, T.A., Lipinsky, E.S. (Eds.), CRC Hand Book of Bio Solar Resources, vol. 2. CRC Press, pp. 147–155.

Kresovich, S., Henderlong, P.R., 1984. Agronomic potential of sorghum as a raw material for ethanol production in Central Ohio. Energ. Agric. 3, 145–153.

National Conference on Prospects of Ethanol as Sustainable Fuel, 2003. Technical Papers, APITCO Ltd, Hyderabad, September, 2003.

Rao, M.J.V., 1985. Techniques for Screening Sorghums for Resistance to Striga. Information Bulletin No. 20, ICRISAT. International Crops Research Institute for the Semi-Arid Tropics, Patancheru, AP, India.

Schneider, H., Wang, P.Y., Chan, Y.K., Maleszka, R., 1981. Conversion of xylose into ethanol by the yeast *Pachysolen tannophilus*. Biotechnol. Lett. 3, 89–92.

Seetharama, N., Bidinger, F.R., Rao, K.N., Gill, K.S., Mulgund, M., 1987. Effect of pattern and severity of moisture deficit stress on stalk rot incidence in Sorghum. Use of line source irrigation technique, and the effect of time of inoculation. Field Crops Res. 289–308.

Seetharama, N., Dayakar Rao, B., Ratnavathi, C.V., Parwez, S., Mathew, B., Bharath Kumar, K.A., 2002. Sorghum: a potential raw material for alcohol production. In: Presented in Seminar on 'Current Development in Alcohol Industry' Conducted by Distillers Association of Maharashtra, 25 August 2002, Mumbai.

Winrock International, 2001. Ethanol from Alternative Sources, Ethanol, vol. 2. Winrock International India, New Delhi.

CHAPTER

5

Sorghum Syrup and Other by Products

C.V. Ratnavathi[1] *and U.D. Chavan*[2]

[1]ICAR-Indian Institute of Millets Research, Rajendranagar, Hyderabad, India [2]Mahatma Phule Krishi Vidyapeeth, Rahuri, Maharashtra, India

OUTLINE

Sorghum Biochemistry: An Industrial Perspective.
DOI: http://dx.doi.org/10.1016/B978-0-12-803157-5.00005-8

Abstract

Sweet sorghum syrup, also referred to as sorghum molasses, sorgho, or sorgo, is made by boiling the sweet juice of the sorghum cane (*Sorghum bicolor*). Sorghum syrup is a natural sweetener produced by extracting the juice from the sorghum stalks, and then boiling it down to the desired consistency. Generally, good quality syrup can be produced from sorghum genotypes with a high percentage of reducing sugars and a low percentage of sucrose in their juice. Nearly 500 kg syrup prepared from the juice of hybrid "Madhura" has been test marketed mainly in Phaltan and Pune. The varieties RSSV-9, RSSV-24, RSSV-45, NSS-221, NSS-104, and SSV-84 gave good quality syrup containing iron, calcium, and potassium. Glucose syrup is produced from purified sorghum starch by saccharification using immobilized enzymes like glucoamylase and glucose isomerase. Sorbitol is also produced from sorghum starch hydrolyzed by chemical hydrogenation in the presence of a nickel catalyst and is used as a sweetener.

5.1 INTRODUCTION

Sweet sorghum syrup, also referred to as sorghum molasses, sorgho, or sorgo, is made by boiling the sweet juice of the sorghum cane (*Sorghum bicolor*, formerly known as *Holcus sorghum*). Sorghum syrup is not derived from the sorghum grain or from sugar cane. And while it is often referred to as molasses, molasses is a by-product of the sugar industry—it is what is left when the granulated white sugar is removed. Sorghum syrup is a natural sweetener produced from the juice extracted from the stalks of sweet sorghum plants.

Mostly grown commercially for making sorghum syrup, the stout, corn-like sorghum plants are handsome and interesting horticultural grasses with plume-like tassels. Members of the Poaceae family, sorghums have been grown for centuries for grain, syrup, brooms, and forage crops. Sorghums like sandy soil or garden loam, and need to be cultivated and kept free of weeds like corn. Sweet sorghum is mostly grown for forage to produce "sorghum syrup" unlike most other sorghum varieties that are grown for grains. Some of the varieties of sorghum can get as tall as 8 to 15 feet, so these larger canes need to be grown in rows at least 6 feet apart. The plants take 120 days to mature.

Sorghum syrup is a sweet sticky syrup, dark amber in color with a thick consistency. It does not have any unpleasant after taste. It can be used as a table sweetener or exchanged for other sweeteners in baking in a one to one ratio. It tastes delicious spread on hot toast or crisp biscuits. Sorghum syrup may crystallize like honey, but liquefies when gently reheated.

5.2 SYRUP PRODUCTION

Sorghum syrup is produced by extracting the juice from the sorghum stalks, and then boiling it down to the desired consistency. Sweet sorghum syrup production offers farmers an excellent opportunity to improve farm income and productivity. Ideally suited for the small landowner with limited capital, this crop requires only 1 to 3 acres. Sweet sorghum yields 800 to 1200 L of syrup per acre, and sorghum syrup sells for Rs. 1000/L. The marketing outlook for sorghum syrup is also very favorable. Almost all the sorghum syrup produced is sold within 2 months of it being processed. Sorghum syrup is generally unavailable from May to October. So, even if syrup production is increased several-fold, a ready market will be available. Sweet sorghum, or "sorgo," is closely related to other sorghum crops. It differs from grain sorghum mainly in that its grain yields are low and its stalks are taller and juicier and have high sugar content. It reproduces by seed and produces tillers, but it has no rhizomes. It is a perennial grass under tropical conditions, but it is winter-killed in areas where frost occurs. Some sweet sorghum varieties are grown for syrup production, while others are grown for forage (silage).

5.3 PROCESSING OF SYRUP

5.3.1 Juice Extraction

All the mills in operation today are old—many with patents dating from before 1900. Replacement parts must often be specially fabricated or cannibalized from other mills. Mills of the types currently used are scarce; consequently, mill repair and replacement continues to be a major concern. The percentage of juice extracted is an important factor in mill operation. The juice extraction rate depends upon the mill speed, the moisture content of the cane, the mill adjustment, and the feeding rate. The rollers must be adjusted to spacings close enough to produce maximum extraction. To set the mill rollers initially, evenly space the feed roller 3/8 inch from the top roller. As a general rule, juice is lost if the bagasse ("pomace" or "chews") contains visible juice and is not broken at the joints as it comes from the mill. Cane ordinarily contains more than 70% water and 10 to 15% fiber, but it is impossible to extract all the juice. With a three-roller power mill, the weight of the juice extracted should be 50 to 60% of the weight of the stalks, unless they are very hard and dry. In smaller mills and with sorghum cane of poor quality, the extraction is frequently 50% or lower. Moreover, early in the harvesting season, when the crop may be less mature, the juice may be low in soluble sugars. In such cases, the yield of juice is proportionately reduced.

When unusually low extraction is being obtained in a small mill, it may be practical to gather up the bagasse and run it through the mill again. Often 50% more juice can be obtained by "double passing" than by putting the stalks through only once. Be sure to check the mill to see how much juice it is extracting. Under normal conditions, an efficient mill will deliver 22 to 25 L of juice from 250 kg of clean stalks. Weigh 250 kg of stalks, run them through the mill, and catch and weigh the juice. If necessary, tighten the rollers to increase the extraction of juice. As a general rule, juice is lost if the crushed stalks contain visible juice and are not broken at the joints as they come from the mill. Many older mills cannot be adjusted tightly enough without breaking to reach this maximum amount of extraction. When adjusting, always tighten the rollers down slowly.

5.4 PRODUCTION OF NATURAL SYRUP FROM SWEET SORGHUM STALK JUICE (NARI, PHALTAN, INDIA) (SMALL SCALE)

Generally, good quality syrup can be produced from sorghum genotypes with a high percentage of reducing sugars and a low percentage of sucrose in their juice, while high quality jaggery production requires the opposite composition. For ethanol production the total sugar content of juice is important, not its composition.

Sweet sorghum syrup production offers farmers an excellent opportunity to improve their income from sorghum crop. The marketing outlook for sorghum syrup is also very favorable, but the processing of sweet sorghum juice is the most critical aspect of making high quality syrup. The yield and quality of sorghum syrup are influenced by the equipment and process used in manufacturing and by the syrup maker's knowledge and skill. A chemical composition of sweet sorghum syrup compared with honey is given in Table 5.1.

5.5 ANALYSIS OF SAMPLE OF MADHURA BY CFTRI, MYSORE AND ITALAB PVT. LTD., MUMBAI

A protocol was developed for the production of natural, chemical-free, quality syrup from the juice of sweet sorghum hybrid "Madhura." Nearly 500 kg syrup prepared from juice of hybrid "Madhura" has

TABLE 5.1 Chemical Composition of Sweet Sorghum Syrup Compared with Honey

	Sweet sorghum syrup	Honey
Calorific value, Cal/g	2.60	3.26
Total soluble solids, % wt	77.00	81.00
Proteins (N X 6.25), % wt	1.65	–
Ash, % wt	3.69	0.59
Calcium (mg/100 g)		
Phosphorous (mg/100 g)	160.00	5.00
Riboflavin (Vitamin B_2) (mg/100 g)	11.00	4.10
Vitamin C (mg/100 g)	10.00	0.06
Nicotinic acid (mg/100 g)	11.50	5.00
Iron (mg/100 g)	153.00	32.00
Sodium (mg/100 g)	0.86	0.59
Potassium (mg/100 g)	86.00	4.70
Sulfur (mg/100 g)	1810.00	90.00
Benzoic acid (mg/100 g)	Not detected	8.00
Added coloring matter (mg/100 g)	Not detected	
Pesticide residues (mg/100 g)	None	
	Not detected	

been test marketed mainly in Phaltan and Pune during last three years. A bottling machine has been used successfully to package the syrup so that its shelf-life is increased (Fig. 5.1). The response of consumers to the colored syrup has been very encouraging. Also the syrup is entirely chemical-free as only natural ingredients such as the aqueous extracts of okra fruits or plants are used for facilitating scum removal. The nutritional quality of syrup was also found to be excellent. When extensive screening of a large number of sweet sorghum genotypes was done, it was found that the varieties RSSV-9, RSSV-24, RSSV-45, NSS-221, NSS-104, and SSV-84 gave good quality syrup. Also the hybrids developed at NARI such as Madhura, NARI-SSH-3, NARI-SSH-15, NARI-SSH-40, and NARI-SSH-21 produced good quality syrup.

Economic analysis of table syrup production from sweet sorghum hybrid "Madhura" was carried out for one hectare during one season for a processor as well as a farmer by considering all the costs of a syrup processing unit as well as raw material and transport. The study revealed that the total cost of production of table syrup from 22.5 t/ha

Crushing of Madhura stripped stalks

Extracted Madhura juice being stained into a settling tank

Syrup preparation from Madhura juice

Temperature recording and scum removal

Bottling of Madhura syrup with filling machine

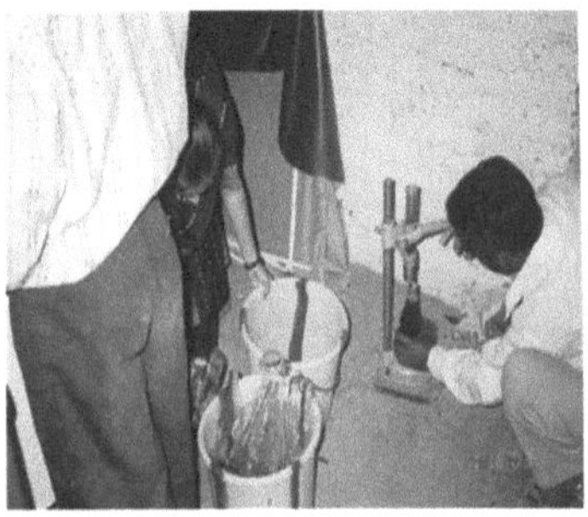

Crown capping of filled bottles of Madhura syrup

FIGURE 5.1 Preparation and bottling of Madhura syrup.

of stripped stalks of Madhura for a farmer is about Rs. 55,000 and about Rs. 64,000 for a processor producing 2000 kg of syrup (9% recovery) in one season (2005-06). If calculated on a per kg basis, the cost of table syrup for a farmer and a processor would be about Rs. 27 and Rs. 31, respectively. The sensitivity analysis of costs based on variable stalk production shows that the costs of syrup production for a farmer as well as a processor could be reduced linearly with a linear increase in stalk yields of sweet sorghum. For example, for 35 t/ha of stripped stalks of sweet sorghum with 9% syrup recovery, the costs would be Rs. 18 and Rs. 6/kg for a farmer preparing table and crude syrup, respectively.

5.6 MANUFACTURING THE SYRUP (LARGE SCALE)

The building where the juice is evaporated to syrup should be a sanitary area suitable for food preparation, meeting good manufacturing practices (GMPs). The area should be covered and screened to keep out filth, pets, birds, rodents, and insects. The floor should be concrete or made of other impervious material that is easily cleaned. It should be graded to drain so it can be washed with water. A pressurized, potable water source for cleaning and hand washing should be available. Hand washing facilities complete with soap, hot water, and single service towels must be available in the processing area.

Buying and storage: Once mature the leaves are stripped from the canes in the field, which is a labor-intensive task. Removing the leaves before cutting increases the yield of syrup and supposedly, the flavor quality of the syrup is superior. Then the canes are cut at ground level with a machete or similar tool and stacked in piles. The seed heads are harvested. The best ripe seed is saved for planting. The nutritious grain is fed to chickens. They need to sit for at least four or five days so the enzyme that turns starch into sugar can have time to work; they can sit for as long as two weeks. Much of this process depends on the weather. The cane is then fed into a sorghum mill or press, which crushes the cane. The juice, which is a light green color, is caught in buckets. The spent cane (which is referred to as pumee) is fed to livestock. The tray for boiling down the syrup is primitive and is designed to be heated over a wood fire so there is a stone chimney. The tray is first filled with water, which is brought to the boil and then the cane juice is poured into the tray. The tray, which is adjusted to be slightly inclined uphill, has little channels and the juice is cooked to make sweet syrup. The syrup is pushed with flat tools uphill and the water evaporates; then the syrup is skimmed. Finally, the sticky syrup is poured or ladled from large containers into jars.

5.7 FILTERING AND SETTLING THE JUICE

Raw green juice, when squeezed from the cane, should be filtered through a coarse screen to filter out the larger pieces of dirt, cane, and debris. The large gears on a horizontal mill may be positioned directly over the expressed juice flowing out of the press. Care should be taken to ensure that grease from the gears does not contaminate the raw juice. This problem can be corrected by placing a strip of sheet metal below the gears to deflect any excess grease. Use only food-grade grease. This lubricant can be obtained from a lubricant/fuel supplier. The juice should be run from the mill through a pipe or hose to the juice-settling tanks, where it is strained through a fine mesh screen, muslin, or fine nylon cloth as it enters the tank. Use two or three tanks, each big enough to hold the juice from 2 or 3 h of milling. Let the juice settle a minimum of 2 h before evaporating. Holding the juice more than 3 to 4 h without refrigeration or without heating (as described later) may cause it to ferment and spoil. The outlet on the settling tank should be at least 1 inch above the bottom of the tank so the settled material is not drawn into the evaporator. A separate opening in the bottom of the tank will be needed to clean out the tank after drain off.

In some operations, settling tanks are heated to a point just below boiling and held for about 2 h. The temperature should not be allowed to go below 160°F as it is held overnight. This holding temperature allows for much of the skimmings to rise to the top and the settlings to precipitate to the bottom. The design of the tank is important, since heat applied directly to the bottom of the tank causes problems in settling. Using this preheat system with a steam-heated operation is ideal. The steam coils are placed on the side of the preheat tank instead of the bottom. Removal of skimmings and settlings in the preheat tank allows contaminates to be removed early in the process and reduces the chances of "boiling-in" excess skimmings during the evaporation step. The earlier you remove the skimmings and insolubles from the juice, the better chance you have of making high-quality, light-colored syrup. Once skimmings are boiled into the syrup, they are almost impossible to remove later in the process.

5.8 EVAPORATING THE JUICE

Evaporators may be batch (kettles) or continuous-flow types. The 12-foot continuous-flow evaporator is the most common. Most pans are the continuous type with dimensions about 3.5 feet by 12 feet by 6 inches, but their size can vary considerably. They are divided into many sections by baffles that extend across the pan and are arranged so the open end of one is opposite the closed ends of those on either side of it. A self-skimming trough may be attached to each side of the evaporator, which greatly reduces the labor for skimming the juice. A "water jacket" in the final six-inch section of the evaporator pan helps moderate the temperature of the finished syrup and helps keep the syrup from scorching. Juice evaporators are made of galvanized iron, black iron, stainless steel, or copper. Use of galvanized iron is highly discouraged when making syrup today. Because cane juice is slightly acidic, it leaches the zinc out of the galvanized metal. These pans are usually made with lead solder, as well. This lead also leaches into the syrup. These pans should not be used today. Black iron pans will make the syrup darker, because high iron levels leach into the syrup. The use of iron is discouraged, although there are no known health hazards at this point.

The ideal metal for evaporators is stainless steel or copper. Both metals transfer heat efficiently. However, stainless steel evaporators need a much more uniform heat source for even heat distribution. Most new pans today are made of stainless steel. Most processors use stainless steel for all containers in the process. The ideal source of heat is steam, although most producers use LP gas. The quality of the syrup depends to a large extent upon the variety of sweet sorghum, the type of soil on which the crop is grown, the type and amounts of fertilizer used, and the kind of growing season. However, it may also be greatly influenced by the equipment and process used in manufacturing and by the skill of the syrup maker.

Skimming: When heat is applied to the juice, much of the starch becomes soluble, but certain proteins and other nonsugar substances begin to coagulate. If allowed to settle, some of this coagulated material rises to the surface of the juice and some sinks to the bottom. The best practice is to remove this material as quickly as possible by skimming as soon as it appears on the surface of the juice.

Success in making syrup depends first on the thoroughness with which the juice is skimmed before it begins to boil rapidly. The agitation of the juice due to active boiling breaks the coagulated material into smaller particles, making them more difficult to remove by skimming than the original mass. This breaking up of coagulated material is commonly referred to by syrup makers as "boiling-in" the impurities. Additional nonsugar substances separate as boiling continues and the juice becomes denser, making it advisable to continue the skimming until the juice has been evaporated to the density of finished syrup, even though careful skimming has been done at the beginning of the evaporation.

Continuous evaporators have many points of superiority. They are constructed in such a way as to produce a quick concentration of the juice to syrup, and, with proper operation, to facilitate efficient skimming. If the pan is operated so the skimmings are properly concentrated and removed, the major function of the operation is to ensure that the syrup is drawn from the pan when the proper density is reached.

The advantages of continuous-type evaporators are:

- The design allows for rapid evaporation, which is essential in making light-colored syrup.
- The syrup is concentrated in a thin layer, thus increasing the rate of boiling and foaming and affording a better opportunity for thorough skimming.
- Heat is applied to the bottom of the evaporator, giving an upward motion to the coagulated material, which makes skimming easier.

The disadvantages are:

- More attention is required to maintain a properly regulated flow of juice.
- There is increased danger of scorching the syrup and altering its color and flavor.
- More careful attention to firing is necessary.
- It requires more skill to obtain uniform syrup density.

In continuous evaporation, a steady stream of juice flows by gravity into the front end of the pan and then flows slowly to the opposite end, at which point it reaches the density of syrup. The juice end of the pan is sufficiently lower than the finishing end, to maintain a juice layer from 2 to 2½ inches deep. This depth should give a layer ¾ to 1½ inches deep (preferably only ¾ to 1 inch) in the finishing end of the evaporator. In other words, the juice end of the evaporator is mounted on the furnace about one inch lower than the syrup end. The evaporator must be level from side to side. If operating during the day only, the evaporator should be kept partly full of water overnight. This can be accomplished by flooding the pan with water after drawing off as much syrup or semisyrup as possible. At the end of the day, the evaporation needs to have reached only the semisyrup stage. If the semisyrup is well skimmed before removal, in the morning it may be put back into the syrup end of the evaporator. The syrup compartment is shut off from the rest of the pan by means of a gate until the evaporator is working well again with fresh juice.

The easiest way to begin the day's operation is to start with juice in the pan over only two-thirds of its length, with either water or semisyrup in the syrup compartment, blocked off with a gate. Transferring juice or syrup of low density to parts of the pan where the syrup is becoming too dense is bad practice, but it is sometimes unavoidable at the start. As soon as syrup is being finished in the back compartment of the evaporator and clean semisyrup is being made from the fresh juice, it is time to permit a continuous flow of juice. The flow of juice ordinarily should be kept as nearly constant as possible without dipping from one compartment to another. Such dipping detracts from the clarity of the syrup, because of the mixing of juice and syrup at different stages of clarification. Mixing of high- and low-density juice usually causes a persistent cloudiness in the finished product. The rate at which juice is run into the evaporator and syrup is run out is now controlled by the syrup maker.

When the evaporator has started to work well, the juice seldom boils in the first compartment, which is the coolest part of the evaporator, unless for some reason the inflow of cold juice is temporarily stopped. This juice has a smooth, relatively cool surface, over which the skimmings form a blanket. This blanket is occasionally removed with a perforated skimmer. If the furnace is properly constructed, the boiling of the juice increases in vigor toward the back end of the pan as far as the section under which the fire is hottest. This causes the skimmings to run counter to the flow of juice to the cooler or front portion of the evaporator.

By the time the juice reaches the hottest part of the pan, which is about 1½ feet beyond the middle, it has been evaporated nearly to semisyrup density and is fairly well cleaned. As the syrup becomes more concentrated, however, additional impurities separate out, and this material also should be carefully removed by skimming. For efficient skimming, a hot fire must be maintained to "roll" the foam.

5.9 CONCENTRATING THE SYRUP

A common problem with continuous evaporators is the difficulty of concentrating the syrup to uniform density. Many operators are able, by experience, to judge fairly accurately the density of syrup while it is still boiling. Some do this by dipping a skimmer or a syrup "rake" into the boiling syrup, holding it up, and noting how the cooling syrup "flakes off." But even an experienced operator is not always consistent with this subjective method.

No amount of experience can take the place of accurate measuring, and the uncertainties of guessing the density can be easily eliminated with the use of a syrup-maker's thermometer. Syrup from different varieties of sorghum boils differently. A good thermometer is more useful in the operation of a shallow evaporator than is the hydrometer, which is often recommended for this purpose.

In using the hydrometer, it is necessary to draw off a cylinder of syrup from the evaporator and float the hydrometer in it. This procedure is troublesome when using a shallow evaporator, because the syrup is very hot and is in such a thin layer that it is not easily dipped out.

The thermometer most suitable for the purpose is one protected by a substantial metal case, with a 10-inch scale graduated from approximately 10–121°C. The bulb should not touch either the bottom or side of the evaporator. By providing a broad metal back piece so that it will stand up, such a thermometer may be kept continuously in the syrup to indicate accurately the point of final evaporation.

When you use a thermometer to test syrup for its density, occasionally check the accuracy of the thermometer by placing it in boiling water and noting the boiling point. Water should boil at 100°C at sea level.

5.10 FINISHING THE SYRUP

The syrup is usually finished at a temperature higher than the boiling point of water if syrup of fairly heavy density is desired, although a slightly different finishing temperature may be used with equal success. Therefore, when using an accurate thermometer at a point 500 feet above sea level, finished syrup would boil at 107.2–107.7°C. Syrup with 78% solids can be produced by concentration until the boiling point of the syrup is 107.7°C. This relationship permits the density of the syrup to be readily determined by simply observing the syrup temperature during boiling.

To enable the juice to evaporate uniformly from the pan, the temperature must be controlled along the pan's length. Boiling of the juice should start about one quarter of the length of the pan from the juice end and remain at a constant temperature until the skimmings are completely removed before reaching the last quarter of the pan. As the syrup density increases, the boiling temperature will gradually increase.

When shutting the pan down, allow the furnace to cool slightly before the syrup is drained off. Remember that no part of the pan should be dry for more than an instant or the syrup's quality may be impaired, and the pan can be damaged. The semisyrup in the center of the pan must be allowed to follow closely behind the syrup being moved to the outlet. Strain the finished syrup through muslin, two thicknesses of cheesecloth, or 45-mesh screen wire as it is run into a suitable container for cooling. Cool the strained syrup obtained from each "stir-off" in a large container to about 60–71.1°C as quickly as possible before putting it in a permanent container of 3.75 L or smaller. The faster you cool the syrup, the less color will develop. Some producers pump the finished syrup through a pipeline that passes through an enclosed cool-water tank to lower the temperature quickly to 60–71.1°C. The syrup is then strained into holding containers for canning. This quick cooling helps maintain the syrup's lighter color, compared to slower cooling methods. If the enzyme isomerase is added to help prevent crystallization, the syrup must be below 65.5°C. Otherwise the enzyme will be inactivated.

The finished syrup is thoroughly sterilized by boiling. If canned above 65.5°C in clean containers, it will not spoil or ferment. The container must be airtight to prevent fermentation or surface mold growth. Thoroughly wash large containers such as metal drums several times with boiling water or steam and dry them thoroughly before adding syrup. It takes from 20–45 L of raw juice to finish 3.75 L of sorghum syrup. The finished syrup will weigh approximately 5.2 kg/L.

5.11 SEMI SYRUP

Since juice cannot be held for long periods of time without spoiling, making semisyrup is an attractive alternative in some cases. Research at the University of Tennessee has shown semisyrup of at least 45°C Brix can be held at 20°C for at least 21 days. A reading in degrees Brix indicates the percentage by weight of sugar in a solution at a specified temperature. Higher concentrations can also be made and the storage time would be lengthened, if stored at the same temperature.

How long you can safely hold semisyrup depends on the sugar concentration, the temperature at which it is held, and the sanitation of the system. Also, if you install a finishing pan in your process to concentrate the semisyrup to finished syrup, the process for obtaining your desired Brix will be more accurate.

5.12 CARE OF THE EVAPORATOR

It is seldom desirable to make syrup day and night; therefore, after each day's run the evaporator must be cleaned. To do so, run water into the pan in the late afternoon and let it follow the syrup as it is finished and drawn off. Leave at least one inch of water in the pan overnight. Never heat a dry pan, or it will warp.

In the morning, scrub out the pans with a nonmetallic abrasive cleaner. A 1–2% nitric acid solution should also be used on the stainless steel pans following the first cleaning. Nitric acid is preferred over sulfuric acid for cleaning stainless steel pans. The alkaline detergents (used the night before) will remove the mineral deposits, and the acid solution (used the next morning) will completely remove salt precipitates from the stainless steel pan. Always wear rubber gloves when using any of these cleaners.

The used solution should be discharged to a covered pit away from trees or streams to prevent injury and pollution. Thoroughly flush out the evaporator with clean water before beginning to cook. Rubber gloves and safety glasses or goggles should be worn when using these cleaning materials.

5.13 GENERAL SANITATION

Sorghum, like any other food product, should be processed under good sanitary conditions. Everyone in the exposed product area must always wear clean clothing and hair restraints. Use GMPs during all processing operations and in the construction of the facility.

Even though bacteria will not generally grow in sorghum syrup, because of its high sugar concentration, yeast and molds will grow in it. Small areas of residual syrup, especially if diluted with water, harbor millions of yeast and mold cells and will increase the chance of syrup spoilage.

All surfaces coming into contact with the juice or finished syrup should be washed, sanitized with i/ ~ imaginary quantity cup sodium hypochlorite (Clorox) 3.75 L of water and allowed to air dry. Fill jars with hot syrup and seal them immediately. This will prevent spoilage. Use food grade products, including the grease used on the mill, for all operations.

5.14 LABELING YOUR CONTAINERS

Each container must be labeled with a true statement identifying the product; that is, sorghum syrup, sorghum and corn syrups, or corn syrup with sorghum syrup. Any optional ingredients or "processing aids" used in the processing of the syrup, such as enzymes, do not have to be labeled. The name and address of the manufacturer or distributor and the contents of the package (volume or net weight) must also be on the label.

5.15 USING ENZYMES FOR PROCESSING SYRUP

5.15.1 Gelling

Depending on the growing season and the variety of sweet sorghum grown, sorghum syrup will occasionally gel when cooked to the required sugar concentration. Gelling results from higher-than normal amounts of starch in the raw juice. It is usually associated with overmature sorghum (as cane matures the sugar converts to starch).

When made from sorghum harvested at the proper stage and allowed to settle over 2 h, most sorghum juice will not contain excessive starch. Occasionally, however, gelling will take place even when the juice is allowed to settle properly.

Managing starch in the juice is one of the most difficult problems in producing consistently high-quality syrup. When the juice is boiled, the starch thickens, as gravies and cream pies thicken, causing the syrup to "gel."

5.15.2 Enzymes

If settling the juice does not remove enough of the starch to prevent gelling, an enzyme that breaks down starch into sugars and dextrins can be used. Several methods can be used successfully if you have a basic understanding of starch and enzymes.

Starch occurs as granules, which are protected from enzyme action because they are insoluble in cold juice. Therefore, starch granules must be ruptured before the enzymes can act on them. Depending on what type they are, the starch granules will rupture when heated to a temperature of 60–82°C. The exact temperature at which the starch granules in sweet sorghum will rupture is unknown. So, the juice should be heated to more than 82°C.

All enzymes are proteins and are destroyed by heat, but some are more heat-stable than others. Of the three types of amylase enzymes (enzymes that will convert starch into sugar or dextrins), two are relatively heat-stable, while the other is easily destroyed by high temperatures. Although any of the enzymes can be used, the heat-stable enzymes are safer since they are less likely to be accidentally destroyed by heat.

Enzymes initiate reactions but are not consumed in them, so only small amounts of the enzyme are necessary. The amount of enzyme needed depends on that particular enzyme's activity, but the recommendations should work for most of the commercial, high-temperature enzymes available. Detailed instructions are available from the enzyme supplier.

5.15.3 Methods to Prevent Gelling

One method to remove starch is to heat the juice, cool it down, and then add the enzyme. First, heat the raw juice to about 99°C for few minutes to rupture the starch granules. After heating, allow the juice to cool to 85°C or below and then add about 30 mL (2 tablespoons) of the amylase enzyme (liquid) per 378 L of juice. Mix the enzyme thoroughly with the juice, either by hand stirring with a paddle or by mechanical agitation, and allow it to settle for an hour or more so that the enzyme will have time to convert the starch to sugar. Then evaporate the juice in the usual manner.

Another way to remove the starch is to add 30 mL (2 tablespoons) of a high-temperature amylase enzyme (liquid) to 378 L of the raw juice and slowly heat the juice during evaporation. The juice should be heated slowly enough so the starch granules rupture and allow the enzymes to convert the starch to sugar before the heat destroys the enzyme. The juice is then evaporated as usual.

A third method and one of the best, for removing the starch is:

- Evaporate the juice to semisyrup.
- Draw the semisyrup off into large containers and let it cool to 85°C.
- Add a high-temperature enzyme (liquid) to the semisyrup. Add the enzyme at the rate of 30 mL (2 tablespoons) of enzyme to the semisyrup produced from 100 gallons of raw juice.
- Let it stand for one or more hours to convert the starch into sugar.
- Finish the semisyrup in the usual manner.

Occasionally, finished syrup will gel even though the juice was allowed to settle thoroughly. In this case, the finished syrup can be reheated to 70–82°C and treated with 30 mL of the enzyme per 3.75 L of the finished syrup and then rebottled. Although reboiling the syrup to destroy the enzyme may be desirable, it is not necessary since the enzyme acts only on the starch in the syrup. This reheating will normally make the syrup darker in color.

5.15.4 Crystallization

Sorghum syrup will often crystallize during storage. Crystallization is frequently a problem for producers who store their syrup through the winter months. Crystallization is usually caused when sucrose is present in concentrations too great to remain dissolved at the storage temperature. You can reduce crystallization by converting some of the sucrose to glucose and fructose (other forms of sugar) with the enzyme invertase. Glucose will also crystallize readily under certain conditions (honey is a good example).

5.15.5 Methods for Using Invertase

Invertase is an enzyme that will be destroyed if it is heated above 65°C. However, it can be added at several points in the process. For processors who make a semisyrup and hold it in a tank for further evaporation, this holding tank is one point in the process when the enzyme may be added. As long as the semisyrup remains below 65°C, the enzyme will continue to break sucrose down into its two monosaccharides (glucose and fructose). Adding the enzyme in the semisyrup tank allows the processor to experiment with how long to allow the enzyme to work to best prevent crystallization. When the semisyrup is evaporated to the final syrup, the enzyme will be inactivated.

Another recommended method is to finish off the syrup in the usual manner and cool it below 65°C. Then add 0.4 kg of invertase per 378 L of finished syrup and bottle. One of the problems with this technique is that eventually all the sucrose will be broken down to glucose and fructose. This breakdown does not change the taste or sweetness of the syrup, but, chemically, the syrup will look similar to corn syrup. On several occasions, the FDA has suspected invertase-treated syrup of being adulterated with corn syrup. Every case, however, was resolved favorably for the producer.

5.15.6 Note About Enzymes

Enzymes are not stable over long periods of time. Depending on the source of the enzyme, liquid enzymes should be stable for 6 months or more and dry enzymes for about 12 months or more. Some suppliers have analyzed enzymes after four years of storage at refrigeration temperatures and found them to retain 75% of their original activity. However, you should not buy a large supply of enzyme at one time, expecting to use it for several years. Store all enzymes in the refrigerator.

5.15.7 Health Benefits of Syrup

Since it contains iron, calcium, and potassium, sorghum is good for you—unlike other liquid sweeteners like high fructose corn syrup. Read labels to be sure that you are buying a pure sorghum product, not one with added sweeteners. Sorghum syrup is sweet, yet it has a little sourness and a slight bitterness; it is not as sweet as maple syrup or honey. Often, sorghum was called "long sweetening" while sugar was called "short sweetening." The best way to eat sorghum syrup is on a hot biscuit or corn bread. It is also good with seasonal fall produce such as squash, pumpkin, apples, pears, persimmons, and pecans. It can used in pies, puddings, cakes—any baked good—and it makes great barbecue sauce, baked beans and chili. Or try some sorghum syrup on this hearty, toothsome corn bread for a wonderful tasty treat! Buy good quality sorghum syrup and store in a cool, dry, dark place in the cupboard at room temperature. It is stable, has a good shelf life, and keeps well for up to a year at room temperature. Pure sorghum syrup is unprocessed and raw.

Nutritional information per serving of sorghum syrup is given in Table 5.2. Sorghum syrup has mostly sucrose and trace amounts of glucose and fructose. It may cause a sudden spike of blood sugar levels, so diabetics have to be cautious and use this in moderation. It is a good source of magnesium. It helps in relieving fatigue, relaxing the muscles, nerves, and blood vessels, thus relieving the symptoms of asthma, migraine headaches, tension and soreness in muscles, etc. It is a good

TABLE 5.2 Nutritional Information per Serving

Serving size	**100 g of sorghum syrup**	
% DAILY REQUIREMENTS		
Total Energy	290 calories	14%
Total energy from carbohydrates	290 calories	
Total fat	0 g	0%
Trans fat	0 g	0%
Cholesterol	0 g	0%
Total carbohydrate	75 g	25%
Dietary fiber	0 g	0%
Sugars	75 g	
Sucrose		
Glucose		
Fructose		
Protein	0 g	0%
MINERALS		
Calcium	150 mg	15%
Iron	3.8 mg	21%
Magnesium	100 mg	25%
Phosphorus	56 mg	6%
Potassium	1000 mg	29%
Sodium	8 mg	0%
Zinc	0.4 mg	3%
Copper	0.1 mg	6%
Manganese	1.5 mg	77%
Selenium	1.7 mg	2%
VITAMINS		
Thiamine	0.1 mg	4%
Riboflavin	0.2 mg	10%
Niacin	0.1 mg	1%
Vitamin B6	0.7 mg	34%

source of iron, which is used in respiration at the cellular level by synthesizing hemoglobin that helps to carry oxygen to cells. It is also a good source of potassium; this along with low amounts of sodium helps to maintain blood pressure and reduces water retention. It is a very good source of manganese, which is an essential cofactor in many enzymatic reactions in the body, especially in the production of energy and antioxidants defenses. It is a moderate source of other minerals such as calcium, phosphorus, zinc, copper, and selenium, which are essential to maintain the metabolic activities of the body, and a moderate source of riboflavin and Vitamin B6, which are essential cofactors in enzymatic reactions of the body.

5.16 ENZYMES

5.16.1 α-Amylase, Glucoamylase, and Glucoisomerase

Unless otherwise stated that the bacterial α-amylase (*Bacillus* sp., EC 3.2.1.1 type-II-A, alpha 1,4-glucan, and glucanohydrolase) was obtained from E. Merck and SISCO. According to the manufacturing company, the α-amylase had an activity of 1500 units/mg protein. Similarly, the glucoamylase (*Aspergillus niger*, EC 3.2.1.3., Exo-1,4-α-glucosidase1,4-α-D-Glucan glucohydrolase) was procured from E. Merck and SISCO which had an activity of 5000 units/gram solid. Glucose isomerase from *Artobacter* spp. was obtained from Microbial Culture Division, National Chemical Laboratory, Pune. This was purified earlier in the laboratory, and has an activity of 12.5 IU/g on dry weight. The purification of the said enzyme was performed using ammonium sulfate fractionation followed by dialysis, gel filtration on Biogel, P-200, Pharmacia double jacketed glass (quartz) column (2.5 × 55 cm) pre-equilibrated with 0.05 M tri-HCl, pH 7.0 containing 0.2 mM $CoCl_2$ which exhibits 125 IU/g dry weight.

5.17 ENZYME ACTIVITY

5.17.1 α-Amylase

One percent solution of soluble sorghum starch in 0.1 M citrate phosphate buffer (pH 6.9) was incubated with the bacterial species *Bacillus* sp., EC 3.2.1.1 type-II-A), diluted with an equal volume of buffer at 37°C for 30 min. The resultant reducing sugars in the hydrolysate were quantitatively estimated for saccharolytic and dextrinizing activities according to the method of Bernfeld (1955) and the dextrinizing activity was determined by the method of Smith and Roe (1949). One unit of α-amylase activity is defined as the amount of enzyme required to release 1 μM of maltose/min under the optimum experimental conditions as described by Tharanathan et al. (1980).

5.17.2 Glucoamylase

The enzyme was incubated with 1% solution of soluble sorghum starch in 0.05 M sodium acetate buffer (pH 4.6) at 60°C for 15 min. The liberated glucose was estimated by the glucose oxidase method of Dahlquist (1964). One unit of enzyme activity is defined as the amount of enzyme required to release 1 μM of glucose/min under the optimum experimental conditions as described by Tharanathan et al. (1980).

5.17.3 Glucoamylase, Transglucocidase, and β-Amylase in Glucoamylase

Glucoamylase	Transglucocidase	β-Amylase
99.5	Not detected	0.001

5.17.4 Glucose Isomerase

The glucose isomerase activity was assayed as per the method of Lloyd et al. (1972). The assay mixture contain suitably diluted enzyme solution (1 mL) or 2 g of presoaked immobilized enzyme was added to 10 mL of 6% aqueous solution of fructose (0.1 M tris-HCl, pH 7.0 containing 0.2 mM, $CoCl_2$) and shaken at 30°C. The glucose was estimated quantitatively by the glucose oxidase method (Dahlquist, 1964), modified by Tharanathan et al. (1980). One unit of enzyme activity is defined as the amount of enzyme required to release one micromole of glucose per minute under the experimental condition. The fructose was quantitatively determined as per the procedure of Wankhede et al. (1979) and Wankhede et al. (1977) using PC, TLC, and carbon celite chromatography.

The GLC of partially methylated aldetol acetates was performed by the method of Bjorndal et al. with a Netal Chromatogrph equipped with FID and fitted with a glass column (0.32 × 152 cm) packed with 3% of ECNSS-M on gas chrom-Q (100–200 mesh) at column temperature of 170°C. The flow rate for nitrogen was 25 mL/min. The fructose and other oligosaccharides were also quantified by GLC as its TMS-derivatives on 3% OV-1 on Gas chrom-Q (100–200 mesh) at column temperature programming (100–220°C).

5.18 STARCH ISOLATION AND METHODS ADOPTED

5.18.1 Proximate Composition of Sorghum Grains

The proximate composition of sorghum grains (ie, moisture, ash, fat, protein, carbohydrate, and crude fiber content) was essentially carried out by standard AOAC (1990) methods.

5.18.2 Isolation and Production Kinetics of Starch From Sorghum Grains

Sorghum grains (50 g) were cleaned and healthy grains were steeped in water (1:2, w/v) for 10 h with a few mL of toluene. In addition, the mercuric chloride was added in very low concentration (0.001 M) to arrest α-amylase activity during steeping. The steeped grains were washed thoroughly with water and then subjected to homogenization. The resultant slurry was filtered through muslin cloth followed by sieving through a standard mesh sieve (150 μ). The upper supernatant was siphoned off and the crude starch was collected by centrifugation at 4000 × g for 10 min. The residue was left after the crude starch extraction was further processed as above until most of the starch was recovered from it (until a negative test with I_2-KI reagent). The crude starch was further purified by suspension in distilled water (1:5) and the sodium chloride (NaCl) was added to achieve the final concentration to 0.1 M to which toluene was added (3:1, v/v). It was then kept for shaking for 1–2 h and the denatured protein was removed by centrifugation. Thus, purified starch was suspended in water and recovered by centrifugation. Then, it was filtered through a Buchner funnel (with acetone wash) and air-dried.

5.18.2.1 Starch Isolation

The isolation and purification of starch from sorghum grains were achieved essentially by the method of Wankhede et al. (1990). In a typical experiment, Sorghum grains (50 g) were cleaned and, healthy grains were steeped in water (1:2, w/v) for 10 h with a few mL of toluene. In addition, the mercuric chloride was added in very low concentration (0.001 M) to arrest the α-amylase activity during steeping. The steeped grains were washed thoroughly with water and then subjected to homogenization. The resultant slurry was filtered through muslin cloth followed by sieving through standard mesh sieve (150). The upper supernatant was siphoned off and the crude starch was separated by centrifugation at 4000 gx for 10 min. The residue left after the crude starch extraction was further processed as above until most of the starch recovered from it (till negative test with I_2-KI reagent). The crude starch was further purified by suspending in distilled water (1:5, w/v) and the sodium chloride (NaCl) was added to achieve the final concentration to 0.1 M to which toluene was added (3:1, v/v). It was then kept for shaking for 1–2 h and the denatured protein was removed by centrifugation. Thus, purified starch was suspended in water and recovered by centrifugation. Then, it was filtered through a Buchner funnel (with acetone wash) and air-dried.

5.18.2.1.1 Standardization of Suitable Process for the Extraction and Recovery of Starch From Molded and Blackened Sorghum Grains

5.18.2.1.1.1 EXTRACTION AND RECOVERY OF STARCH The black sorghum grains were soaked in distilled water from 24 to 168 h separately at ambient temperature (27 ± 2°C) and 60°C. The soaking water was changed after every 24 h. The soaking grains were blended with water (1:2, w/v) in a laboratory blended for 2 min. The blending was repeated for two more minutes after a pause of 1 min. The slurry was screened through a double-layered muslin cloth. The residue on the muslin cloth was washed with additional water to solubilize and filter out the remaining starch. The filtrate was centrifuged at 8000 × g for 30 min. The supernatants were discarded and the residue was suspended in 0.2% sodium hydroxide solution to solubilize the proteins. The contents were allowed to stand for 3 to 4 h to settle the starch. The yellowish upper sodium hydroxide layer was decanted and the settled starch was resuspended in sodium hydroxide until the sodium hydroxide layer became colorless. The settled starch was then washed with water and the contents were screened through the 0.045 mm sieve. The filterate was centrifuged and the starch obtained was recorded as residue in centrifuge tubes and dried at ambient temperature (Fig. 5.2).

A maximum absorption of water was observed in the grins soaked at 60°C for 96 h. However, the highest starch yield was obtained when the grains were soaked at 60°C for 144 h. The dilute alkali (0.2%) was found to be a superior grain soaking and residue washing medium than the dilute acids or water. A significant higher starch yield was obtained with the use of 0.045 mm sieve (59.3%) as a slurry filtration device as compared to that of muslin cloth (33.8%).

The starch obtained with muslin cloth exhibited slightly higher purity but significantly lower recovery. A soaking of grains for 144 h at 60°C in 0.2% NaOH followed by grinding in water and filtration of the slurry through 0.045 mm sieve, centrifugation of the filtrate at 8000 × g for 30 min, washing of the residue in 0.2% NaOH followed by water washing, centrifugation, and drying of the residue has been standardized to obtain a maximum of 62.2% bright starch from molded and blackened sorghum. The product contained 91.6% starch (Fig. 5.3).

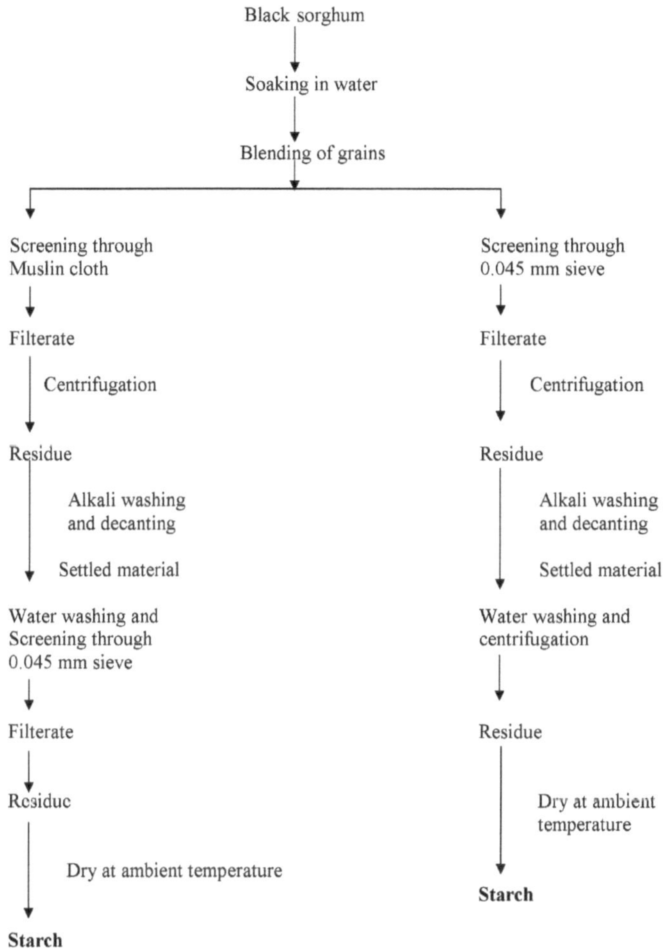

FIGURE 5.2 Standardization of starch extraction processes from sorghum.

A process to get white starch from the mold-infected sorghum with black or red/brown discoloration is described here. It involves soaking of grains in 0.2% solution of H_2SO_4 or NaOH (1:2, w/v) for 12 h at 60°C, followed by blending in the same extract for 2–3 min and filtration of slurry through 0.045 mm sieve (Fig. 5.4). The extracted starch is recovered by centrifugation of the filtrate at 8000 × g for 20 min, washing the starch-pellet with water to remove residual acid/alkali, and drying the product at 40°C. Sodium hydroxide was suitable for extraction of starch from grains with both black and red/brown discoloration while H_2SO_4 was suitable for only blackened grains. The physico-chemical properties of sorghum starch are given in Tables 5.3–5.12 and Figs. 5.5 and 5.6.

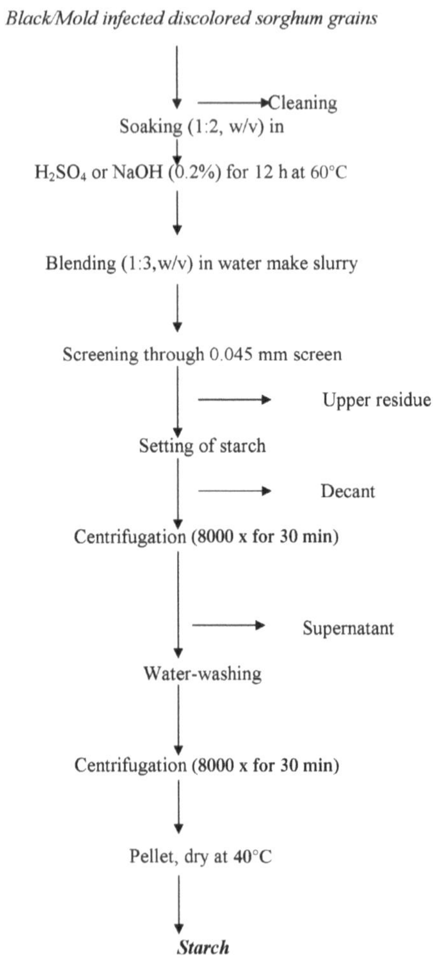

FIGURE 5.3 Standardized process for extraction of starch from black/mold infected discolored sorghum grains.

5.18.3 Standardization of Starch Production From Sorghum Grains

In the present investigation attempts have been made to standardize various parameters, that is, soaking time and temperature.

5.18.4 Effect of Soaking Temperature on the Yield of Starch

The effect of soaking temperature on the extent of release of starch from sorghum grains was studied. The soaking time was kept constant at 20 h and the temperature of the soaking medium was varied from 30, 40, 50, 55, and 60°C.

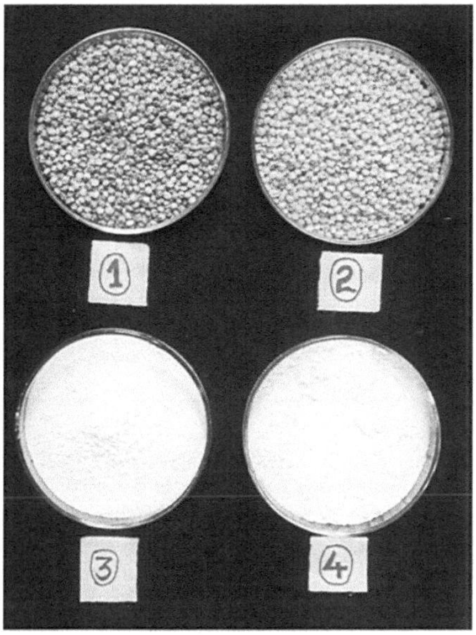

1.Molded grains 2. Sound grains,
3. Starch from 1 and 4. Starch from 2

FIGURE 5.4 Starch from molded and sound grain.

TABLE 5.3 Water Absorption by Mold Infected Blackened Sorghum

Grain soaking temperature (°C)	Weight gain after soaking of grains (h)			
	6	12	24	Mean
30	24.8 ± 0.6	31.9 ± 0.7	38.7 ± 0.8	31.8
40	27.2 ± 0.7	33.6 ± 0.6	40.1 ± 0.4	33.7
50	37.7 ± 0.6	36.4 ± 0.8	42.6 ± 0.7	38.9
60	39.9 ± 0.7	42.9 ± 0.6	43.0 ± 0.4	41.9
70	41.1 ± 0.4	43.0 ± 0.5	43.1 ± 0.6	42.4
Mean	33.1	37.6	41.5	37.7

TABLE 5.4 Effect of Soaking Medium on Extraction and Recovery of Starch From Black Sorghum

Grain soaking medium (%)	Extracted starch (%)
NaOH, 0.1	52.00 ± 2.4
H_2SO_4, 0.1	23.38 ± 1.8
Lactic acid, 0.1	34.4 ± 1.9
Water	33.8 ± 2.1

TABLE 5.5 Effect of Different Soaking Periods and Extraction Media on Yield of Starch From Black Sorghum

Soaking and extraction medium	Starch yield (%) After soaking grains (h)		
	6	12	24
Water	45.7 ± 1.6	52.2 ± 1.8	53.0 ± 1.7
Lactic acid, 0.2%	49.4 ± 1.9	56.7 ± 1.5	57.2 ± 1.5
Potassium metabisulfite, 0.2%	50.0 ± 1.8	59.2 ± 1.6	60.0 ± 1.7
Sodium hydroxide, 0.2%	52.8 ± 1.7	64.0 ± 1.4	64.3 ± 1.6
Sulfuric acid, 0.2%	53.7 ± 1.9	63.8 ± 1.8	64.1 ± 1.5

TABLE 5.6 Effect of Increasing Concentration of Alkali as Soaking Medium on Starch Extraction and Recovery

NaOH concentration (%)	Starch recovery (%)
0.1	52.0 ± 2.4
0.2	54.4 ± 2.6
0.5	54.8 ± 2.8

TABLE 5.7 Effect of Screening Method on Recovery of Starch

Screening method	Starch recovery (%)
Muslin cloth	33.8 ± 1.6
Sieve (0.045 mm)	59.3 ± 2.7

TABLE 5.8 Effect of Period and Temperature of Soaking of Grains on the Extraction and Recovery of Starch From Black Sorghum Grains

Soaking period (h)	Starch recovery (%) at	
	27 ± 2°C	60°C
24	17.9 ± 0.50	18.4 ± 0.90
48	20.2 ± 1.30	20.8 ± 1.10
72	23.7 ± 0.80	24.7 ± 0.80
96	27.2 ± 1.60	27.6 ± 1.20
120	28.5 ± 0.80	32.1 ± 1.60
144	32.0 ± 0.95	33.8 ± 0.85
168	32.4 ± 0.80	34.0 ± 0.60

TABLE 5.9 The Yield, Content, and Recovery of Starch by Different Methods of Extraction

Sr. no.	Type of method used			Percent starch in black grain	Starch product from grain (%)	Actual starch content in extracted product (%)	Starch recovery on initial grain basis (%)
	SM	BM	RW				
1.	W	W	N	64.7	33.8 ± 1.20	96.0 ± 1.59	50.2 ± 1.60
2.	N	W	N	64.7	52.0 ± 1.80	93.8 ± 1.65	75.4 ± 1.75
3.	N	N	W	64.7	48.3 ± 1.90	94.7 ± 1.57	70.7 ± 1.70
4.	W	W	N	64.7	59.3 ± 1.80	92.0 ± 1.83	84.3 ± 1.81
5.	N	W	N	64.7	62.2 ± 1.95	91.6 ± 1.92	88.0 ± 1.79
6.	N	N	W	64.7	61.8 ± 1.10	92.2 ± 1.50	88.1 ± 1.67

Method 1 to 3 with muslin cloth, Method 4 to 6 with 0.045 mm sieve. *SM*, soaking medium; *BM*, blending medium; *RW*, residue washing medium; *W*, water; *N*, NaOH (0.2%).

TABLE 5.10 Chemical Composition of Starch Extracted by Different Method

Starch extraction method	Starch (%)	Moisture (%)	Protein (%)	Total sugars (%)	Total ash (%)
1.	96.0 ± 1.59	4.5 ± 0.33	0.47 ± 0.10	0.47 ± 0.02	0.50 ± 0.03
2.	93.8 ± 1.65	4.0 ± 0.35	1.03 ± 0.09	0.45 ± 0.01	0.25 ± 0.02
3.	94.7 ± 1.60	5.0 ± 0.29	1.24 ± 0.08	0.43 ± 0.03	0.10 ± 0.04
4.	92.0 ± 1.57	4.5 ± 0.25	1.06 ± 0.06	0.41 ± 0.05	0.39 ± 0.02
5.	91.6 ± 1.92	4.8 ± 0.30	1.23 ± 0.07	0.45 ± 0.04	0.40 ± 0.03
6.	92.2 ± 1.50	4.8 ± 0.25	1.66 ± 0.06	0.43 ± 0.03	0.42 ± 0.01

Method 1 to 3 with muslin cloth, Method 4 to 6 with 0.045 mm sieve. Methods: 1. WWN, 2. NWN, 3. NNW, 4. WWN, 5. NWN, 6. NNW. *SM*, soaking medium; *BM*, blending medium; *RW*, residue washing medium; *W*, water; *N*, NaOH (0.2%).

5.18.5 Effect of Soaking Time on the Yield of Starch Sorghum Grains

In the present study, soaking time as a function of liberation of starch from grains was investigated. Unless otherwise mentioned the isolation and purification was essentially done by the method mentioned. Soaking time was varied, that is, 6, 9, 12, 16, 20, and 24 h. The yield of starch was calculated to optimize the soaking time.

TABLE 5.11 Functional Properties of Extracted Starch From Black Sorghum

Starch extraction method	Water absorption capacity (mL/g)	Oil absorption capacity (mL/g)	Gelatinization temperature (°C)
1.	0.90 ± 0.06	0.99 ± 0.05	78.6 ± 1.2
2.	1.12 ± 0.07	1.18 ± 0.06	77.2 ± 1.1
3.	1.20 ± 0.06	1.11 ± 0.07	78.0 ± 1.01
4.	0.92 ± 0.07	1.32 ± 0.04	80.2 ± 1.10
5.	0.93 ± 0.05	0.99 ± 0.06	76.8 ± 1.30
6.	0.95 ± 0.06	1.01 ± 0.05	78.3 ± 1.15

Method 1 to 3 with muslin cloth, Method 4 to 6 with 0.045 mm sieve. Methods: 1. WWN, 2. NWN, 3. NNW, 4.WWN, 5. NWN, 6. NNW. *SM*, soaking medium; *BM*, blending medium; *RW*, residue washing medium; *W*, water; *N*, NaOH (0.2%).

TABLE 5.12 Physicochemical Properties of Starch Extracted From Black or Red/Brown Discolored Sorghum

Parameter	Whole grains	H_2SO_4 extracted starch	NaOH extracted starch
Moisture, %	9.5	3.4	4.0
Starch, %	67.6	94.5	94.0
Protein, %	9.5	0.7	0.8
Ash, %	1.7	0.3	0.4
Fat, %	2.4	0.9	0.9
Fiber, %	2.2	–	–
GT, °C	–	77.5	78.3
Color	Black	White	White
Appearance	Red/brown	Pinkish	White

GT, gelatinization temperature.

5.18.6 Standardization of Production of Glucose Syrup From Sorghum Starch by Acid Hydrolysis

In the present investigation, the standardization of the production of glucose syrup from starch at various concentration (15, 25, and 35%, w/v) as a function of acid concentration (ie, 5, 10, 15, and 20%), various temperature (ie, 50, 60, 70, 80, and 90°C) for different time intervals (ie, 2, 4, 6, 8, 10, and 12 h) was performed.

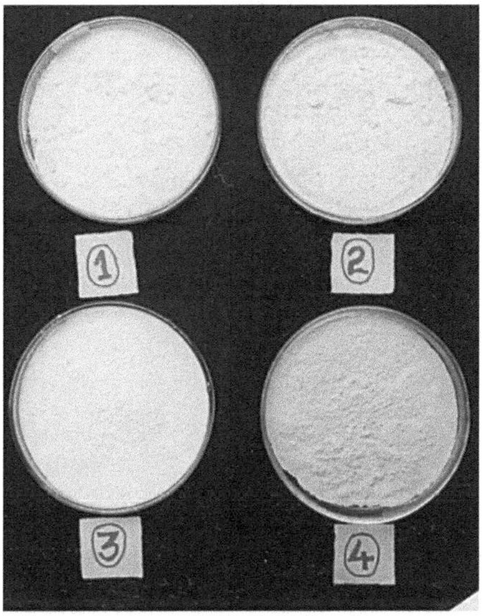

FIGURE 5.5 The physical appearance of starch obtained from four different soaking media (1. NaOH, 0.1%; 2. $H_2 SO_4$, 0.1%; 3. Lactic acid 0.1%; 4. Water).

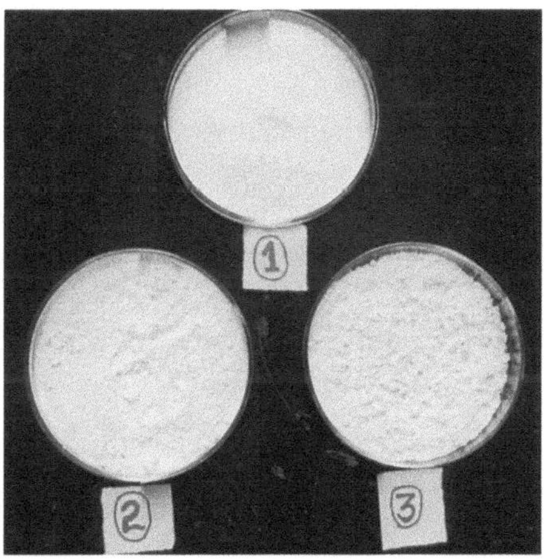

FIGURE 5.6 The physical appearance of starch obtained by using three levels of dilute alkali (1. 0.1; 2. 0.2; and 3. 0.5%).

5.18.7 Standardization of Dextrinization of Sorghum Starch Using α-Amylase

In the present investigation, there was a standardization of dextrinization of sorghum starch at different concentrations (ie, 15, 25, and 35%, w/v) as a function of α-amylolytic dextrinization, using various concentration of α-amylase (60, 65, and 70 units).

5.18.8 Standardization of Production of Glucose Syrup From Sorghum Starch Using Acid Hydrolysis Followed by α-Amylolytic Dextrinization

In the present study, partially dextrinized sorghum starch using optimum conditions of acid hydrolysis followed by α-amylatic dextrinization as a function of various enzyme concentrations was investigated.

5.18.9 Saccharification of Starch

In the present investigation, the saccharification of the liquefied starch was carried out using glucoamylase under the experimental conditions.

5.18.10 Preparation of Immobilized Glucoamylase on DEAE-Cellulose

In a typical experiment, the immobilization of glucoamylase on DEAE-Cellulose was performed as per the procedure described by Anonymous (1979). 53.0 g of a dry glucoamylase preparation (free of trans-glucosylase activity and negligible β-amylase activity) having a glucoamylase activity of 115 U/g was incorporated into 3.8 L of deionized water. The mixture was stirred for 30 min and filter aid added. The mixture was filtered, the filter cake washed, the filtrate and washings combined, and the pH of the combined solutions adjusted to 4.5 using 4 N HCL. 13.3 g DEAE Cellulose (Whatman DE 23) was added, the mixture stirred for 60 min at 60°C, and then filtered and the filter cake washed with deionized water. The recovered moist filter cake had a glucoamylase activity of 99.50 U/g. The moist filter cake is referred to as immobilized glucoamylase.

5.18.11 Effect of Glucoamylase Concentration on the Release of Glucose From Liquefied Starch

In the present study, glucoamylase concentration as a function of glucose syrup production was investigated (Fig. 5.7). Only enzyme concentration was varied, that is, 100, 115, and 130 units. The yield of glucose syrup was calculated to optimize the enzyme concentration.

5.19 ESTIMATION OF GLUCOSE

5.19.1 Paper Chromatography

5.19.1.1 Qualitative Estimation of Sugars

A Whatman paper (3 mm) was taken and the baseline was drawn with the help of a scale, 13 cm above the bottom of paper. The paper was

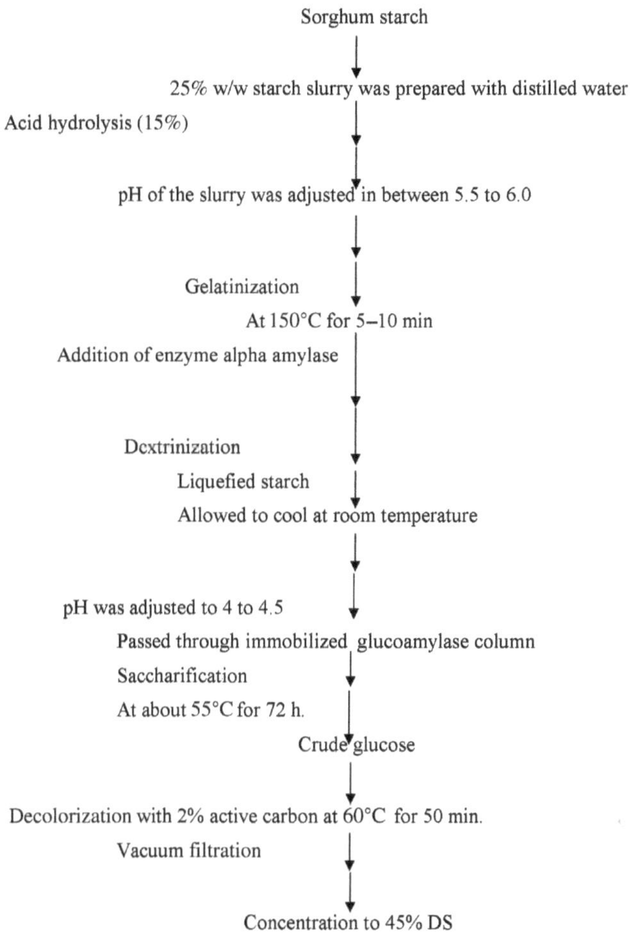

FIGURE 5.7 Enzymatic hydrolysis of sorghum starch to glucose syrup.

divided as per the sample numbers. The moving end of the paper was serrated to increase the flow rate of solvent. About 4 to 10 μL of sample was applied on the baseline with the help of capillary tubes. Simultaneously one standard was also applied on the baseline. The paper was kept in chromatography chamber and the solvent was applied in the boat. It was run for 18–20 h, followed by drying of the paper in air. It was then sprayed with the following reagents in sequence.

1. Saturated $AgNO_3$—Acetone solution-$AgNO_3$ was dissolved in 0.1 mL distilled water to form the saturated solution. This solution was added in 100 mL acetone. The precipitate formed was dissolved by adding little quantity of distilled water.

2. 0.1 M NaOH in methanol.
3. 0.5 M Sodium thiosulfate solution.

The sugars were detected on the paper chromatogram on the basis of their RF values. The solvent used for the above method was prepared as follows.

Butanol: Acetic acid: Distilled water (4:1:5, v/v) were taken in a separating funnel. It was mixed well for 15 to 20 min. After keeping them in standing position, two layers were formed. The lower layer was discarded and the upper layer was used as solvent for paper chromatography.

5.19.1.2 Quantitative Estimation of Glucose

Whatman paper (3 mm) was divided in two parts. One part was used for the samples for color development and another part for the same samples without color development. Paper was serrated to increase the slow rate. About 4 μL sample was applied on the base line of paper with the help of capillary tubes in the part used for color development. But for the part which was not sprayed with detecting reagent, the sample of known quantity (0.5 mL) was spotted with the help of micropipettes on the base line. The paper was kept in the chromatography chamber and the solvent was allowed to flow. Then the paper was removed and dried in the current of hot air. Half of the paper was sprayed with the detecting reagent prepared as above. After spraying, spots were developed on the paper. The spots which were parallel to the spot of standard sugars were chosen and the parallel area to these spots on the unsprayed paper was cut down. The prepared squares according to the samples applied were then divided into then strips with the help of scissors. These strips were dissolved in 3 mL of 70% ethanol, from which 1 mL of solvent was taken and color was developed as in Phenol H_2SO_4 method. The intensity of the color was read on the spectrophotometer at 480 nm. The xylose content was estimated from the calibration curve.

5.19.2 Thin Layer Chromatography

The quantitative thin layer chromatography was performed as per the methods of Wankhede and Tharanathan (1976), Tharanathan et al. (1976), and Wankhede et al. (1977) on precoated silica gel plastic sheets (Polygram Sil-G) with the solvent system:

1. 1-Butanol:Ethanol:Water (3:1:1) and
2. 1-Butanol:Pyridine:Water (6:4:3)

5.19.3 Gas Liquid Chromatography

Gas liquid chromatography was carried out by the method of Björndal et al. (1970) with a Netel chromatograph equipped with a flame ionization detector, fitted with a glass column (0.32 × 150 cm) packed with 3% of ECNSS-M on Gas Chrom-W (100–200 mesh), at a column temperature of 170°C. The flow rate of nitrogen gas was 28 mL/min. The retention time (T_R) of alditol acetates are given relative to 1, 5-di-O-acetyl-2,3,4,6-tetra-O-methyl-D-glucitol.

5.19.4 Carbon-Celite Chromatography

It was performed according to the method of Whistler and Durso (1950) as adopted and described by Wankhede and Tharanathan (1976). Animal charcoal was treated with concentrated hydrochloric acid followed by washing with distilled water thoroughly on a Buchner funnel. It was then activated at 120°C, overnight. The mixture of activated charcoal and Celite (2:1, w/w) was suspended in distilled water and a column (31.0 × cm) was packed with the above materials. The concentrated solution of sugars was placed on the column. It was washed with distilled water to elute the monosaccharides, while the oligosaccharides were desorbed by successive elution with ethanol (5–25%, v/v).

5.19.5 Preparation of Immobilized Glucose Isomerase on DEAE Cellulose

In the present study, the immobilization of glucose isomerase on DEAE cellulose was carried out as under 50 g of dry glucose isomerase having activity 125 unit/g dry weight was incorporated into 4 L of deionized water. The resultant mixture was subjected to stirring for 20 min. and the filter aid was added. The resultant mixture was filtered and washed and the filtrate and washing was combined and the pH of the combined solution was adjusted to 7.00 containing 0.05 M $CoCl_2$; 15 g of DEAE cellulose was added followed by stirring at 60°C for 60 min. It was then filtered and the filter cake washed with deionized water. The resultant moist filter cake had a glucose isomerase activity of 125 unit/g dry weight. The resultant moist filter cake used as a source of immobilized glucose isomerase.

5.19.6 Effect of Glucose Isomerase Concentration on the Release of Fructose From Glucose Syrup

In the present study, glucose isomerase as a function of fructose syrup production was investigated (Figs. 5.8 and 5.9). In this investigation, the concentration of enzyme was varied, that is, 125, 130, and 140 units. The yield of fructose syrup was calculated to optimize enzyme concentration.

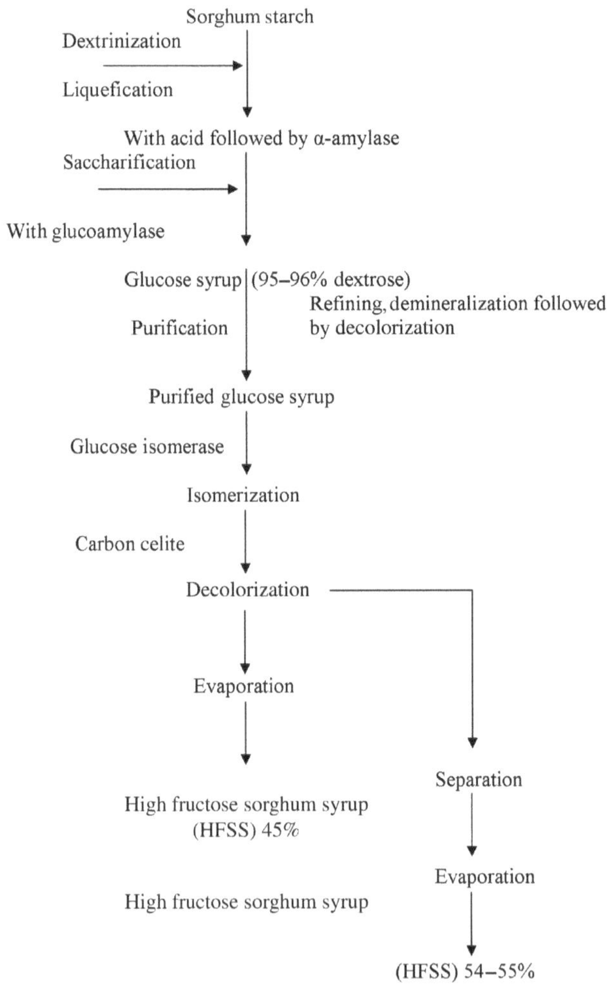

FIGURE 5.8 Production of high fructose syrup from dextrinized sorghum starch using immobilized glucose isomerase (*Artobacter* spp.) at optimum experimental conditions.

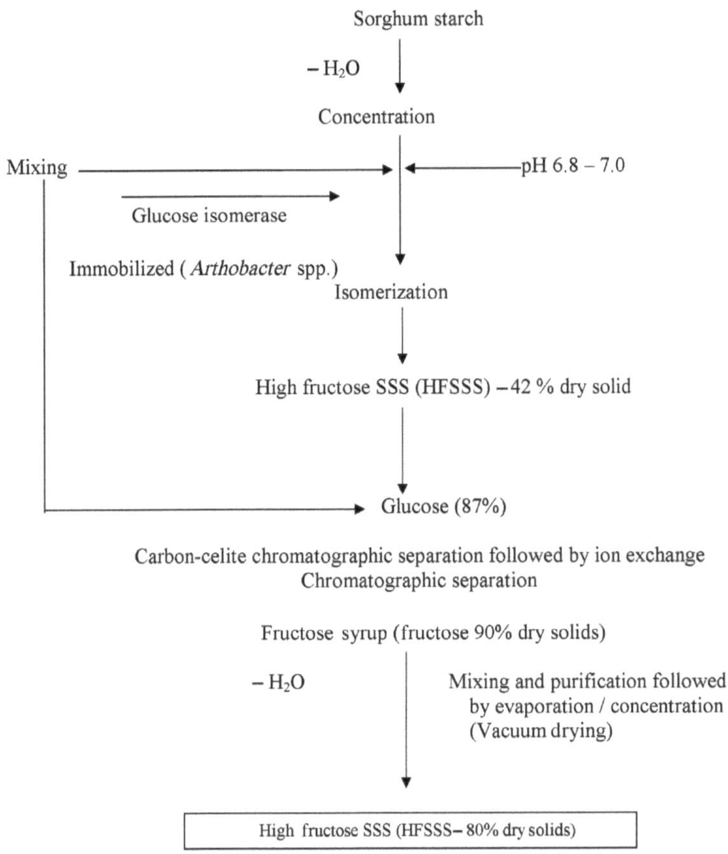

FIGURE 5.9 High glucose syrup from dextrinized sorghum starch syrup (DE 95, 40% solids).

5.19.7 Production of High Fructose Syrup From Rain Damaged Sorghum Grains

5.19.7.1 Chemical Composition of Sorghum Grains

The grain flour of sorghum genotype CSH 16 was analyzed quantitatively for its chemical composition. The results (Table 5.13) revealed that the starch content was found to be 70.30% in the grain sample. It is also clear from the results that sorghum grain contains a slightly lower amount of protein and fat.

The results obtained in this investigation are in good conformity with the results reported by Ganorkar (2001), Kulkarni (2000), and Neucere and Sumrell (1980). However possibilities of differences in composition of sorghum grains may be due to environmental stress,

TABLE 5.13 Proximate Composition of Sorghum Grains[a]

S. no.	Parameters (%)	CSH 16
1.	Moisture	8.70
2.	Protein	8.92
3.	Starch	70.30
4.	Soluble sugars	2.85
5.	Crude fiber	2.73
6.	Crude fat	3.59
7.	Ash	1.30

[a]*Unless otherwise mentioned, the values represent the average of three determinations.*

geographical differences, and cultivation practices. Nevertheless there is good agreement with regards to starch content among the reports claimed by different investigators.

5.19.8 Isolation and Standardization of Starch From Sorghum Grains

Starch was isolated from sorghum grains by a wet-milling process under optimized condition.

5.19.8.1 Chemical Composition of Starch

Chemical composition of sorghum starch obtained from CSH 16 is presented in Table 5.14. The data indicated that the carbohydrate content was 95.3% of starch. After exhaustive purification the protein content of sorghum starches could be reduced to 1.2% which was comparatively higher as compared to the values reported by other workers (Subramaniam and Jambhunathan, 1981; Wankhede and Umadevi, 1982; and Ganorkar, 2001).

5.19.9 Studies on Standardization of Dextrinization of Starch With Acid

The results indicated that the starch was liquefied/dextrinized by using acid under experimental conditions.

TABLE 5.14 Proximate Composition and Physico-Chemical Properties of Sorghum Starch[a]

S. no.	Parameters	Percent content
1.	Moisture	7.2
2.	Protein (%)	1.2
3.	Fat (%)	1.1
4.	Total carbohydrate (%)	95.3
5.	Ash (%)	0.28
6.	Crude fiber	0.82
7.	Yield	68.0
8.	Starch granule characteristics shape of granules	Polygonal to round
9.	Color of starch powder	Bright white

[a]Proximate analysis was carried out in moisture free samples. The values represent the average of three determinations.

TABLE 5.15 Effect of Acid Concentration on the Dextrinization of Starch by Acid[a]

Starch (g)	Acid concentration (% v/v)	Hydrolysis (%)	Dextrinized Product (g)
25	5	12.00	3.00
25	10	36.00	9.00
25	15	45.00	11.25
25	20	44.00	11.00

[a]Each value represents average of three determinations. Unless otherwise mentioned, Acid hydrolysis of starch was carried out with hydrochloric acid at temperature of 50°C for period of 10 h.

5.19.9.1 *Effect of Acid Concentration on the Dextrinization of Starch by Acid*

It is clear from the results reported in Table 5.15 that when the acid concentration is increased from 5 to 20%, hydrolysis is also increased.

5.19.9.2 *Effect of Starch Concentration on the Dextrinization of Starch by Acid*

It is revealed from the results (Table 5.16) that the rate of liquefaction of starch is optimum at starch concentration of 25%.

TABLE 5.16 Effect of Starch Concentration on the Dextrinization of Starch by Acid[a]

Starch (%)	Acid concentration (% v/v)	Hydrolysis (%)	Dextrinized product (g)
15	15	23	3.45
25	15	44	11.00
35	15	27	9.45

[a]*Each value represents average of three determinations. Unless otherwise mentioned, acid hydrolysis of starch was carried out with 15% hydrochloric acid at temperature of 50°C for period of 10 h.*

TABLE 5.17 Effect of Temperature on the Dextrinization of Starch by Acid[a]

Starch (g)	Temperature (°C)	Hydrolysis (%)	Dextrinized product (g)
25	40	42.25	10.55
25	50	44.00	11.00
25	60	40.50	10.10
25	70	41.25	10.30
25	80	40.00	10.00
25	90	42.00	10.50

[a]*Each value represents average of three determinations. Unless otherwise mentioned, acid hydrolysis of starch was carried out with 15% hydrochloric acid for period of 10 h.*

5.19.9.3 *Effect of Temperature on the Dextrinization of Starch by Acid*

It is evident from the results reported in Table 5.17, that the dextrinization of the starch was decreased as the temperature of hydrolysis increased. The maximum hydrolysis was observed at 50°C.

5.19.9.4 *Effect of Time on the Dextrinization of Starch by Acid*

It can be from the results (Table 5.18) that as the time of dextrinization is increased the percentage of hydrolysis of sorghum starch is found to increase. The time of 10 h was found to be optimum for the highest dextrinization of starch.

TABLE 5.18 Effect of Time on the Dextrinization of Starch by Acid[a]

Starch (g)	Time (h)	Hydrolysis (%)	Dextrinized product (g)
25	2	6	1.50
25	4	13	3.12
25	6	22	5.50
25	8	30	7.50
25	10	44	11
25	12	36	9.0

[a]*Each value represents average of three determination. Unless otherwise mentioned, acid hydrolysis of starch was carried out with 15% hydrochloric acid at temperature of 50°C. Unless otherwise indicated, D-glucose, maltose, and α-dextrins were detected in the acid hydrolysates of starch.*

TABLE 5.19 Effect of Starch Concentration on the Dextrinization of Starch by α-Amylase[a]

Starch (%)	Enzyme concentration (units)	Hydrolysis (%)	Dextrinized product (g)
20	60	27	5.40
25	60	43	10.75
30	60	30	9.00

Each value represents average of three determinations.
[a]*Unless otherwise mentioned enzyme concentration, time, and temperature are 60 units, 48 h, and 70°C, respectively, kept constant throughout study.*

5.19.10 Studies on Standardization of Dextrinization of Starch by α-Amylase

5.19.10.1 Effect of Starch Concentration on the Dextrinization of Starch by α-Amylase

It is evident from data given in Table 5.19 that as the concentration of starch increased from 20 to 25%, the percentage of hydrolysis of starch was also increased. It is also observed that a further increase in concentration showed an adverse effect on hydrolysis. Therefore, 25% starch was found to be optimum for dextrinization of starch. These results are in close agreement with the results reported by Kshirsagar (1986).

Enzyme hydrolysis of starch was carried out with α-amylase under optimum experimental conditions. Unless otherwise mentioned, D-glucose, maltose, malto-oligosaccharides along with α-dextrins were detected in the hydrolysates

TABLE 5.20 Effect of α-Amylase Concentration on the Dextrinization of Starch by α-Amylase[a]

Starch (g)	Enzyme Concentration (Units)	Hydrolysis (%)	Dextrinized product (g)
25	60	33	8.25
25	65	43	10.75
25	70	38	9.50

Each value represents average of three determinations.
[a]*Unless otherwise mentioned starch concentration, time, and temperature are 25%, 48 h, and 70°C, respectively, kept constant throughout study.*

TABLE 5.21 Effect of Temperature on Dextrinization of Starch by α-Amylase[a]

S. no.	Starch	Temperature	Hydrolysis	Dextrinized product (g)
1.	25	50	33.2	8.30
2.	25	60	41.55	10.37
3.	25	70	43	10.75
4.	25	80	42.8	10.70
5.	25	90	40.5	10.12

[a]*Unless otherwise mentioned starch concentration, time, and enzyme concentration are 25%, 48 h, and 65 units, respectively, kept constant throughout study.*

5.19.10.2 Effect of α-Amylase Concentration on the Dextrinization of Starch by α-Amylase

It is evident from data given in Table 5.20 that, as the concentration of enzyme increased from 60 to 65 units the percentage of hydrolysis of starch was also increased. It is also observed that a further increase in concentration showed an adverse effect on hydrolysis. Therefore 65 units of enzyme was used in the present study. These results are in close agreement with the results reported by Kshirsagar (1986).

Enzyme hydrolysis of starch was carried out with α-amylase under optimum experimental conditions (Tables 5.21–5.23).

The data pertaining to effect of pH on dextrinization of starch by α-amylase was depicted in Table 5.23. The effect of pH ranging from 4.5 to 7.0 on the dextrinization of starch by α-amylase was studied. As the pH was increased from 4.5 to 6.5% hydrolysis increased. However, at pH 7.0 the percentage of hydrolysis was found to decrease. Therefore, a pH of 6.5 was found to be optimum for the dextrinization of starch by α-amylase.

TABLE 5.22 Effect of Time on Dextrinization of Starch by α-Amylase[a]

S. no.	Starch	Time	Hydrolysis (%)	Dextrinized product (g)
1.	25	12	28.6	7.15
2.	25	24	32.7	8.17
3.	25	36	41.5	10.37
4.	25	48	43.0	10.75
5.	25	60	42.5	10.62
6.	25	72	41.3	10.32

[a] *Unless otherwise mentioned starch concentration, temperature, and enzyme concentrations are 25%, 70°C, and 65 units, respectively, kept constant throughout study.*

TABLE 5.23 Effect of pH on Dextrinization of Starch by α-Amylase[a]

S. no.	Starch	pH	Hydrolysis (%)	Dextrinized product (g)
1.	25	4.5	32.8	8.20
2.	25	5.0	34.4	8.60
3.	25	5.5	40.3	10.07
4.	25	6.0	42.5	10.62
5.	25	6.5	43.0	10.75
6.	25	7.0	42.8	10.70

[a] *Unless otherwise mentioned starch concentration, temperature, enzyme concentration, and time are 25%, 70°C, 65 units, and 48 h, respectively, kept constant throughout study.*

5.19.11 Standardization of Saccharification of Starch

It is well known that starch hydrolysis was undertaken historically using acid. However, it has been experienced that this possesses some disadvantages such as corrosion of the reactor, poor yield of hydrolysis, formation of unwanted sugars and sugar derivatives possessing bitter taste, etc. Moreover, the problems of caramelization, pyrodextrins formation, and odd taste, especially salty, cannot be ruled out exclusively in the acid hydrolysis method.

Therefore, attempts were made to hydrolyze starch with an optimum concentration of acid at mild conditions followed by α-amylase hydrolysis to obtain maltodextrin (dextrinization). The resultant dextrinized starch syrup was subjected to saccharification (ie, sweet taste) using immobilized glucoamylase.

TABLE 5.24 Effect of Percent Dry Solids of Dextrin Syrup of Sorghum Starch on Saccharification by Glucoamylase[a]

S. no.	Percent dry solids of dextrin syrup of sorghum starch[b]	Enzyme concentration (Units)	Saccharification (%)	Glucose (g)
1.	20	115	80.2	16.04
2.	25	115	94.5	23.62
3.	30	115	92.1	22.63

[a]*Unless otherwise mentioned enzyme concentration, temperature, and time are 115 units, 70°C, and 10 h, respectively, kept constant throughout study.*
[b]*The Dextrose equivalent (DE) of 20, 25, and 30% are 9.2, 11.5, and 13.8, respectively.*

TABLE 5.25 Effect of Enzyme Concentration on Saccharification of Sorghum Starch by Glucoamylase[a]

S. no.	Percent dry solids of dextrin syrup of sorghum starch[b]	Enzyme concentration (units)	Saccharification (%)	Glucose (g)
1.	25	110	90.5	22.62
2.	25	115	94.5	23.72
3.	25	120	92.4	23.01

[a]*Unless otherwise mentioned percent dry solids of dextrin syrup of sorghum starch, temperature, and time are 25%, 70°C, and 10 h, respectively, kept constant throughout study.*
[b]*The Dextrose equivalent (DE) = 11.5.*

5.19.11.1 *Effect of Percent Dry Solids of Dextrin Syrup of Sorghum Starch on Saccharification by Glucoamylase*

Sincere attempts were made to standardize the percentage of dry solids of dextrin syrup from sorghum starch on saccharification by glucoamylase. It is evident from the data given in Table 5.24 that as the sorghum starch dextrin syrup (Dry solids %) increased from 20–25%, the percentage of saccharification was increased from 80.2 to 94.5%. However, at 30% dry solids the percentage of saccharification was found to decrease. Therefore, 25% dry solids was found to be optimum.

5.19.11.2 *Effect of Enzyme Concentration on Saccharification of Sorghum Starch by Glucoamylase*

The effect of enzyme concentration on saccharification of sorghum starch by glucoamylase from 110 to 120 units was studied and the results are presented in Table 5.25. The data revealed that as the

TABLE 5.26 Production of Glucose Syrup by Saccharification Using Native Gluco-Amylase at Optimum Condition[a]

S. No.	Sorghum starch dextrin syrup (dry solids percentage)[b]	Enzyme concentration (units)	Saccharrification (%)	Glucose (g)
1.	25	115	94.5	23.62

[a]*Unless otherwise mentioned percent dry solids of dextrin syrup of sorghum starch, time, temperature, and enzyme concentration are 25%, 10 h, 60°C, and 115 units, respectively, kept constant throughout study.*
[b]*The Dextrose equivalent (DE) = 11.5.*

concentration of enzyme increased from 110 to 115 units, the percentage of saccharification also increased. However, as the enzyme concentration exceeds 115 units, it adversely effects the percentage of saccharification. So, an enzyme concentration of 115 units was found to be optimum.

5.19.11.3 *Effect of Time on Saccharification of Sorghum Starch by Glucoamylase*

The data pertaining to effect of time ranging from 2 to 12 h on saccharification of sorghum starch by glucoamylase was presented in Table 5.26. As the time increased the percentage of saccharification also increased. But after 10 h the percentage of saccharification was found to decrease from 94.5 to 93.62%. Therefore, a time of 10 h was found to be optimal for saccharification of starch by glucoamylase.

5.19.11.4 *Effect of Temperature on Saccharification of Sorghum Starch by Glucoamylase*

Sincere efforts were made to optimize the temperature for saccharification of sorghum starch by glucoamylase. The effect of temperature ranging from 50% to 90°C was studied (Table 5.27). As the temperature increased from 50% to 70°C there was an increasing percentage of saccharification. But at elevated temperature (80–90°C), the percentage of saccharification was found to decrease. Therefore, a temperature of 70°C was found to be optimum. Production of glucose syrup by saccharification using native glucoamylase at optimum condition (Table 5.28).

Attempts were made to undertake the optimum production of glucose-syrup by saccharification employing purified glucoamylase at optimum experimental conditions. It is evident from the results that the sorghum starch dextrin syrup to the extent of 25% dry solids employing glucoamylase at optimum enzyme activity (115 units). The yield of saccharification (ie, dextrose yield) was found to be 94.5% on a laboratory scale in a batch type operation.

TABLE 5.27 Effect of Temperature on Saccharification of Sorghum Starch by Glucoamylase[a]

S. no.	Percent dry solids of dextrin syrup of sorghum starch[b]	Temperature (°C)	Saccharification (%)	Glucose (g)
1.	25	50	82.3	20.57
2.	25	60	94.3	23.63
3.	25	70	94.5	23.62
4.	25	80	92.2	23.05
5.	25	90	88.16	22.04

[a]*Unless otherwise mentioned percent dry solids of dextrin syrup of sorghum starch, time, and enzyme concentration are 25%, 10 h, and 115 units, respectively, kept constant throughout study.*
[b]*The Dextrose equivalent (DE) = 11.5.*

TABLE 5.28 Effect of Time on Saccharification of Sorghum Starch by Glucoamylase[a]

S. no.	Percent dry solids of dextrin syrup of sorghum starch[b]	Time (h)	Saccharification (%)	Glucose (g)
1.	25	2	52.1	13.025
2.	25	4	65.6	16.40
3.	25	6	80.2	20.05
4.	25	8	89.6	22.40
5.	25	10	94.5	23.62
6.	25	12	93.62	23.40

[a]*Unless otherwise mentioned percent dry solids of dextrin syrup of sorghum starch, temperature, and enzyme concentration are 25%, 70°C, and 115 units, respectively, kept constant throughout study.*
[b]*The Dextrose equivalent (DE) = 11.5.*

In this investigation the production of glucose syrup (ie, saccharification) was undertaken using immobilized glucoamylase (*A. niger*) at optimum experimental condition at various time intervals (Table 5.29). It can been seen from experimental data presented in Table 5.29 that the production of glucose-syrup was higher to the extent of 94.5% saccharification as compared to native glucoamylase. However, it has been observed that as the time of saccharification increased from 24–120 h, there was a decline trend in the saccharification (%) of dextrinzed sorghum starch syrup. This may be attributed to the fact that the resultant decline may be due to the lowering effect of the activity on α–1,6 linkages of the sorghum starch dextrins. The decline

TABLE 5.29 Variation in the Production of Glucose Syrup on the Saccharification Using Immobilized Glucoamylase at Optimum Experimental Condition[a]

S. no.	Period (h)	Glucoamylase (units)[b]	Saccharification (%)	Glucose (g)
1.	0	115	94.5	23.56
2.	24	115	93.0	23.25
3.	48	100	92.8	23.20
4.	72	100	92.8	23.20
5.	96	98.5	85	21.25
6.	120	70	80	20.00

[a] *Unless otherwise mentioned percent dry solids of dextrin syrup of sorghum starch, time, temperature, and enzyme concentration are 25%, 10 h, 70°C, and 115 units, respectively, kept constant throughout study.*
[b] *Unless otherwise mentioned the production of glucose syrup was undertaken with batch type laboratory scale using pharmacia double jacketed glass (quartz) column without peristaltic pump equipped with other accessories.*

trend in the saccharifying activity at least to some extent can be correlated to the formation of transglycosidation products especially isomaltose.

5.20 STARCH AND MALTODEXTRIN FROM SORGHUM

The starch prepared from black sorghum grain has a vital role as a potential raw material for the production of maltodextrin and it may be explored for industrial or commercial exploitation in different food products as a fat replacement to reduce the calorific value. The metabolism of maltodextrins is similar to starch. Food products for diabetic patients can be prepared by using maltodextrins as a fat replacement. Maltodextrins give rise to hypoglycemic activity and they are also a promising food substance for heart patients because it acts as a very good hypocholesterolemic agent.

The maximum amount of maltodextrin was liberated at 4.5% concentration of hydrochloric acid hydrolysis process at 120 min interval, that is, 95 and 92% for PVK 801 and CSH 9 sorghum varieties and their dextrose equivalents were 9 and 8, respectively. The maltodextrin contained 98 and 97% total carbohydrate in PVK 801 and CSH 9, respectively (Table 5.30). The results of rheological attributes, particularly the pasting behavior of maltodextrin, revealed that the viscosity was found to be maximum at 40°C and then decreased as the temperature increased. The microscopic examination of maltodextrin granules revealed that they are mostly irregular-shaped particle aggregates of

TABLE 5.30 Chemical Composition and Physical Parameters of Sorghum Starch

	Content (%)	
Parameter	**PVK-801**	**CSH-9**
Moisture	9.80	9.50
Protein	1.10	1.29
Fat	0.90	0.95
Ash	0.24	0.45
Crude fiber	0.73	0.67
Total carbohydrate	96.81	95.60
Yield of starch	75.50	72.85
Granule shape	Polygonal to round	Polygonal to round
Granule size	8.0–9.5 μm	8.0–9.1 μm
Color of starch	Bright white	Mild white
Iodine affinity	4.4	3.9
Bulk density	0.59	0.58
Molecular weight		
Amylose (daltan)	2,45,000 + 10,000	2,45,000 + 10,000
Amylopectin (daltan)	$>0.59 \times 10^6$	$>0.59 \times 10^6$

4–5 μm in size (Table 5.31). The results indicated that good quality cookies and toffees could be prepared using maltodextrin with 3% replacement of shortening. Maltodextrins are metabolized similar to starch. Thus, fat replacers like maltodextrin can be recommended for use in food products for diabetics, as it exhibits hypoglycemic activity. Moreover, it will be a very promising for heart patients also as it acts as a very good hypocholesterolemic agent. The comparative characteristics of starch and maltodextrin of sorghum are given in Table 5.32 and Fig. 5.10.

5.21 GLUCOSE AND HIGH FRUCTOSE SYRUP

5.21.1 Production of Glucose Syrup by Saccharification Using Native Glucoamylase at Optimum Condition

Attempts were made in the NATP project to undertake the optimum production of glucose-syrup by saccharification employing purified

TABLE 5.31 Physico-Chemical Properties of Sorghum Starch and Maltodextrin

Constituent	Starch	Maltodextrin
Moisture (%)	9.50–9.80	7.5–8.0
Protein (%)	1.10–1.29	0.05–0.07
Fat (%)	0.90–0.95	0.07–0.08
Ash (%)	0.24–0.45	0.6–0.7
Fiber (%)	0.67–0.73	0.4–0.5
Total carbohydrates (%)	95.60–96.81	97–98
Yield (%)	73–76	–
Granule shape	Round, polygonal	Irregular
Granule size (μm)	8.0–9.5	3.0–5.0
Color of granule	Bright white	Mild white
Iodine affinity	3.9–4.4	–
Bulk density	0.58–0.59	0.57–0.59
Molecular weight of amylose (daltan)	2,45,000	1,75,000
Molecular weight of amylopectin (daltan)	$>0.59 \times 10^6$	$>0.59 \times 10^6$

TABLE 5.32 Physico-Chemical Constituents of Maltodextrin

	Content (%)	
Parameter	PVK-801	CSH-9
Moisture	8.0	7.6
Protein	0.05	0.07
Carbohydrate	98.0	97.0
Fat	0.07	0.08
Ash	0.6	0.7
Bulk density	0.579	0.575
Granule shape	Irregular	Irregular
Granule size	3–5 μm	4–5 μm
Granule color	Bright white	White
Molecular weight		
Amylose (daltan)	1,75,000 + 5000	1,75,000 + 5000
Amylopectin (daltan)	$>0.59 \times 10^6$	$>0.59 \times 10^6$
Dextrose equivalent	9	8

Sorghum grains

↓

Cleaning (removal of foreign matters)

↓

Steeping in water (20 h, at 50°C and 4.5 pH)
(750 ppm KMS, 0.002 M $HgCl_2$, few drops of toluene in water)

↓

Washing with clean water

↓

Blending/preparation of slurry (1:3, w/v, seed : water)

↓

Separation of starch
(Squeezing of slurry through muslin cloth and then passing through 150 m sieve)

↓

Protein denaturation
(Add 1.5 g NaCl and 5 mLToluene in slurry)

↓

Shaking for 1 h

↓

Centrifugation at 4200 RPM for 15 min.

↓

Isolation of starch

↓

Drying (vacuum drying)

↓

Air drying for over night

↓

Starch

↓

Acid hydrolysis
(4.5 % HCl conc. 50°C temp. and 120 min time)
Vacuum drying

↓

Isolation → Maltodextrin

FIGURE 5.10 Flow diagrams for the isolation and purification of starch and maltodextrin.

glucoamylase at optimum experimental conditions. It is evident from the results that the sorghum starch dextrin syrup to the extent of 25% dry solids was used employing glucoamylase at the optimum enzyme activity (115 units) to produce glucose syrup. The yield of saccharification (ie, dextrose yield) was found to be 94.5% on a laboratory scale in a batch type operation.

The production of glucose syrup employing immobilized glucoamylase (*A. niger*) at optimum conditions was undertaken experimentally (Table 5.26). It indicates that the production of glucose-syrup was

TABLE 5.33 Selection of Promising Genotypes of Sweet Sorghum for the Production of High Fructose Syrup

S. no.	Genotype	Total sugar (%)	Nonreducing sugar (%)
1.	RSSV-9	18.8	15.76
2.	Keller	18.5	16.30

higher to the extent of 94.5% saccharification as compared to native glucoamylase. However, it has been observed that as the time of saccharification increased from 24 to 120 h, there was a decline in the saccharification (%) of dextrinized sorghum starch syrup. This may be attributed to the fact that the resultant decline may be due to the lowering effect of the activity on α-1, 6 linkages of the sorghum starch dextrins. The decline trend in the saccharifying activity at least to some extent can be correlated to the formation of transglycosidation products, especially isomaltose.

The optimal conditions for the isomerization were found to be 125 units of enzyme concentration at 65°C with pH 7.0.

5.21.2 Production of High Fructose Syrup From Sweet Sorghum Juice

Attempts have been made to identify the promising and high yielding juice genotypes of sweet sorghum. The highest and most significant juice extraction was recorded in RSSV-9 and the Brix of up to tune of 16.70% was recorded in genotype Keller (Table 5.33). Similarly it has been reported that the maximum value (3.56) and higher total sugar content and nonreducing sugar content was reported in RSSV-9 and Keller, respectively.

5.21.3 Production of High Fructose Syrup From Juice of Sweet Sorghum Genotype

It is well known that the sweet sorghum juice contains glucose, fructose, and sucrose. The hydrolysis of sucrose can be achieved by acid hydrolysis. However, it has been observed that this has several disadvantages such as caramelized color formation, inhibition by product formation, corrosion of the reactor, etc. Moreover, the involvement of costly refining processes, relatively low degree of hydrolysis, and usages of highly purified chemical as a starting material for hydrolysis will lead to an uneconomical and costly process. Therefore attempts were made on a laboratory scale as a trial using invertase enzyme.

TABLE 5.34 Production of HFS Using Native Invertase, a Batch Type on Laboratory Scale at 65°C for 48 h and pH 4.5

S. no.	Genotype	Initial fructose	Final fructose
1.	RSSV-9	3.15	9.32
2.	Keller	3.26	9.78

TABLE 5.35 Variation in the Production of Glucose Syrup on the Saccharification Using Immobilized Glucoamylase During Recycling at Optimum Experimental Conditions[a]

	Saccharification (%) Period (h)[b]		Glucose (g) Period (h)	
Usage of recycled immobilized glucoamylase	**0**	**24**	**0**	**24**
I-cycle	94.5	93.0	23.62	23.25
II-cycle	80.3	64.8	20.07	16.20

[a]*Unless otherwise mentioned percent dry solids of dextrin syrup of sorghum starch, time, temperature, and enzyme concentration are 25%, 10 h, 70°C, and 53 g, respectively, kept constant throughout study.*
[b]*Unless otherwise mentioned the production of glucose syrup was undertaken with batch type laboratory scale using pharmacia doubble jacketed glass (quartz) column without peristaltic pump equipped with other accessories.*

In the typical experiment, 250 mL of juice of the above varieties were subjected to invertase hydrolysis in the Pharmacia double jacketed glass (quartz) columns (2.5 × 55 cm) at a sugar concentration of 40% at 65°C for 48 h. The results revealed that the fructose concentration was found to increase considerably in the resultant reaction mixture. The result (Table 5.34) showed considerable increase in the fructose content at laboratory scale.

5.21.4 Variation in the Production of Glucose Syrup on the Saccharification Using Immobilized Glucoamylase During Recycling at Optimum Experimental Conditions

The effect of using recycled glucoamylase on the production of glucose-syrup was studied (Table 5.35). It is observed from the data that the optimum production of glucose-syrup at 0 and 24 h was found to be 94.5 and 93.0% respectively in first cycle. However, it is evident from the result that the saccharifying activity of sorghum dextrinized syrup was found to be comparatively lower during the second cycle. This may be

TABLE 5.36 Effect of Enzyme Concentration on Isomerization of Glucose Syrup by Glucose Isomerase[a]

S. no.	Glucose syrup (%)	Enzyme concentration (units)	Glucose (%)	Malto-oligosaccharides (%)	Fructose (%)
1.	40	115	51.62	6.58	41.80
2.	40	120	50.84	6.76	42.40
3.	40	125	47.50	7.48	45.00
4.	40	130	48.38	7.12	44.50

[a]*Unless otherwise mentioned percent glucose syrup, pH, and temperature are 40%, 7.0, and 65°C. These were kept constant throughout study.*

attributed to the lowering of glucoamylase activity during the cycling operation. The possibility of inhibition by the desired other products (sugar derivatives) due to transglucocidation cannot be ruled out.

5.22 STUDIES ON STANDARDIZATION OF ISOMERIZATION OF GLUCOSE SYRUP

5.22.1 Effect of Enzyme Concentration on Isomerization of Glucose Syrup by Glucose Isomerase

Sincere efforts were made to study effect of enzyme concentration on isomerization of glucose syrup by glucose isomerase. The data related with the same is depicted in Table 5.36. It is evident from Table 5.36 that as the enzyme concentration increases from 115 to 125 units there is an increase in the percentage of fructose. However, at an enzyme concentration of 130 units there was decline trend in the percentage of fructose. Therefore 125 units of enzyme concentration was found to be optimized for the isomerization of glucose syrup by glucose isomerase.

5.22.2 Effect of Temperature on Isomerization of Glucose Syrup by Glucose Isomerase

The effect of temperature ranging from 45 to 85°C on the isomerization of glucose syrup by glucose isomerase was studied. The related data are presented in Table 5.37. From the data it can be clearly summarized that as the temperature increased from 45 to 65°C there was increase in the percentage of fructose. However, at elevated temperature (75–85°C), there was decline in the percentage of fructose. Therefore, a temperature of 65°C was found to be optimal.

TABLE 5.37 Effect of Temperature on Isomerization of Glucose Syrup by Glucose Isomerase[a]

S. no.	Glucose syrup (%)	Temperature	Glucose (%)	Malto-oligosaccharides (%)	Fructose (%)
1.	40	45	50.62	6.58	42.80
2.	40	55	47.12	8.08	44.80
3.	40	65	47.53	7.45	45.00
4.	40	75	49.18	7.48	43.28

[a]*Unless otherwise mentioned percent glucose syrup, pH, are 40%, 7.0. These were kept constant throughout study.*

TABLE 5.38 Effect of pH on Isomerization of Glucose Syrup by Glucose Isomerase[a]

S. no.	Glucose syrup (%)	pH	Glucose (%)	Malto-oligosaccharides (%)	Fructose (%)
1.	40	5	50.13	7.07	42.8
2.	40	6	48.38	7.12	44.5
3.	40	7	47.50	7.48	45.0
4.	40	8	51.75	4.45	43.8
5.	40	9	52.02	5.48	42.5

[a]*Unless otherwise mentioned percent glucose syrup, temperature, are 40%, 65°C. These were kept constant throughout study.*

5.22.3 Effect of pH on Isomerization of Glucose Syrup by Glucose Isomerase

The effect of pH ranging from 5 to 7.5 on isomerization of glucose syrup by glucose isomerase was investigated. The data obtained is presented in Table 5.38. It is evident from the Table that as pH increased from 5 to 7 there was increasing percentage of fructose. But a further increase in pH, adversely affected the percentage of fructose. Therefore, pH 7 was found to be the optimal condition.

5.22.4 Studies on the Production of Sorbitol From Purified Sorghum Starch Glucose

The results on sorbitol production indicated that the glucose was converted to sorbitol using Raney Nickel catalyst under experimental conditions. It is evident from data that the conversion of glucose to sorbitol was found to the extent of 90%. The hydrogenation kinetics was investigated using various parameters. Hydrogenation of purified sorghum starch glucose (50% in 200 mL) was carried out in the presence of a Nickel catalyst (4%) and hydrogen gas (pressure 400 psi) at a temperature of 140°C.

The sorbitol salt obtained from hydrogenation of glucose was analyzed and found that it had various impurities such as Nickel catalyst, color, organic acid, and breakdown products of glucose. Removal of all these impurities is necessary for the recovery of pure sorbitol with a good yield. Raney Nickel catalyst is removed from the solution by filtration and the other organic impurities are removed by passing hydrogenated syrup successively through columns of a strong acid cation exchange resin and a weak base anion exchange resin. The pure sorbitol crystals in stable form were obtained by cooling sorbitol syrup of 68 Brix from 25 to 15°C within 30 min with stirring.

5.22.5 Production of Sorbitol From Sorghum Starch

Sorbitol solution appears to be a clear, colorless, and syrup liquid with a pleasantly sweet taste. The solution is essentially odorless. The sorbitol content of the solution was found to be 66% (Table 5.39). Study of the composition of the prepared solution revealed that the solution contains 0.21% reducing sugars and 0.70% total sugars. In order to assess the importance and feasibility in food and other industries, the viscosity profile of sorbitol solution is considered as one of the

TABLE 5.39 Biochemical Properties of Sorbitol Solution

Parameter	Observations
Appearance	Clear, colorless, syrup liquid
Color	Watery white
Taste	Pleasantly sweet
Odor	Essentially odorless
Assay (%) D-sorbitol	66.0
Reducing sugars (%)	0.21
Total sugars (%)	0.70
Residues on ignition (%)	0.10
Water (%)	29.5
pH (10% solution)	6.5

TABLE 5.40 Viscosity Profile for Sorbitol Syrup

	Apparent viscosity (Cps)			
	Shear rate (S^1)			
Temperature (°C)	**4.51**	**34.73**	**97.41**	**451**
20	160	147	132	98
30	110	105	88	71
40	80	72	66	49
50	65	58	48	34
60	55	47	32	20

TABLE 5.41 Functional Properties of Sorbitol Solution

Parameter	Observations
Refractive index (20°C)	1.45
Relative sweetness (10% solution)	60:100
Cooling effect	In 6 s
Thermal stability	160–180°C
Flashpoint	Above 300°C
Specific gravity at 25°C	1.28

important parameters. Sorbitol, being a monosaccharide polyol, contributes little to the viscosity of the solution to which it is added. The viscosity of sorbitol solution over a wide range of shear rates and temperatures has been carried out by using Haake's Rotoviscometer (RV 20). The flow behavior of sorbitol solution at various temperatures (20–26°C) was investigated. The results indicated that the viscosity of the sorbitol solution linearly decreases with the increase in the temperature and shear rate (Table 5.40).

Various functional properties of sorbitol were extensively studied. The refractive index of the solution (at 20°C) was found to be 1.45. The relative sweetness of the 10% solution was compared with the sweetness of standard sucrose. The relative sweetness of the solution was found to be 60:100. The data on the cooling effect of the solution revealed that the solution gives a cooling effect within six seconds of putting it in mouth. The solution was found to be thermally stable up to 160–180°C (Table 5.41). The flashpoint of the sorbitol solution was found to be 300°C and the specific gravity of the solution was 1.28 at 25°C. The non-carcinogenic properties of sorbitol are making it increasing accepted as a sweetener and the insulin-independent mechanism of sorbitol make it an ideal sweetener for diabetics. Sorbitol produced from sorghum starch hydrolysis by chemical hydrogenation in the presence of nickel catalyst was used as sweetener. The milk chocolate prepared using sucrose scored a higher value for all the quality attributes than the milk chocolates prepared by using sorbitol and sucrose in combination. Milk chocolate manufactured from sorbitol resulted in a low-viscosity chocolate base, which was corrected by using additives (starch or guar gum 2%).

5.22.6 Properties of Sorbitol

Solubility: The concentration of saturated solution of sorbitol at 40°C is 75% (w/w).

Relative Sweetness: 10% aqueous solution of sorbitol at 25% is 60% relative to sucrose. *Hygroscopicity*: Sorbitol powder is not very hygroscopic. It absorbs less than 10% water at relative humidity from 40 to 75%.

Cooling Effect: When Sorbitol (70%) dissolved in mouth saliva it gives cooling effect in 6 seconds.

References

Anonymous, 1979. In: Johnson, J.C. (Ed.), Immobilized Enzymes Preparation and Engineering Recent Advances. Noyes Data Corporation Publishers, New Jersey, p. 310.

AOAC, 1990. Official Methods of Analysis. Association of Official Analytical Chemists, Washington, DC.

Bernfeld, P., 1955. Amylase, alpha and beta. In: Colowick, S.P., Kaplan, N.O. (Eds.), Methods in Enzymology, vol. 1. Academic Press, New York, pp. 149–158.

Björndal, H., Hellerqvist, C.G., Lindberg, B., Svensson, S., 1970. Gas-Flüssigkeits-Chromatographie und Massenspektrometrie bei der Methylierungsanalyse von Polysacchariden. Angew. Chem. Int. Ed. Engl. 9, 610–619.

Dahlquist, A., 1964. Method for assay of intestinal disaccharidases. Anal. Biochem. 7, 18.

Ganorkar, P.M., 2001. Studies on Production, Functional and Rheological Attributes of Sorghum (*Sorghum bicolor* L. Moench. Var. PVK-801) Starch Carboxymethyl Starch and Their Utilization in Foodstuffs. M.Tech. thesis (Food Sciences). Marathwada Agricultural University, Parbhani.

Kshirsagar, S.S., 1986. Studies on Suitability of CSH 5 Sorghum Starch for Enzymatic Production of Glucose. M.Tech. thesis (Food Science) submitted to Marathwada Agricultural University, Parbhani.

Kulkarni, A.S., 2000. Studies on production, functional and Rheological attributes of sorghum (*Sorghum bicolor* L. Moench. Var. PVK-801) starch maltodextrin and their utilization in low calorie foodstuff. M.Tech. thesis (Food Sciences). Marathwada Agricultural University, Parbhani.

Lloyd, N.E., Khaleeluddin, K., Lamnn, N.R., 1972. Automated method for the determination of D-glucose isomerase activity. Cereal Chem. 49, 544.

Neucere, N.J., Sumrell, G., 1980. Chemical composition of different varieties of grain sorghum. J. Agric. Food Chem. 28, 19–21.

Smith, B.W., Roe, T.H., 1949. A photometric method for the determination of alpha amylase in blood and urine, with use of the starch iodine colour. J. Biol. Chem. 179, 53–59.

Subramaniam, V., Jambhunathan, R., 1981. In: Proceedings of International Symposium on Sorghum Quality ICRISAT, India, pp. 251.

Tharanathan, R.N., Parmahans, S.V., Wankhede, D.B., 1980. Amylolytic susceptibility of native groundnut and ragi starch granules as viewed by scanning electron microscopy. Starke. 32 (5), 158–161.

Tharanathan, R.N., Wankhede, D.B., Raghvendra Rao, M.R., 1976. Mono- and oligosaccharide composition of groundnut (*Arachis hypogea*). J. Food Sci. 41, 715.

Wankhede, D.B., Rathi, S.S., Gunjal, B.B., Patil, H.B., Walade, S.G., Rodge, A.B., et al., 1990. Studies on isolation and charecterization of starch from pearl millet (*Pennisetum americanum* (L.) Leeke.) grains. Carbohydr. Polym. 13, 17–28.

Wankhede, D.B., Saroja, R., Raghvendra Rao, M.R., 1977. New starches: preparation and properties of Starch and its fractions of two varieties in Groundnuts (*Arachis hypogea*). Starch. 29, 223–228.

Wankhede, D.B., Shehahaz, A., Raghvendra Rao, M.R., 1979. Preparation and physicochemical properties of starches and their fractions from finger millet (*Eleusine coracana*) and foxtail millet (*Setaria italica*). Starch. 31 (5), 153–159.

Wankhede, D.B., Tharanathan, R.N., 1976. Sesame (*Sesamum indicum*) carbohydrates. J. Agric. Food Chem. 21, 655–659.

Wankhede, D.B., Umadevi, S., 1982. Preparation and some of the physicochemical properties of polydexrtins from ragi, wheat, and rice starches. Starch. 34, 162–165.

Whistler, R.L., Durso, D.F., 1950. J. Am. Chem. Soc. 72, 677.

Further Reading

Anonymous, 1994. Glucose syrup enhancers. Food Ind. 47 (3), 16–17.

Birch, G.G., 1981. The chemistry and properties of glucose syrup. Developments in food Carbohydrates.

Panse, V.G., Sukhatme, P.V., 1985. Statistical Methods for Agricultural Workers. second ed. ICAR, New Delhi.

Wankhede, D.B., Deshpande, H.W., Gunjal, B.B., Bhosale, M.B., Patil, H.B., Gahilod, A.T., et al., 1989. Studies on physico-chemical pasting characteristics and amylolytic susceptibility of starch from sorghum (*Sorghum bicolor* L. Moench.). Starch/Strake. 41 (4), 123–127.

CHAPTER

6

Sorghum Processing and Utilization

C.V. *Ratnavathi*

ICAR-Indian Institute of Millets Research, Rajendranagar, Hyderabad, India

OUTLINE

Abstract

Sorghum traditionally has been used in many places for animal feed—nearly 40% of the global sorghum crop is now being used for human consumption. Sorghum has gained lot of importance as a human food in India in recent years, due to its gluten-free nature, antioxidant properties, and low glycemic index which mean that is favorable as a diabetic cereal. Innovative processing methods of sorghum grain are required to enable it to reach the urban market in different forms. This chapter deals with the processing methods of grain sorghum. Primary processing includes grading, cleaning, destoning, dehulling, grading, and polishing of the grain to improve the appearance of the grain and also the market price of the grain.

Sorghum Biochemistry: An Industrial Perspective.
DOI: http://dx.doi.org/10.1016/B978-0-12-803157-5.00006-X

Secondary processing consists of the preparation of flakes, semolina, and pasta from sorghum to make the different food products that can be made readily available for the urban consumer.

Sorghum is one of the oldest known grains, originating in Africa and India, where it is commonly used in a variety of foods. While sorghum traditionally has been used in many places for animal feed, nearly 40% of the global sorghum crop is now being used for human consumption. Around the world, it is used in food products including porridges, breads, cookies, tortillas, and extruded commercial products (Arvin Donley, 2013). Growers in the United States are increasing their interest in producing human food products from sorghum due to the development of white sorghum. Sorghum has gained a lot of importance as a human food in India also in recent years due to its gluten free nature, antioxidant properties, and low glycemic index, which means that it is favorable as a diabetic cereal.

The processing of sorghum industrially has not gained much importance. Sorghum is generally made into flour and is used for various food preparations. In the sorghum growing regions of Africa and Asia the grain is made into flour by hand pounding. Traditionally, water was added to the grain by up to 30–40% and the grain was decorticated by hand pounding with wooden pestles in a mortar for 10–15 min. Bran was immediately separated from the endosperm material by winnowing with a traditional basket winnower. The decorticated grain was then conditioned to 25–30% moisture and pounded into a meal. All the milled portions of a batch were blended together and spread on jute bags in the open sun to dry for 18–24 h.

6.1 ABRASIVE DECORTICATION AND HAMMER MILLING

The first type of milling procedure adopted for sorghum after hand pounding was abrasive decortication and hammer milling where decortication is carried out through abrasion and by a hammer mill—either flour or grits/semolina is usually made. Later it was commercialized in Botswana. Milling equipment comprised a Rural Industries Innovation Centre (RIIC, Kanye, Botswana) PRL-type dehuller and hammer mill. A cleaned dry 10 kg batch of each sorghum type was fed into the barrel of the dehuller through a hopper fitted with a flow regulator. The bran was progressively abraded off and removed by means of a cyclone fan. The grain was decorticated to the operator's satisfaction. The decorticating time for each batch ranged from 3 to 8 min, depending on the sorghum type. The grain was then milled using a hammer mill fitted with a 2.0 mm opening screen (Rohrbach et al., 2000; Rooney and Waniska, 2000). There has also been an intensive effort to introduce RIIC decorticators in Andhra Pradesh. Decortication improved the quality of the flour from sorghum and millets so that it could be used in new ways (Geervani and Vimala, 1993).

6.2 ROLLER MILLING

Roller mill is an advanced milling technology developed over abrasive decortications and hammer milling. The details of the commercial roller mill (RM) (Maximill, Kroonstad, South Africa) with two pairs of fluted rolls and a rated throughput of 500 kg/h used was well explained by Kebakile et al. (2008). The top rolls (coarse break rolls) had 8 flutes/25 mm and the bottom rolls (fine break rolls) had 22 flutes/25 mm. All roll pairs operated at differentials of 1.5:1. Five kg batches of clean grain were tempered to 16% moisture for 15 min in sealed plastic buckets at ambient conditions, stirring at intervals of 5 min to uniformly distribute the added water. The tempered grain was roller milled immediately using top and bottom roller gaps of 0.80 and 0.30 mm, respectively. The milled stock was separated on vibrating sieves of mesh sizes 1.00, 0.850, 0.710, and 0.710 mm arranged in descending order. The first two sieves retained the bran, while the last two separated the meal. The extraction of flour is higher than in a hammer mill.

6.3 GRAIN PROCESSING

Though sorghum is nutritionally superior, its consumption has been decreasing gradually due to the nonavailability of processed clean grain and its products in the urban markets. To increase sorghum consumption among the urban population, the development of processing technologies and also the availability of semiprocessed products are a prerequisite. As a step towards this, under the World Bank aided project 20 pieces of machinery for different processes were procured and retrofitted at ICAR-Indian Institute of Millets Research, Hyderabad. Grain processing is of two types: (1) Primary processing; (2) Secondary processing.

6.4 PRIMARY PROCESSING

Grading: This is the processing of bulk farm produce to clean grain through grading and cleaning by a cleaner-cum-grading machine. This machine grades grain based on the size of the grain and separates it

FIGURE 6.1 Grader in operation.

TABLE 6.1 Percent Recovery of Grain in Primary Processing (Grading machine)

	Genotype	Total grain taken	Bold grain recovered	Percent recovery of clean grain
1.	C-43	15.0	13.8	92.00
2.	M35-1	15.0	13.1	89.33
3.	LPJ	15.0	14.5	96.60

Data includes the mean of three replications.

into three sizes (Fig. 6.1): bigger size, medium size, and small size and it also separates unwanted particles. The grain thus separated is collected directly into gunny bags and can be sent to the market. This helps the farmer in getting the correct price for the grain as per its quality. The bulk operation of these processes can be done mechanically (Table 6.1).

Destoning: Destoning (the removal of stones from bulk farm produce) is also another primary processing method of the grain and is carried out by a destoner (Figs. 6.2–6.5). Destoning is the removal of waste, stones, and glumes from the grain, which is very much necessary as it improves the storage of the grain and consumer acceptability for usage. This is also a kind of value addition to the grain. The percent recovery of the grain from the bulk lots either from the farm or market usually ranges from 89.33 to 96.60%.

Dehulling or pearling of grain sorghum: It is essential to polish/pearl the grain prior to milling into rawa or flour in a pearling machine. In this process the outer surface of the grain is polished. It may not

FIGURE 6.2 Destoner in operation.

FIGURE 6.3 Stones separated in upper chamber.

only remove the bitterness associated with the pericarp, but also reduces insoluble fiber and dirt and microorganisms present on the naked grain, and also improves the appearance of the grain. The mold and fungi that attack the grain grown during the *Kharif* season (rainy) can blacken the grain. This blackening usually leads to a low market price. Though the grain productivity is greater compared to the postrainy season, the producer ends up with less income due to this blackening problem. Dehulling or pearling is an abrasion technique used to polish the grain. The pearling machine shown in Fig. 6.6 is a modified version of earlier decorticators. Here the grain after polishing is collected on a tray where the bran that is removed is separated by a shaking operation. The bran is also separated during the process and is collected in a container. This bran can be mixed into cattle feed. The recovery of the grain is 95–98% in pearling (Fig. 6.7).

FIGURE 6.4 Destoner in operation.

FIGURE 6.5 Cleaned grain.

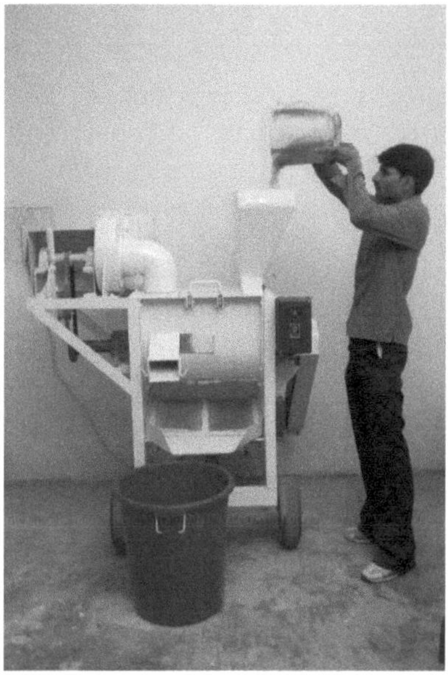

FIGURE 6.6 Dehuller in operation.

FIGURE 6.7 Polished grain recovered.

6.5 SECONDARY PROCESSING

For the preparation of any food product the grain should be in the form of flour or in a form that can be conveniently used. Grain was processed into flour, coarse rawa, fine rawa, flakes, or pops through the respective machines. The quality of the products was evaluated by using these products to prepare various foods that are commonly made from rice and wheat. Sorghum flour can be made using a grinding mill called a pulverizer (Figs. 6.8–6.10) and the flour is sieved to make it fine. The fine flour is used in *roti, cake,* and *biscuit* preparations. Sorghum semolina, locally called *rawa* or *suji*, can be made in two forms with two different particle sizes, coarse1.18 mm and fine 0.6 mm, respectively. Both forms of rawa can commercially be made available for market. Rawa recovery will be 50–55% in the case of coarse rawa and 45–50% in the case of fine rawa and the remaining 50% is flour. Rawa is used for preparing Indian culinary preparations such as *idli, upma* and *kesari,* and *laddu*. Fine rawa is made with a Brabender semolina mill (Fig. 6.10). The rawa made in a Brabender semolina machine is made entirely from the endosperm and the pericarp is removed as bran separately.

Sorghum flakes (beaten sorghum): Sorghum grain can be processed into flakes or beaten sorghum using the flaking machine. The flaking machine comprises of two parts. One is a roaster which is used for roasting the grain. The other part is a roller flaker used for making the

FIGURE 6.8 Pulverizer in operation.

FIGURE 6.9 Coarse semolina prepared.

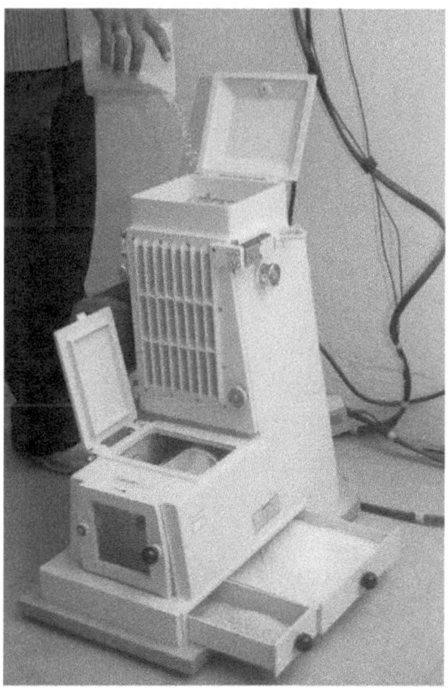

FIGURE 6.10 Brabender fine semolina machine.

TABLE 6.2 Effect of Processing on the Chemical Quality of the Product

Sample		Grain	Flakes	Parboiled rawa
M35-1	Starch	63.4	63.35	64.7
C-43		64.5	60.35	62.15
M35-1	Protein	11.09	11.99	11.82
C-43		10.07	10.3	10.68
M35-1	Fat	2.02	2.42	2.48
C-43		2.6	2.58	2.44

FIGURE 6.11 Grain being put in roaster.

flakes. The flaking machine unit is ideally purposed for rice and the same machine was used for sorghum grain by optimizing the temperature and moisture. The appearance, quality, and taste of these products was found to be acceptable and on a par with rice, wheat, and maize (Table 6.2).

Preparation of flakes: Sorghum grain is soaked for 12–18 h in water and the grain is drained and air dried in the shade. The grain, after drying, is put in the roaster (Fig. 6.11) and roasted for 2 min at the standardized temperature which was set already. After removal from the roaster the grain (Fig. 6.12) is slowly entered into the roller flaker

FIGURE 6.12 Grain taken out after roasting.

FIGURE 6.13 Grain being put in roller mill.

for flaking (Figs. 6.13–6.14). The pericarp on the grain is separated and removed. The grain is pressed into thin flakes within 15 min. The flakes are made in batches and the capacity in continuous mode is estimated to 5 kg/h. The mill is farmer friendly—farmers can make their own flakes and market the product.

FIGURE 6.14 Grain rolled in to flakes.

TABLE 6.3 Recovery of Pops From Grain

S. no.	Genotypes	Weight after soaking	grain weight before popping	Popping ability (%)
1.	Dagadi	505	200	60.4
2.	**CSV 216R**	**420**	**100**	**76.2**
3.	M35-1	560	280	50.0
4.	DSV 5	570	170	70.2
5.	**PKV kranti**	**500**	**105**	**79.0**
6.	CSV 22	440	200	54.6
7.	**CSV 18**	**515**	**80**	**84.5**

Preparation of pops: Usually pops are prepared from special sorghum grain prescribed for pops. But, like maize, sorghum grain have the ability to pop after preconditioning of the grain. This process of preconditioning was standardized and normal released *Kharif* cultivars like CSH 16 also can be used for making pops. The grain was soaked for 6–8 h and the grain was air dried in the shade on a cloth. The grain is then put in a smart pop machine and pops are made. The recovery percentage of pops among seven cultivars was measured in three replications. Four genotypes were superior having more than 70% pops and three genotypes, CSV 18 (84.5%), PKV *kranti* (79%), and CSV 216R (76.2%), are identified as being the best for popping and are the superior genotypes (Table 6.3; Fig. 6.15).

FIGURE 6.15 Pops made from superior genotypes.

6.6 PRODUCTION OF PROCESSED FOODS THROUGH MACHINES

Roti cannot be made using a roller with sorghum dough due to the absence of gluten; hence dough is made sticky using hot water. However, rolling of roti needs skill and traditionally it is done by hand tapping on a flat stone. It is not only time-consuming, but if not done correctly the *rotis* will be thick and unpalatable. To overcome this, a machine was developed to facilitate rolling wet *rotis* commercially. The electric machine consists of a rotary revolving, circular flattening plate with an upper rolling pin (Fig. 6.16a and b). A foot operated lever applies pressure gradually on the rotating dough, flattening it uniformly. A thin *roti* is formed between two plastic sheets which is removed carefully and baked on a hot pan/tawa. Using this machine 40–45 rotis can be made in one hour. Thirty five thin *rotis* each weighing 25 gm can be made per kg of flour using the machine. The wastage of flour is also minimized. This machine is ideal for small hotels, canteens, and hostels.

Multigrain roti testing: A big automatic roti, unleavened pan cake (*chapathi*) machine (Figs. 6.17–6.18) which is similar to a tortilla machine was used experimentally for sorghum roti preparation. Roti

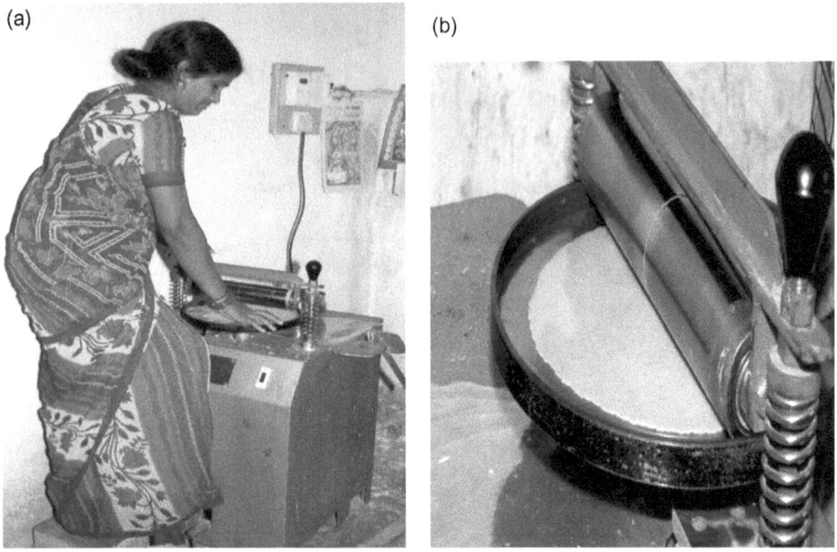

FIGURE 6.16 (a) Roti machine. (b) Rolled roti.

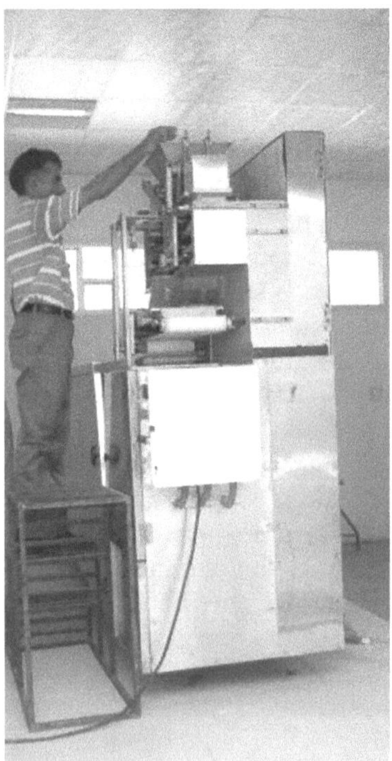

FIGURE 6.17 Automatic roti machine in operation.

FIGURE 6.18 Roti being prepared.

from sorghum was tested using dough which was blended initially with wheat since the dough from sorghum is not sticky as it is gluten free. Sorghum and wheat could be blended in 60:40 ratio and subsequently multigrain atta, including wheat, bajra, ragi, and soybean, was tried. Before the dough was tested on the machine a preliminary experiment was carried out on *roti* quality analysis from different dough made from different combinations of flours of sorghum, wheat, finger millet, pearl millet, and soybean. The range of starch in the combinations varied from 51.9 to 63.3 and fat ranged from 1.8 to 2.67. The percent protein ranged from 11.3 to 14.16. The dough has good viscosity and the mean diameter of the *roti* made from composite flours was 24.16 cm. as compared to *Rabi* sorghum grain flour which was 22.39 cm. The composite flour with 50% wheat was suitable to prepare *rotis* using the big *roti* machine.

Extruded products from sorghum: Sorghum is known to have low digestibility and the digestibility is known to be decreased after cooking. However, sorghum digestibility is increased by processes such as fermentation, flaking, extrusion, and popping. Extrusion (Fig. 6.19) is a very ideal process where the nutrient availability will be enhanced and the product can be an ideal breakfast and snack food. Since sorghum is nonsticky, a single screw extruded product could be successfully made with a blend of maida and sorghum semolina. Two kinds of single

FIGURE 6.19 Single screw extruder.

FIGURE 6.20 Vermicelli.

FIGURE 6.21 Conchiglie.

screw cold extruded pasta products, vermicelli (Fig. 6.20) and conchiglie (Fig. 6.21), could be made with this combination. This is a semiprocessed product and ready-to-cook breakfast cereal from sorghum also has a very good potential niche market in urban areas.

References

Donley, A., 2013. Sorghum milling study http://www.world-grain.com/News/News%20Home/Features/2013/7/Sorghum%20milling%20study.aspx?cck=1.

Geervani, P., Vimala, V., 1993. Operational Research on Development of Sorghum Food Enterprises for Alternative Uses and Supplementary Feeding. College of Home Science, Andhra Pradesh Agricultural University, Hyderabad, India.

Kebakile, M.M, Taylor, J.R.N., Rooney, L.W., 2008. Roller milling – an alternative dry milling process for sorghum. In: Alternative Cereal Processing Technologies, Lobatse, Botswana, November 4–6, 2008. http://projekt.sik.se/traditionalgrains/participants/Abstract_CV_Sweden/Martin_Kebakile.htm.

Rohrbach, D., Mupanda, K., Seleka, T., 2000. Commercialization of Sorghum Milling in Botswana: Trends and Prospects. Working Paper Series No. 6. Socioeconomics and Policy Program, International Crops Research Institute for the Semi-Arid Tropics, Bulawayo, Zimbabwe, 24 pp.

Rooney, L.W., Waniska, R.D., 2000. Sorghum food and industrial utilization. In: Smith, C.W., Frederiksen, R.A. (Eds.), Sorghum, Origin, History, Technology and Production, pp. 689–729. Common dry milling procedures like hand pounding abrasive dehulling, hammer mill use and roller milling and grain conditioned roller milling.

Index

Note: Page numbers followed by "*f*" and "*t*" refer to figures and tables, respectively.

L

M

N

O

P

R

T

FSC
www.fsc.org
MIX
Papier | Fördert gute Waldnutzung
FSC® C083411

Kolibri 360 GmbH
Ferdinand-Jühlke-Straße 7
99095 Erfurt, Deutschland
produktsicherheit@kolibri360.de